White Dwarf Atmospheres and Circumstellar Environments

Edited by
D. W. Hoard

Related Titles

Kwok, S.

Organic Matter in the Universe

2011

ISBN: 978-3-527-40986-0

Barnes, R. (ed.)

Formation and Evolution of Exoplanets

2010

ISBN: 978-3-527-40896-2

Shore, S. N.

Astrophysical Hydrodynamics
An Introduction

2007

ISBN: 978-3-527-40669-2

Shaw, A. M.

Astrochemistry
From Astronomy to Astrobiology

2006

ISBN: 978-0-470-09137-1

Stahler, S. W., Palla, F.

The Formation of Stars

2004

ISBN: 978-3-527-40559-6

Spitzer, L.

Physical Processes in the Interstellar Medium

1998

ISBN: 978-0-471-29335-4

Shapiro, S. L., Teukolsky, S. A.

Black Holes, White Dwarfs and Neutron Stars
The Physics of Compact Objects

1983

ISBN: 978-0-471-87317-4

White Dwarf Atmospheres and Circumstellar Environments

Edited by
D. W. Hoard

WILEY-VCH Verlag GmbH & Co. KGaA

The Editor

Dr. D. W. Hoard
Spitzer Science Center
California Institute of Technology
Pasadena, USA

Library of Congress Card No.: applied for

British Library Cataloguing-in-Publication Data:
A catalogue record for this book is available from the British Library.

Bibliographic information published by the Deutsche Nationalbibliothek
The Deutsche Nationalbibliothek lists this publication in the Deutsche Nationalbibliografie; detailed bibliographic data are available on the Internet at http://dnb.d-nb.de.

Typesetting le-tex publishing services GmbH, Leipzig
Cover Design Adam-Design, Weinheim
Printing and Binding Fabulous Printers Pte Ltd, Singapore

Printed in Singapore
Printed on acid-free paper

ISBN Print 978-3-527-41031-6

ISBN oBook 978-3-527-63657-0
ISBN ePDF 978-3-527-63659-4
ISBN ePub 978-3-527-63658-7
ISBN Mobi 978-3-527-63660-0

Where are the stars, pristine as great ideas? Behind clouds the heavens saturate with luminous dust...

("Where Are The Stars Pristine" from Palladium by Alice Fulton. Copyright 1987 by Alice Fulton. Used with permission of the University of Illinois Press.)

Contents

Preface

White dwarf stars play a key role in a wide variety of astrophysically important scenarios. They not only provide a glimpse into the distant future of our own Sun, but are the evolutionary endpoints of the majority population of low mass stars in the Galaxy. As relic cores of normal stars, white dwarfs provide insights into stellar evolution, and expose material created during a stellar lifetime of nuclear burning to direct examination. Binary stars containing white dwarfs are linked to the chemical enrichment of the interstellar medium (via nova explosions), and are laboratories to probe the processes of mass transfer and accretion that power the central engines of quasars and govern the formation of stars and planetary systems. Type Ia supernovae, used as standard candles for measuring cosmological distances, are believed to result from accretion onto white dwarfs and/or white dwarf-white dwarf collisions.

In many ways, white dwarfs are relatively "simple" and well understood objects: partially crystallized balls of mostly electron-degenerate carbon and oxygen, with the mass of a sun packed into the volume of an earth.[1] They do not produce any new energy, but slowly radiate away the trapped energy of billions of years of nuclear fusion. We can "listen" to the ringing of acoustic waves in their interiors, and produce physically realistic model spectra that are almost indistinguishable from the real thing. However, in recent years, new discoveries have made it clear that the immediate environments of white dwarfs, from their photospheres out, can be – and often are – as interesting as the white dwarfs themselves. The flotsam and detritus surrounding white dwarfs, largely undetectable or overlooked during most

1) This is an extreme physical situation that was eloquently described by Arthur Stanley Eddington in a lecture transcribed in his 1927 book *Stars and Atoms* (Oxford: Clarendon Press, p. 50), as "a density much transcending our terrestrial experience" and, somewhat more pithily, "a tight squeeze". Eddington also relates the initial reaction to the inferred physical properties of the second known white dwarf, Sirius B: "We learn about the stars by receiving and interpreting the messages which their light brings to us. The message of the Companion of Sirius when it was decoded ran: 'I am composed of material 3000 times denser than anything you have ever come across; a ton of my material would be a little nugget that you could put in a matchbox.' What reply can one make to such a message? The reply which most of us made in 1914 was – 'Shut up. Don't talk nonsense.'".

of the last 100 years of astronomical observations of white dwarfs, turn out to have their own tales to tell about the past, present, and future of these objects.

During approximately the last decade, advances in observational techniques and detector technology have opened up new regimes of wavelength and sensitivity to the study of white dwarfs. In particular, satellite observatories such as the Hubble and Spitzer Space Telescopes have enabled dramatic new discoveries about white dwarfs. For example, observations with Spitzer have shown that the presence of dusty debris disks around white dwarfs, which can often only be detected in the mid-infrared, is fairly common. Meanwhile, ever more sophisticated model atmosphere calculations have enabled the increasingly realistic generation of white dwarf synthetic spectra that can be used as diagnostic comparisons with observations.

The tale of Subrahmanyan Chandrasekhar is well known among those astronomers who study white dwarfs: as a 19-year-old student, on a long sea voyage to England in 1930 to begin his graduate studies at Cambridge University, he whiled away his time by modifying a theory proposed by his soon-to-be graduate advisor, the British astronomer Ralph Fowler, to include special relativistic effects. By combining quantum mechanics and Einsteinian relativity, Chandra determined that the mass of a star that can end its life as a white dwarf (and, hence, the mass of a white dwarf itself) has an upper limit, which is now named in his honor. In part for this work, Chandra was awarded the Nobel Prize in Physics in 1983 (shared with William A. Fowler) "for his theoretical studies of the physical processes of importance to the structure and evolution of the stars."[2] The preparation of this book coincided with the 100th anniversary of both the birth of Chandra and the classification of the first white dwarf, 40 Eridani B – auspicious omens for a book about white dwarf stars.

The chapters in this book will *not* present detailed treatises on the formation or internal structure of white dwarfs themselves, topics which have been extensively covered in other works (from Chandra onward). Nor will they focus on white dwarfs as members of detached or interacting binary star systems. Instead, the focus of this book is shifted somewhat away from the white dwarfs themselves, and onto the relatively new and fascinating topics of peculiar atmospheric compositions and the dust, nebulae, and (potentially) planets that surround white dwarfs. We will start in the geometrically thin, nondegenerate atmospheres of the white dwarfs and move outward into their circumstellar environs.

Acknowledgments

The cover of this book shows the spectral energy distribution of the dusty white dwarf GD 16 from the optical to the mid-infrared, illustrating the infrared excess, along with a depiction of the white dwarf and its circumstellar disk. Robert Hurt (Spitzer Science Center) kindly made some minor modifications to the original

2) http://nobelprize.org (15 June 2011)

image from the Spitzer press release sig09-002 by Jay Farihi[3] for use on this cover. The original image is courtesy of NASA/JPL-Caltech.

The rapid pace of advances in the understanding of white dwarf stars and their circumstellar environments, especially over the course of the last ten years, would not have been possible without the contributions of many theorists and researchers, as well as the availability of data from numerous surveys and data archives. On behalf of myself and the other authors, I would like to acknowledge the following facilities that have been of particular (although not exclusive) usefulness in exploring the atmospheres and circumstellar environments of white dwarfs as discussed in this book (listed in order by wavelength regime, from short to long):

1. The NASA-CNES-CSA Far Ultraviolet Spectroscopic Explorer, FUSE, which was operated for NASA by the Johns Hopkins University under NASA contract NAS5-32985.
2. The NASA Galaxy Evolution Explorer, GALEX, which is operated for NASA by the California Institute of Technology under NASA contract NAS5-98034.
3. The International Ultraviolet Explorer satellite, which was a collaboration among three groups: NASA, the European Space Agency (ESA), and the United Kingdom's Science and Engineering Research Council (SERC; now called the Particle Physics and Astronomy Research Council, PPARC).
4. The NASA/ESA Hubble Space Telescope, which is operated by the Association of Universities for Research in Astronomy, Inc., under NASA contract NAS 5-26555.
5. The Digitized Sky Surveys, which were produced at the Space Telescope Science Institute under US Government grant NAG W-2166. The images of these surveys are based on photographic data obtained using the Oschin Schmidt Telescope on Palomar Mountain and the UK Schmidt Telescope. The plates were processed into the present compressed digital form with the permission of these institutions. The National Geographic Society – Palomar Observatory Sky Atlas (POSS-I) was made by the California Institute of Technology with grants from the National Geographic Society. The Second Palomar Observatory Sky Survey (POSS-II) was made by the California Institute of Technology with funds from the National Science Foundation, the National Geographic Society, the Sloan Foundation, the Samuel Oschin Foundation, and the Eastman Kodak Corporation. The Oschin Schmidt Telescope is operated by the California Institute of Technology and Palomar Observatory. The UK Schmidt Telescope was operated by the Royal Observatory Edinburgh, with funding from the UK Science and Engineering Research Council (later the UK Particle Physics and Astronomy Research Council), until 1988 June, and thereafter by the Anglo-Australian Observatory. The blue plates of the southern Sky Atlas and its Equatorial Extension (together known as the SERC-J), as well as the Equatorial Red

3) http://www.spitzer.caltech.edu/images/
2054-sig09-002-Emission-from-the-White-Dwarf-System-GD-16 (9 May 2011)

(ER), and the Second Epoch [red] Survey (SES) were all taken with the UK Schmidt.

6. The Sloan Digital Sky Surveys, SDSS and SDSS-II. Funding for the SDSS and SDSS-II was provided by the Alfred P. Sloan Foundation, the Participating Institutions, the National Science Foundation, the US Department of Energy, the National Aeronautics and Space Administration, the Japanese Monbukagakusho, the Max Planck Society, and the Higher Education Funding Council for England. The SDSS was managed by the Astrophysical Research Consortium for the Participating Institutions.
7. The Two Micron All Sky Survey, 2MASS, which was a joint project of the University of Massachusetts and the Infrared Processing and Analysis Center/California Institute of Technology, funded by the National Aeronautics and Space Administration and the National Science Foundation.
8. The Spitzer Space Telescope, which is operated by the Jet Propulsion Laboratory, California Institute of Technology, under a contract with NASA.

Pasadena, California, 2011 *D. W. Hoard*

List of Contributors

You-Hua Chu
Astronomy Department
University of Illinois at
Urbana-Champaign
1002 W. Green Street
Urbana, IL 61801
USA

John H. Debes
NASA's Goddard Space Flight Center
Greenbelt, MD 20771
USA

Rosanne Di Stefano
Harvard-Smithsonian Center for
Astrophysics
60 Garden Street
Cambridge, MA 02138
USA

Patrick Dufour
Département de Physique
Université de Montréal
C.P. 6128
Succursale Centre-Ville
Montréal, QC H3C 3J7
Canada

Jay Farihi
Department of Physics and Astronomy
University of Leicester
Leicester, LE1 7RH
United Kingdom

Mukremin Kilic
Harvard-Smithsonian Center for
Astrophysics
60 Garden Street
Cambridge, MA 02138
USA

Edward M. Sion
Department of Astronomy and
Astrophysics
Villanova University
Villanova, PA 19085
USA

1
Hot White Dwarfs

Edward M. Sion

1.1
Introduction

Research on hot white dwarfs during the past 30 years has greatly expanded, as many new discoveries and the new questions they raise have emerged from increasingly larger, deeper surveys conducted with multimeter class ground-based telescopes, the International Ultraviolet Explorer (IUE), the Hubble Space Telescope (HST), the Extreme Ultraviolet Explorer (EUVE), and the Far Ultraviolet Spectroscopic Explorer (FUSE). This review will focus on white dwarfs ranging in temperature from 20 000 up to 200 000 K and higher, which are the hottest white dwarf stars known. Since the mid-twentieth century, the earliest spectroscopic surveys of white dwarf candidates from the proper motion selected samples of Willem Luyten and Henry Giclas were carried out by Jesse Greenstein, Olin Eggen, James Liebert, Richard Green, and others. The selection criteria employed in many of these surveys did not reveal a large number of hot white dwarfs because the surveys lacked ultraviolet sensitivity and also missed objects with low flux levels in the optical. Nevertheless, the earliest surveys quickly revealed that white dwarfs divide into two basic composition groups with hydrogen-rich (the DA stars) and helium-rich atmospheric compositions (the DB and other non-DA stars). The origin of this dichotomy still represents a major unsolved problem in stellar evolution, although theoretical advances in late stellar evolution made starting in the 1980s, as well as advances in modeling envelope physical processes and mass loss, have shed important new light on this puzzle (Chayer *et al.*, 1995; Fontaine and Michaud, 1979; Iben, 1984; Iben *et al.*, 1983; Schoenberner, 1983; Unglaub and Bues, 1998, 2000; Vauclair *et al.*, 1979).

The spectroscopic properties of white dwarfs are determined by a host of physical processes which control and/or modify the flow of elements and, hence, surface abundances in high gravity atmospheres: convective dredge-up, mixing and dilution, accretion of gas and dust from the interstellar medium and debris disks, gravitational and thermal diffusion, radiative forces, mass loss due to wind outflow and episodic mass ejection, late nuclear shell burning and late thermal pulses, ro-

White Dwarf Atmospheres and Circumstellar Environments. First Edition. Edited by D. W. Hoard.

tation, magnetic fields, and possible composition relics of prior pre-white dwarf evolutionary states. Virtually all of these processes and factors may operate in hot white dwarfs, leading to the wide variety of observed spectroscopic phenomena and spectral evolution.

The basic thrust of research on hot white dwarfs is three-fold: (i) to elucidate the evolutionary links between the white dwarfs and their pre-white dwarf progenitors, whether from the asymptotic giant branch (AGB), the extended horizontal branch, stellar mergers, or binary evolution; (ii) to understand the physics of the different envelope processes operating in hot white dwarfs as they cool; and (iii) to disentangle and elucidate the relationships between the different spectroscopic subclasses and hybrid subclasses of hot white dwarfs as spectral evolution proceeds. This includes the source of photospheric metals, the chemical species observed, and the measured surface abundances in hot degenerates.The evolutionary significance of certain observed ion species is complicated by the role of radiative forces and weak winds in levitating and ejecting elements at temperatures $> 20\,000$ K.

1.2
Remarks on the Spectroscopic Classification of Hot White Dwarfs

Before discussing the hot white dwarfs, a brief discussion of their spectroscopic classification is appropriate. The non-DA stars fall into six subclasses, including the PG 1159 stars ($75\,000 \text{ K} < T_{\text{eff}} < 200\,000$ K); DO ($45\,000 \text{ K} < T_{\text{eff}} < 120\,000$ K), and DB ($12\,000 \text{ K} < T_{\text{eff}} < 45\,000$ K). The remaining three subclasses, which are too cool ($T_{\text{eff}} < 12\,000$ K) to show helium, are DC (pure continuum, no lines), DQ (helium-dominated but with carbon molecular bands and/or atomic carbon dominating the optical spectrum), and DZ (helium dominated with lines of accreted metals and sometimes H). The three cool spectral subgroups, as well as the PG 1159 stars, are discussed elsewhere in this volume (see Chapters 2 and 3 by M. Kilic and P. Dufour), and are not covered here. For convenience, Table 1.1 lists the various primary white dwarf spectroscopic classification symbols, any of which can also be assigned as a secondary symbol to form hybrid spectral classes. For example, the identified classes of hot, hybrid composition degenerates are predominantly DBA, DAB, DAO, and DOA.

The DA stars are much easier to classify because the Balmer lines of hydrogen are detectable across a vast range of T_{eff}, from 4000 up to 120 000 K and higher. Figure 1.1 shows representative optical spectra of several DA white dwarfs. Hot DA stars that contain detectable helium are classified as DAO if He II is present and DAB if He I is present. Because of the importance of temperature as a direct luminosity and age indicator in white dwarfs, and the fact that white dwarfs span enormous ranges of T_{eff} (e. g., the H-rich white dwarfs span a temperature range from 4500 to 170 000 K!), a temperature index was introduced by Sion *et al.* (1983) defined as $10 \times \theta_{\text{eff}}$ ($= 50\,400 \text{ K}/T_{\text{eff}}$). Thus, for the hot DA and non-DA stars, their spectral types can be expressed in half-integer steps as a function of temperature; for example, the DA sequence extends from DA.5, DA1, DA1.5, DA2, DA2.5, ...,

Table 1.1 Definition of primary spectroscopic classification symbols.

Spectral type	Characteristics
DA	Only Balmer lines; no He I or metals present
DB	He I lines; no H or metals present
DC	Continuous spectrum, no lines deeper than 5% in any part of the electromagnetic spectrum
DO	He II strong; He I or H present
DZ	Metal lines only; no H or He lines
DQ	Carbon features, either atomic or molecular in any part of the electromagnetic spectrum
P	magnetic white dwarfs with detectable polarization
H	magnetic white dwarfs without detectable polarization
X	peculiar or unclassifiable spectrum
E	emission lines are present
?	uncertain assigned classification; a colon (:) may also be used
V	optional symbol to denote variability

DA13. A DA2 star has a temperature in the range 22 400–28 800 K, while a DA2.5 has T_{eff} in the range 18 327–22 400 K. Similarly, the sequence of DB stars extends from DB2, DB2.5, DB3, DB3.5, and so on. A DB2 star has a temperature in the range 22 400–28 800 K, while a DB2.5 has T_{eff} in the range 18 327–22 400 K. Figure 1.2 shows representative optical spectra of several DB white dwarfs. For the hot DA stars ($T_{eff} > 20\,000$ K), the temperature index ranges are given in Table 1.2. A similar range of temperature index is defined for the hot non-DA stars.

The spectroscopic appearance of the DO and DB subclasses is determined by the ionization balance of He I and He II. The DO white dwarfs show a pure He II spectrum at the hot end, and a mixed He I and He II spectrum at the cool end. However, the hottest non-DA stars are problematic to classify because (1) many are planetary nebula nuclei and isolated post-AGB stars sharing the hallmark spectroscopic characteristics of the PG 1159 degenerates, though with gravities lower than $\log g = 7$,

Table 1.2 Temperature index ranges for hot DA stars.

Spectral type	T_{eff} Range (K)	$10 \times \theta_{eff}$ Range
DA.25	200 000	–
DA.5	100 800	–
DA1	40 320–67 200	1.25–0.75
DA1.5	28 800–40 320	1.75–1.25
DA2	22 400–28 800	2.25–1.75
DA2.5	18 327–22 400	2.75–2.25

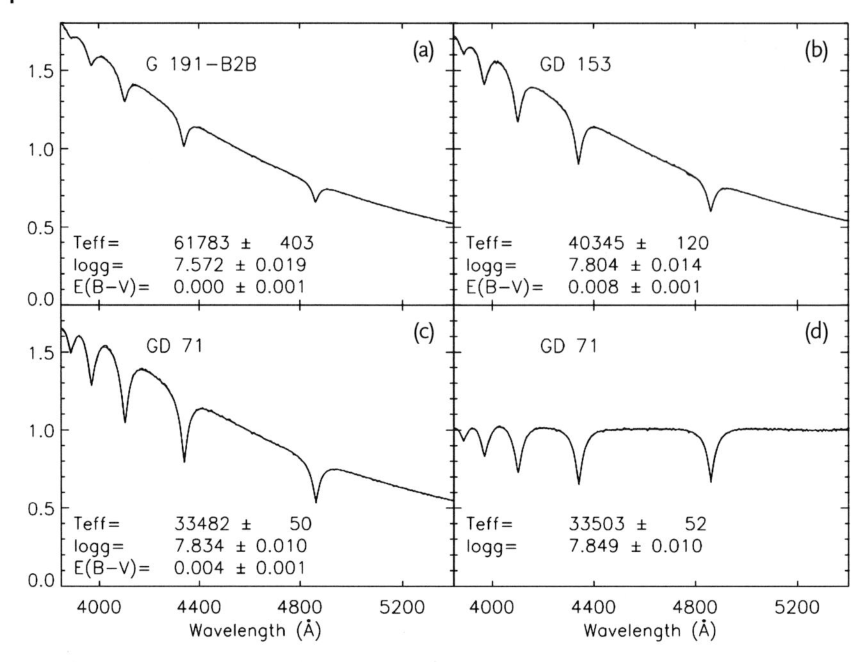

Figure 1.1 Spectra and model fits of three hot DA white dwarfs, (a) G191-B2B, (b) GD 53, and (c) GD 71, that are used as primary flux calibration standards for HST. Absorption lines of the hydrogen Balmer series are prominent in the spectrum of each star. The fluxes are in f_λ units, normalized to have a median value of one in the range 3850–5400 Å (a–c). Panel (d) shows the model fit for GD 71 when the continuum shape is rectified. Note that the spectra and model fits are essentially indistinguishable in the plots. From Allende Prieto *et al.* (2009), reproduced with the permission of Wiley-Blackwell.

which is traditionally adopted as the minimum defining gravity for classification as a white dwarf star (versus a high gravity subdwarf; Greenstein and Sargent, 1974); and (2) the assignment of the primary spectral class for a white dwarf is determined by the element represented by the strongest absorption features in the optical spectrum. However, by this criterion, the PG 1159 stars, for which either oxygen (e. g., O VI) or carbon (e. g., C IV) are the strongest optical lines (with He II features weaker), should be classified as DZQO or DQZO depending upon whether O VI or C IV are strongest, respectively. Since many of these objects have atmospheric compositions which are not completely helium-dominated (cf., Werner and Heber, 1991; Werner *et al.*, 1991), it is inappropriate to assign spectral type DO on the Sion *et al.* (1983) scheme since the primary O-symbol denotes a helium-dominated composition. Hence, the primary spectroscopic type is adopted here as the atom or ion with the strongest absorption features in the *optical* spectrum, where applicable (for example, it is possible that the strongest absorption feature may lie in the ultraviolet). This is the scheme used for classifying the hottest degenerates.

In practice, a degenerate classification is withheld for any PG 1159 star with $\log g < 7$. However, these objects are designated PG 1159 as given by Werner

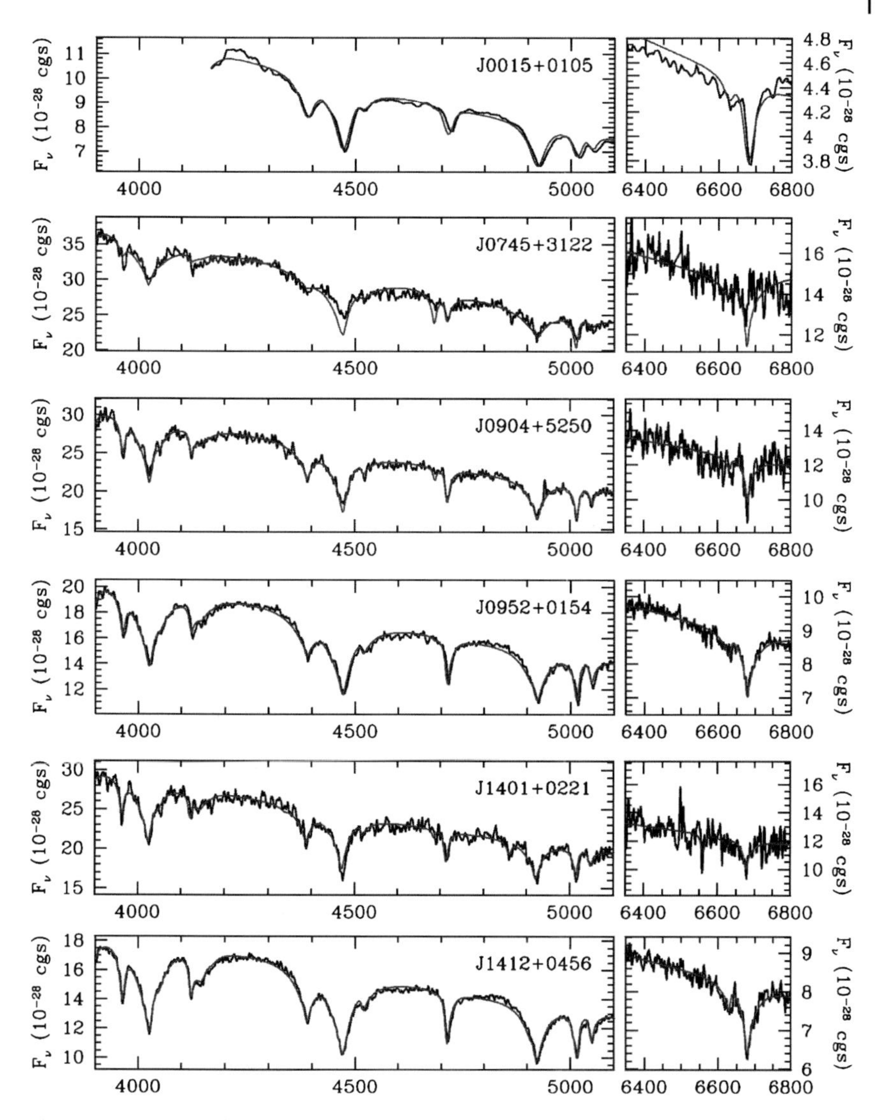

Figure 1.2 Spectra and model fits (smooth lines) of six hot DB white dwarfs, with temperatures in the range 30 000–45 000 K. The prominent absorption features in each spectrum are lines of He I. From Eisenstein *et al.* (2006b), reproduced by permission of the AAS.

and Heber (1991) and Werner *et al.* (1991), and subsequently used by Napiwotzki and Schoenberner (1991) and Dreizler *et al.* (1995). These designations are: E for emission, lgE for low gravity with strong central emission, A for absorption, E_p for emission/peculiar, and E_H, lgE_H or A_H for hybrid PG 1159 stars that have detectable hydrogen. The temperature index would differentiate the hot C-He-O stars from the well-known, much cooler DZ and DQ degenerates below 10 000 K. For example, PG 1159 itself ($\log g = 7$, $T_{\rm eff} = 110\,000$ K, C IV absorption as the strongest optical lines) would be classified as DQZO.4. The obvious disadvantage is the inevitable confusion with the cool DQ and DZ degenerates in cases for which the temperature index is missing or there are no He II absorption features (e. g., H 1504; Nousek *et al.*, 1986). In a case like H1504, where no helium is detected, $T_{\rm eff} = 170\,000$ K, and $\log g = 7$, the classification DZQ.3 is assigned.

The hot DQ stars are an entirely new subclass of hot white dwarfs. Unlike the previously known, and cooler, DQ white dwarfs which have helium-dominated atmospheres, the hot DQ stars have atmospheres that are dominated by carbon! These hot DQs probably evolve from a different progenitor channel than the cool DQs. While the temperature index can be used to distinguish these C-dominated objects from the cooler He-dominated DQ stars, their uniqueness suggests a special classification designation as "hot DQ" stars defined by the dominant presence of C II features in their optical spectra. There is certainly a precedent for having a special designation for this unique class of C-dominated white dwarfs. The PG 1159 stars merited their own special designation, rather than classifying them as DZ, DQ, or DO based upon whether O, C, or He features dominated their optical spectra. The designation PG 1159 is widely used to distinguish these unique, exotic objects from helium-dominated DO stars at lower temperatures. See Chapter 3 by P. Dufour for details of the hot DQ white dwarfs.

1.3
The Hot DA Stars

The total number of hot ($T_{\rm eff} > 20\,000$ K) DA white dwarfs has increased enormously, largely due to the Sloan Digital Sky Survey (SDSS; Eisenstein *et al.*, 2006a; York *et al.*, 2000) but also to smaller surveys such as the Supernova Progenitor SurveY (SPY) project (Koester *et al.*, 2001; Napiwotzki *et al.*, 2003), and the Hamburg–Schmidt (Hagen *et al.*, 1995; Homeier, 2003) and Montreal–Cambridge–Tololo surveys (Demers *et al.*, 1986; Lamontange *et al.*, 1997). Follow-up high quality ground-based spectroscopy of survey objects yield large samples of hot DA white dwarfs with precise temperatures and gravities. Only one of many such examples is the Koester *et al.* (2009) analysis of 615 DA white dwarfs from the SPY project.

When the Balmer lines are extremely weak due to ionization, it is difficult to determine accurate temperatures for the hottest DA white dwarfs. While the Lyman series can be used to measure $T_{\rm eff}$ and estimate $\log g$ in the far-ultraviolet with Orfeus (e. g., Dupuis *et al.*, 1998), the Hopkins Ultraviolet Telescope (HUT; e. g., Kruk *et al.*, 1997), and FUSE (Sahnow *et al.*, 2000), there are two different widely known discrepancies that plague the reliable determination of hot DA physical parameters: (1) an inability to fit all of the Balmer lines simultaneously with consistent atmospheric parameters (the so-called Napiwotzki effect; cf. Gianninas *et al.*, 2010, and references therein); and (2) the disagreement between the parameters derived from fitting optical spectra and those derived from fitting far-ultraviolet spectra (e. g., Finley *et al.*, 1997, and references therein). The Napiwotzki effect has been resolved by adding metals (not detected in the optical spectra) to the model atmospheres, which provides a mild back-warming effect. The fact that the analysis of far-ultraviolet spectra from the FUSE archive reveals a correlation between higher metallic abundances and instances of the Balmer line problem strongly supports this scenario (Gianninas *et al.*, 2010). However, the disagreement between parameters derived from optical and far-ultraviolet spectra remains.

A large fraction of the hot DA stars observed in the extreme- and far-ultraviolet have revealed trace abundances of numerous heavy elements which are presumably radiatively levitated against downward diffusion by radiative forces (Chayer *et al.*, 1995). Extreme-ultraviolet observations of hot DA white dwarfs have been particularly effective in revealing levitated trace metals in their atmospheres (Finley *et al.*, 1997). This occurs because the extreme-ultraviolet flux of a hot DA star can be strongly suppressed due to both the low opacity of the residual neutral hydrogen shortward of 300 Å, and the strong continuum absorption and heavy line blanketing in that same extreme-ultraviolet wavelength range due to any trace metal ion constituents that may be present in the photosphere. Finley *et al.* (1997) point out an extensive literature on extreme-ultraviolet analyses of hot DA white dwarfs including, for example, work by Kahn *et al.* (1984), Petre *et al.* (1986), Jordan *et al.* (1987), Paerels and Heise (1989), Barstow *et al.* (1993), Finley *et al.* (1993), Jordan *et al.* (1994), Vennes *et al.* (1994), Vennes *et al.* (1996), Wolff *et al.* (1996), and Marsh *et al.* (1997).

Absorption features due to C, N, O, Si, Fe, and Ni have been seen, and in one object, PG 1342+444, absorption lines of O VI are detected which had previously only been seen in the subluminous Wolf–Rayet planetary nebula nuclei and the PG 1159 stars (Barstow *et al.*, 2002). On the other hand, there is also a sizable number of hot DA stars that appear to be metal deficient since radiative forces theory (Chayer *et al.*, 1995) predicts that any metals present should be levitated.

There are now well over 100 known DA white dwarfs with $T_{\rm eff} > 60\,000$ K (e. g., Bergeron *et al.*, 1994; Eisenstein *et al.*, 2006a; Finley *et al.*, 1997; Homeier *et al.*, 1998; Kidder, 1991; Koester *et al.*, 2001, 2009; Liebert *et al.*, 2005; Limoges and Bergeron, 2010; Marsh *et al.*, 1997; Vennes, 1999; Vennes *et al.*, 1997; Voss, 2006). Table 1.3 lists a selection of these objects with the most reliable temperature estimates hotter than 70 000 K. Noteworthy among this sample are (i) a uniquely massive hot DA white dwarf, WD 0440–038, with $T_{\rm eff}$ estimates in the range 65 000–72 000 K and $\log g$ in the range 8.4–9.1, corresponding to a mass of $M_{\rm wd} \approx 0.9{-}1.3 M_\odot$ (Dupuis *et al.*, 2002; Finley *et al.*, 1997; Koester *et al.*, 2009; Vennes *et al.*, 1997); and (ii) the fact that the DA stars reach temperatures above 100 000 K, with the hottest currently known DA star, WD 0948+534, at $T_{\rm eff} \gtrsim 130\,000$ K (Liebert *et al.*, 2005). Thus, the hottest DA stars overlap in temperature with the hot DO stars and the PG 1159 stars. This overlap further strengthens the case for the existence of a separate evolutionary channel for the hot DA stars extending up to the domain of the H-rich central stars of planetary nebulae.

The large increase in the number of hot DA stars found in the SDSS has spawned new insights into the luminosity function of hot white dwarfs. A study by Krzesiński *et al.* (2009) used the new SDSS sample of hotter, fainter DA stars (Eisenstein *et*

Table 1.3 A selection of the hottest DA white dwarfs (T_{eff} > 70 000 K). This list was compiled from the literature with input and advice from J. Holberg (private communication) to select the most reliable temperature estimates obtained from the highest quality optical and far-ultraviolet spectra.

WD name	T_{eff} (K)	log g	Reference
0556−375[a]	$70\,275^{+8505}_{-3875}$	$7.37^{+0.13}_{-0.24}$	1
1248−278	70 798 ± 1305	7.23 ± 0.06	2, 3
1312−253[a]	71 153 ± 906	7.06 ± 0.04	2, 3
1312−253[a]	71 234 ± 891	7.02 ± 0.03	2, 3
0440−038[a]	71 545 ± 805	8.538	4
2244+031	72 000 ± 1600	7.78 ± 0.08	5
0440−038[a]	72 340 ± 2319	8.772 ± 0.085	6
0102−185	72 370 ± 1793	7.16 ± 0.08	7
0556−375[a]	72 800 ± 1800	7.58 ± 0.13	8
1547+015[a]	72 978 ± 2554	7.628 ± 0.111	6
1342+443[a]	74 130 ± 2202	7.84 ± 0.11	9[b]
1201−049	74 798 ± 944	7.475	4
1312−253[a]	75 463 ± 825	7.682 ± 0.027	10, 4
1547+015[a]	75 585 ± 561	7.612	4
0158−227	75 758 ± 1128	7.386	4
0630+200	75 792 ± 751	8.398 ± 0.050	6
1827+778	75 800 ± 610	7.68 ± 0.03	5
0616−084	76 320 ± 200	8.05 ± 0.15	11
0111−381	76 857 ± 746	7.367	4
1749+717	76 900 ± 550	7.56 ± 0.03	5
1622+323	77 166 ± 1839	7.838 ± 0.082	6
0441+467[a]	77 300 ± 3400	7.31 ± 0.14	12
0939+262	77 300 ± 1400	7.78 ± 0.06	12
0229−481	77 421 ± 2550	7.549 ± 0.064	13
1547+015[a]	78 101 ± 3409	7.49	9[b]
1342+443[a]	78 700 ± 2700	7.82 ± 0.11	12
1253+378[a]	79 900 ± 3600	6.61 ± 0.20	12
1253+378[a]	79 900 ± 3765	6.61 ± 0.21	9
2146−433	81 638 ± 1245	7.994	4
0441+467[a]	83 800 ± 1700	7.17 ± 0.13	12
1305−017	85 773 ± 992	7.800	4
0345+006	86 850 ± 3544	7.08 ± 0.12	7
1738+669[a]	$88\,010^{+2390}_{-2610}$	$7.79^{+0.14}_{-0.09}$	1
0500−156	94 488 ± 112	7.214	4
1738+669[a]	95 324 ± 1217	7.864 ± 0.035	6
0615+655	98 000 ± 5500	7.07 ± 0.15	5
2246+066	98 000 ± 1700	7.04 ± 0.05	5
0950+139	108 390 ± 16 786	7.39 ± 0.38	9
0948+534	136 762 ± 9805	7.222	9[b]

a Duplicate listing with independent parameter estimates.

b Revised parameters provided by J. Holberg (private communication).

1 Marsh *et al.* (1997); 2 Kidder (1991); 3 Koester *et al.* (2001); 4 Voss (2006); 5 Homeier *et al.* (1998); 6 Finley *et al.* (1997); 7 Limoges and Bergeron (2010); 8 Vennes *et al.* (1997); 9 Liebert *et al.* (2005); 10 Koester *et al.* (2009); 11 Vennes (1999); 12 Bergeron *et al.* (1994); 13 Bragaglia *et al.* (1995)

al., 2006a) to derive a hot white dwarf luminosity function that extends to the most luminous, hottest white dwarfs to date. This luminosity function encompasses DA and non-DA stars over the temperature range $\sim 25\,000\,\mathrm{K} > T_{\mathrm{eff}} > \sim 120\,000\,\mathrm{K}$. The cool end of their luminosity function connects with the hot end of previously determined SDSS white dwarf luminosity functions. By constructing separate DA and non-DA luminosity functions for the first time, Krzesinski *et al.* (2009) noted distinct differences between them. For example, they found a sudden drop in the non-DA luminosity function near $M_{\mathrm{bol}} = 2$, which they interpret as a transition of non-DA atmospheres into DA atmospheres during white dwarf evolution. The transition would occur as trace amounts of hydrogen float to the surface and give rise to H-features in the optical spectrum.

It is well known that the hottest DA stars are cooler than the hottest non-DA stars (Fontaine and Wesemael, 1987; Sion, 1986). Unless some fraction of the PG 1159 degenerates undergoes spectral evolution into hot DA stars when previously "hidden" hydrogen floats up to their surface, then the progenitors of the hottest DA stars represent a separate evolutionary channel. This would be contrary to the single channel scenario of Fontaine and Wesemael (1987). In the Hertzsprung–Russell (H–R) diagram, the hottest DA white dwarfs appear to connect up with the H-rich central stars of planetary nebulae. There is very likely a direct evolutionary connection.

1.3.1 DAO White Dwarfs

The DAO stars are hot DA stars that exhibit ionized helium in their optical spectra. They are characterized by surface temperatures in the range $50\,000\,\mathrm{K} < T_{\mathrm{eff}} < 100\,000\,\mathrm{K}$, low gravities with $\log g < 7.5$, and $\log[\mathrm{He/H}] = -2$ by number (Bergeron *et al.*, 1994; Good *et al.*, 2004, 2005; Napiwotzki, 1999). Searches for binarity as a means of explaining the presence of He II in a DA star as a composite spectrum have been largely negative (Good *et al.*, 2005). There are six DAOs in binaries including four out of 12 low mass DAOs that are in binaries. Thus, the majority of the low mass DAOs appear to *not* be in binaries and, hence, may be the descendants of extended horizontal branch progenitors. Because pure DA and DB stars have monoelemental atmospheres, the hybrid composition DAO stars offer insights into white dwarf spectral evolution.

There is typically a large discrepancy in the T_{eff} values determined for DAO white dwarfs from optical Balmer line spectra compared to temperatures derived from Lyman line spectra obtained with FUSE (see Table 1.4). This behavior echoes the Balmer line versus Lyman line temperature discrepancy noted by Barstow *et al.* (2003) for the hot ($T_{\mathrm{eff}} > 50\,000\,\mathrm{K}$) DA white dwarfs. The spectra of DAO stars are generally better fitted with models utilizing homogeneous, rather than stratified, atmospheres, which led Good *et al.* (2004) to suggest the possibility that homogeneous atmospheres are a better approximation of reality in low gravity DAO white dwarfs.

Table 1.4 The hottest DAO white dwarfs[a].

Name	Effective temperature (K) Balmer lines	Lyman lines
Abell 7	66 955	99 227
HS 0505+0112	63 227	120 000
PuWe 1	74 218	109 150
RE 0720−318	54 011	54 060
Ton 320	63 735	99 007
PG 0834+500	56 470	120 000
Abell 31	74 726	93 887
HS 1136+6646	61 787	120 000
Feige 55	53 948	77 514
PG 1210+533	46 338	46 226
LB2	60 294	87 662
HZ 34	75 693	87 004
Abell 39	72 451	87 965
RE 2013+400	47 610	50 487
DeHt 5	57 493	59 851
GD 561	64 354	75 627

a Data used in this table are from Good *et al.* (2004).

Virtually all of the DAO stars reveal heavy element absorption lines in their spectra. The metal abundances in the DAO stars appear to be generally higher than the metal abundances in their DA counterparts, although Good *et al.* (2004) noted little difference between the metal abundances of DAO stars and DA stars. Moreover, Good *et al.* (2005) found that the metal abundances of the lower gravity, lower temperature DAO stars and the higher gravity DAO stars differed little from each other despite the fact that these two groups of DAO white dwarfs presumably have arisen from different progenitor channels (see below). Good *et al.* (2005) also noted that the metal abundances of the DAO white dwarfs were not markedly different from those of the hot DA stars with metals. The observed DAO metal abundances were compared with the theoretical predictions of Chayer *et al.* (1995) as well as with the wind mass loss calculations of Unglaub and Bues (1998, 2000). Except for Si, none of the metal abundances exceeded the theoretical prediction after taking into account radiative levitation and weak mass loss.

The low masses of DAO white dwarfs would point toward evolution from stars that did not have sufficient mass to ignite core helium burning on the horizontal branch. Hence, the lower mass DAO stars cannot be the descendants of post-AGB evolution. It is possible that they are the descendants of the sdB and sdO subdwarf stars on the extended horizontal branch. On the other hand, the more massive DAO stars with $M_{\rm wd} > 0.5 M_\odot$ appear to represent an evolutionary channel connecting them to the AGB stars since DAO white dwarfs more massive than $0.5 M_\odot$ would have been massive enough for helium shell burning on the horizontal branch. In this scenario, the DAO stars could also be the progeny of H-rich planetary nebula central stars or even hybrid PG 1159 stars (containing some H) in which a "hidden" reservoir of hydrogen floats up to the surface. Recently, Gianninas *et al.* (2010) contended that the post-extreme horizontal branch evolution is no longer needed to explain the evolution for the majority of the DAO stars, and that the presence of metals might drive a weak stellar wind, which, in turn, could explain the presence of helium in DAO white dwarfs. Nevertheless, it is still not possible to definitively establish these different potential evolutionary links.

1.4 The PG 1159 Stars

The most exciting stellar discovery of the Palomar Green survey was a class of extremely hot, high luminosity degenerate objects known as the 1159 stars (Green *et al.*, 1986). Subsequently, large surveys (Palomar Green; Hamburg–Schmidt; Hamburg ESO, Wisotzki *et al.*, 1996, and, most recently, the SDSS) have uncovered the majority of the known PG 1159 stars.[1] The PG 1159 stars reveal spectra typically devoid of hydrogen. Instead, they are dominated by He II and highly ionized, high excitation carbon, especially a broad absorption trough in the region of 4670 Å comprising He II λ4686 and several high excitation C IV lines (Green *et al.*, 1986). Absorption lines of O VI were also detected in the optical ultraviolet (Sion *et al.*, 1985). All of these high excitation features implied very high effective temperatures. However, the first determinations of accurate surface temperatures and the first chemical abundances for these extremely hot objects had to await the development of a new generation of Non-Local Thermodynamic Equilibrium (NLTE) atmosphere codes with the requisite atomic data for millions of transitions (by Klaus Werner and his students; discussed below). Meanwhile, the SDSS made it possible to significantly increase the number of known PG 1159 stars.

Synthetic spectral analyses require studies of lines of successive ionization stages in order to evaluate the ionization equilibrium of a given chemical element. This provides a sensitive indicator of the effective temperature. Since stars with $T_{\rm eff}$ as high as 100 000 K have their Planckian peaks in the extreme ultraviolet wavelength range and the features are highly ionized, most of the metal lines are found in the ultraviolet range. Werner and his colleagues require high signal-to-noise and high resolution ultraviolet spectra for their NLTE analyses of white dwarfs. They utilized a number of spacecraft instruments including the Faint Object Spectrograph, Goddard High Resolution Spectrograph, Space Telescope Imaging Spectrograph, and Cosmic Origins Spectrograph onboard HST in order to acquire spectra of sufficient quality. They carried out state-of-the-art analyses of the hottest (pre-)white dwarfs

1) The most recent and the only discovery within the last 10 years, besides those from the SDSS, is HE 1429−1209, which was discovered as a white dwarf candidate in the Hamburg ESO survey and confirmed as a PG 1159 star by the SPY project (Werner *et al.*, 2004).

by means of NLTE model atmospheres, including the metal-line blanketing of all elements from hydrogen to nickel (Rauch and Werner, 2010).

The spectral analysis of the PG 1159 stars revealed a range of temperatures of $T_{\rm eff} = 75\,000$–$200\,000$ K and gravities of $\log g = 5.5$–8.0 (Dreizler *et al.*, 1994; Rauch *et al.*, 1991; Werner *et al.*, 1996). Prior to the SDSS, only 28 PG 1159 stars were known. From the empirically derived ranges of their parameters, it was obvious that the position of some of the PG 1159 stars (specifically the lower gravity members, subtype lgE Werner, 1992) in the H–R diagram overlapped the hot central stars of planetary nebulae. The remainder of the PG 1159 stars are more compact objects with higher (white dwarf) surface gravities (subtype A or E). Thus, the PG 1159 stars appear to be evolutionary transition objects between the hottest post-AGB and white dwarf phases. Figure 1.3 shows a summary plot of log g versus log $T_{\rm eff}$ from the most recent analyses of the PG 1159 and hot DO white dwarfs. For comparison, evolutionary tracks by Althaus *et al.* (2009) are also plotted.

Until very recently, Fe lines had never been detected in PG 1159 stars even though Fe lines are seen in some hot DA, DO, and DAO white dwarfs. In the far-ultraviolet, features due to Fe VII would be the expected indicators for the presence of Fe, but only if the effective temperature is not so high that the population of Fe VII is far too depleted by ionization. Using the plethora of far-ultraviolet

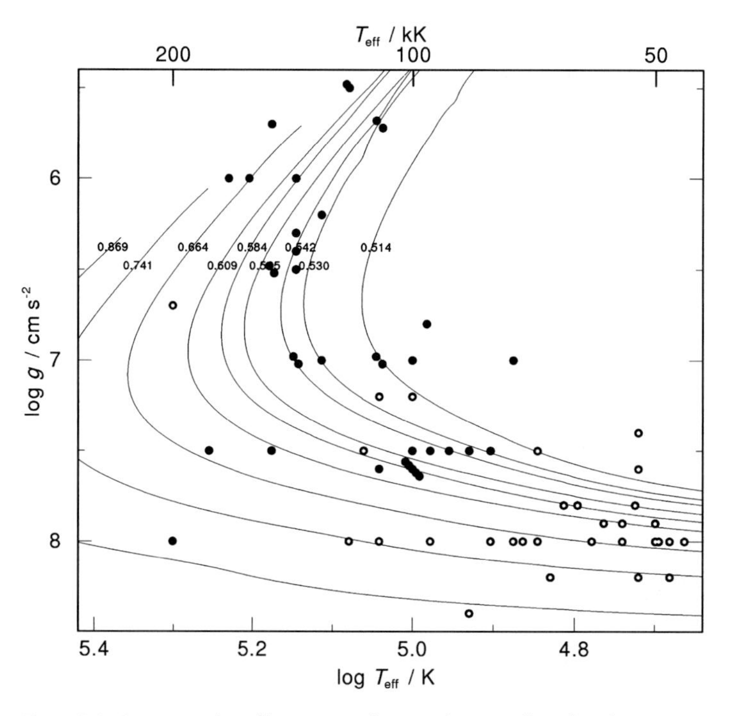

Figure 1.3 Summary plot of log g versus log $T_{\rm eff}$ showing all analyzed PG 1159 stars (filled circles) and hot DO white dwarfs (unfilled circles), with DO white dwarf cooling tracks by Althaus *et al.* (2009).

transitions lying in the FUSE range, Werner *et al.* (2010) have detected Fe features not only in the cooler PG 1159 stars with $T_{\text{eff}} < 150\,000$ K, but also among objects in the hottest subset of PG 1159 stars with temperatures between 150 000 K and 200 000 K, including RX J2117.1+3412, K116, NGC 246, H1504+65, and Longmore 4 (which has revealed evidence of episodic mass ejection, as shown in Figure 1.4; Werner *et al.*, 1992). Among the cooler subset of PG 1159 stars, Fe VIII lines are detected at solar abundance in FUSE spectra, while in the hotter subset, Fe X is the detected species and the analysis of abundances are in progress (Werner, 2010; Werner *et al.*, 1992). Their analyses yielded a solar iron abundance for these stars. These hottest objects are among the most massive PG 1159 stars (0.71–0.82$M_\odot$), while those objects revealing the strongest Fe deficiency are associated with a lower mass range (0.53–0.56$M_\odot$). Nonetheless, the evolutionary significance, if any, of the presence of Fe in solar abundance in some PG 1159 stars versus PG 1159 stars that appear to be Fe-deficient remains unclear.

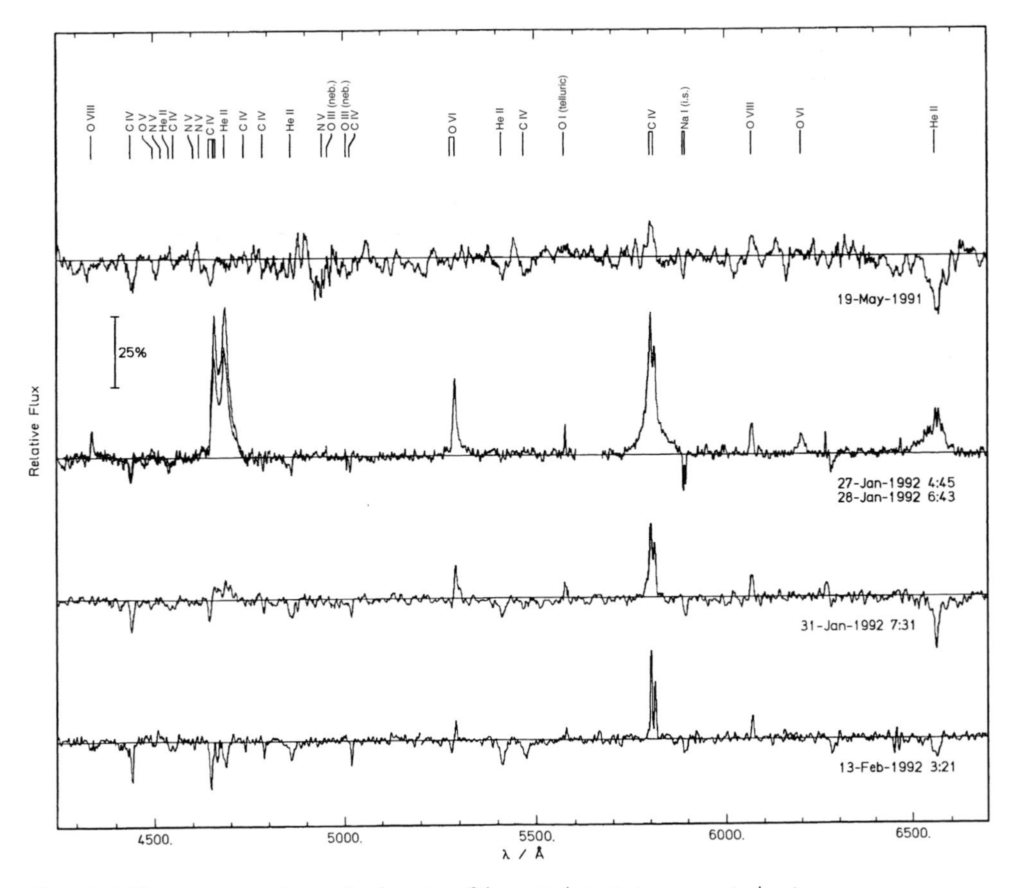

Figure 1.4 Time sequence of normalized spectra of the central star in Longmore 4, showing (from top to bottom) the appearance and rapid decline of a dramatic emission line phase linked to a mass ejection event. From Werner *et al.* (1992), reproduced with permission. © ESO.

It is now widely believed that the hydrogen-deficiency in extremely hot post-AGB stars of spectral class PG 1159 is probably caused by a (very) late helium-shell flash or an AGB final thermal pulse (Iben, 1984) that consumes the hydrogen envelope, exposing the usually-hidden intershell region. Thus, the photospheric elemental abundances of these white dwarfs offer insights into the details of nuclear burning and mixing processes in the precursor AGB stars. Werner *et al.* (2008) compared predicted elemental abundances to those determined by quantitative spectral analyses performed with advanced NLTE model atmospheres. A good qualitative and quantitative agreement is found for many elemental species (He, C, N, O, Ne, F, Si, Ar), but discrepancies for others (P, S, Fe) point at shortcomings in stellar evolution models for AGB stars. PG 1159 stars appear to be the direct progeny of [WC] Wolf–Rayet stars (Werner *et al.*, 2007, 2008), a possibility first suggested by Sion *et al.* (1985) when the same high excitation O VI absorption features detected in the PG 1159 stars were also seen in the optical ultraviolet spectra of O VI planetary nebula nuclei.

1.5 DO White Dwarfs

The DO white dwarfs are hot helium-rich white dwarfs that populate the white dwarf cooling sequence from the hot beginning ($T_{eff} = 120\,000$ K) down to 45 000 K. The optical spectra of hot DO stars covering this range of T_{eff}, and newly discovered in the SDSS, are displayed in Figure 1.5.

At $T_{eff} < 45\,000$ K, the helium-rich cooling sequence is interrupted by the so-called "DB gap" (Liebert *et al.*, 1986) in which, prior to the SDSS, no helium-rich white dwarfs were found down to effective temperatures of 30 000 K. Dreizler and Werner (1996, 1997) determined effective temperatures and surface gravities of all 18 known DO white dwarfs. The DO stars have a nearly pure helium atmosphere, with only relatively few showing weak metal lines. Their surface composition is controlled by the competition between gravitational diffusion and radiative levitation as well as possible weak mass loss. The initial comparisons between theoretical predictions for white dwarfs and the observed abundances were obtained by Fontaine and Michaud (1979) and Vauclair *et al.* (1979), who took into account radiative levitation. These processes are also invoked to explain the transition from the possible progenitors, the hydrogen deficient, C- and O-rich PG 1159 stars (see Dreizler and Heber, 1998, as well as Section 1.4 above, for an introduction to these objects). In this scenario, heavier elements in the atmosphere are depleted and only traces of metals can be kept in the helium rich atmosphere as long as the radiative levitation is sufficiently strong due to high effective temperatures. Finally, at the hot end of the DB gap, enough of the remaining traces of hydrogen floats up to cover the helium rich envelope with a thin hydrogen rich layer. Even though this scenario is able to qualitatively explain the evolution of hot helium rich white dwarfs, these processes are far from understood quantitatively. For example, observed metal abundances (see Table 1.5) do not fit theoretical predictions (Chayer

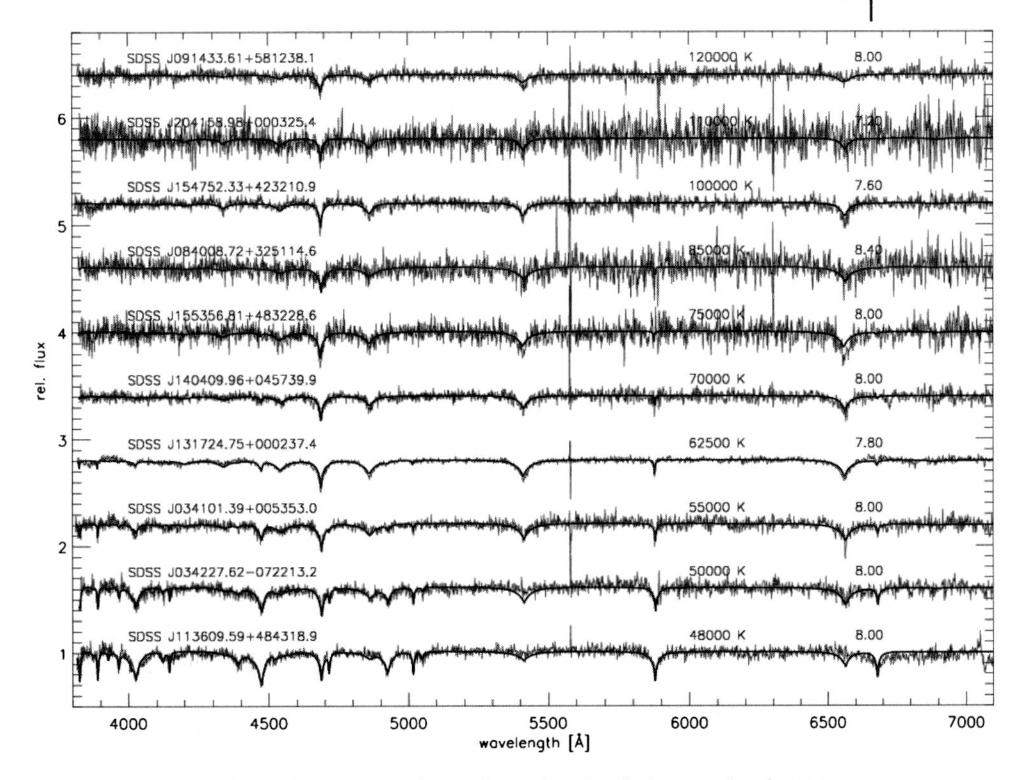

Figure 1.5 Normalized optical spectra (gray lines) of DO white dwarfs discovered in the SDSS, along with model atmospheres (black lines), ordered by decreasing effective temperature. From Hügelmeyer *et al.* (2005), reproduced with permission. © ESO.

Table 1.5 DO white dwarf parameters[a].

Star	T_{eff} (kK)	log *g*	Metal abundances[b]				
			C	N	O	Fe	Ni
PG 1034+001	100	7.5	−5.0	−3.2	−4.1	−5.0	< −5.0
PG 0108+101	95	7.5	−2.0	< −7.3	−3.3	−4.3	−4.3
RE 0503−289	70	7.5	−2.3	−4.8	−3.3	< −6.0	−5.0
HS 0111+0012	65	7.8	−3.0	< −8.5		< −6.0	< −6.0
HZ 21	53	7.8	< −6.0	−5.0	< −6.0		
HD 149 499 B	50	8.0	< −6.0	< −6.0	< −6.0		

a Data used in this table are from Dreizler (1999).
b Logarithm of number ratios relative to He from homogeneous model atmosphere fits.

et al., 1995; Dreizler and Werner, 1996, 1997). Whether this is due to the rather poor observational data, lack of adequate models, or a fundamental problem with the transition scenario itself has so far not been determined.

From the analysis of PG 1159 stars, it is known that the nitrogen abundance varies at least by three orders of magnitude (Dreizler and Heber, 1998). Are the PG 1159 stars the progenitors of the DO white dwarfs? Since nitrogen is destroyed in triple alpha reactions, it is not surprising that the PG 1159 stars are generally nitrogen-poor. Hence, it is possible that those DO stars with low nitrogen abundance could be the direct descendants of PG 1159 stars with low nitrogen abundance.

Among the DO white dwarfs are exotic objects with ultra-high excitation ion (uhei) lines in their optical spectra. These objects were first discovered prior to the SDSS. They show absorption lines of O VII and N VII around 3888 Å, at 5673, and around 6086 as well as O VIII at 4340 , 4658 , and 6064/6068 Å. In a new SDSS uhei DO white dwarf, SDSS J025403.75+005854.4, features due to Ne IX are also present (Krzesiński *et al.*, 2004). The He II lines in these uhei DO white dwarfs are very strong and cannot be fitted with the latest NLTE models. The effective temperature required to excite the detected ions exceeds 500 000 K. The evolutionary track of a massive ($1.2 M_{\odot}$) post-AGB remnant carries it to effective temperatures as high as 700 000 K on very short time scales (Paczyński, 1970). However, Werner *et al.* (1995) have argued that the uhei features in DO white dwarfs cannot be photospheric because the He II lines would fade completely at such high temperatures. Instead, they proposed an hypothesis of optically thick and hot stellar winds, based on the triangular shape of the line profile. Furthermore, the lines are blue-shifted, which favors the assumption of an expanding envelope.

Related to these uhei DO stars is KPD 0005+5106, one of the hottest known DO stars. Werner *et al.* (2007, 2008) discovered highly ionized photospheric metals, such as Ne VIII and Ca X, requiring extremely high temperature, much higher than even previous analyses that yielded $T_{\rm eff} \sim 120\,000$ K. More recently, Wassermann *et al.* (2010) reported the detection of Si VII, S VII, and Fe X in this star. The NLTE analysis of the metals and the He II line profiles yield $T_{\rm eff} = 200\,000$ K with $\log g = 6.7$. The abundances of metals are in the range of 0.7 to 4.3 times solar with an upper limit to any hydrogen present of < 0.034 solar. Remarkably, these new analyses of KPD 0005+5106 reveal it to be much hotter than the next hottest DO white dwarf. In the H–R diagram, it stands alone, far to the left and at higher luminosity than any of the other DO stars. At its high $T_{\rm eff}$ and luminosity, it is likely that the chemical abundances are probably affected by a stellar wind. Thus, diffusion and radiative levitation may not be important factors in controlling the surface abundances. Furthermore, Wassermann *et al.* (2010) found that the chemical abundances of KPD 0005+5106 closely resemble the abundances seen in R Coronae Borealis stars. Since the R Cor Bor objects are widely held to be the product of binary mergers (e. g., Han, 1998; Webbink, 1984), it may imply that KPB 0005+5106 is itself the product of such a merger and, hence, is the evolved progeny of an R Cor Bor giant. If true, such a connection would imply that the surface abundances of KPD 0005+5106 are chemical relics of the progenitor giant and, thus, not controlled by diffusion and radiative levitation. If this interpretation is correct, then KPD 0005+5106 would represent a new evolutionary channel producing DO white dwarfs distinct from the evolutionary channel connecting the PG 1159 stars to the DO stars.

1.6 DB White Dwarfs

The DB stars contain helium to a degree of purity not seen in any other astronomical objects. Even at high signal-to-noise ratio, many spectra only exhibit the absorption lines of He I. If the DB star has accreted metals from a debris disk, comets, or the interstellar medium, then they are classified DBZ, or DBAZ if hydrogen is present (see below). Some hot DB stars exhibit atomic carbon in the far ultraviolet. The best example is the pulsating hot DB star GD 358 (Sion *et al.*, 1989). It remains unclear if the DBA stars have accreted their hydrogen or if it is primordial and a result of convective mixing. The DB white dwarfs by consensus are the progeny of extremely hydrogen deficient post-AGB stars (e. g., see Althaus *et al.*, 2005, 2009, and references therein). The DB cooling sequence extends from the hottest DB stars like GD 358 and PG 0112+122, down to the cooler DB white dwarfs (below 20 000 K), and extending downward to 12 000 K, at which point envelope convection has deepened substantially and the He I lines become undetectable. The distribution by number of the coolest DO stars to the hottest DB stars (30 000 to 45 000 K) is interrupted by the DB gap (Liebert *et al.*, 1986). Prior to the SDSS, within this very wide range of temperature, no objects with H-deficient atmospheres were known to exist.

Now, however, the large number of new white dwarfs discovered in the SDSS (Eisenstein *et al.*, 2006a) has led to the firm placement of no less than 26 DB stars within the DB gap (Hügelmeyer and Dreizler, 2009). This raises suspicions that the DB gap was not a real feature of the white dwarf temperature distribution. On the other hand, there is still a deficit of a factor of 2.5 in the DA/non-DA ratio within the gap (Eisenstein *et al.*, 2006b). However, many other objects whose status is questionable (e. g., DAO, DBA, DAB, masquerading composite DA + DB/DO systems) may alter or eventually erase this deficit. Some of these objects are found within the gap, while others are seen near the gap edges. There is also the complicating factor of circumstellar accretion of hydrogen, and the role played by radiative levitation and weak winds in this temperature interval.

If every DA white dwarf evolving through the DB gap turned into a DB, then there should be a significant spike seen in the number of DB stars at the red edge of the DB gap, which is not observed. This is in contrast to the strong signature of convective mixing and dilution that changes (significantly lowers) the DA to non-DA ratio at lower temperatures, $T_{\mathrm{eff}} < 12\,000$ K (Sion, 1984). Even if the transformation of DA stars into DB stars takes place at 30 000 K and lower, when a sufficiently thin hydrogen layer is convectively mixed downward and diluted by the deepening helium convection zone, this only affects the DA stars with extremely low hydrogen layer masses ($< 10^{-15}\,M_{wd}$), which amounts to approximately 10% of all DA stars cooling through the 45 000 to 30 000 K interval (Eisenstein *et al.*, 2006b). This is contrary to the original contention of Fontaine and Wesemael (1987) that all DA stars should have ultra-thin layers ($< 10^{-12}\,M_{wd}$). Rather, it appears that the fraction of hot DA white dwarfs that transform into non-DA white dwarfs is on the order of 10% of all DA stars.

The problem of whether a DB gap exists is complicated by the known existence of several peculiar DAB, DBA, or DAO stars believed to lie in the 30 000–45 000 K range that (1) show evidence of spectrum variability and/or (2) do not fit atmosphere models, whether homogeneous (completely mixed) in H and He throughout the atmosphere or stratified with the hydrogen all in a very thin upper layer. Several other white dwarfs in or near the DB gap also have peculiar spectra. For example, PG 1210+533, with $T_{eff} = 45\,000$ K, exhibits line variability of H, He I, and He II, probably modulated by rotation (Bergeron *et al.*, 1994). Also, GD 323, with $T_{eff} = 30\,000$ K, is a DAB star with a variable spectrum that cannot be fit completely successfully by either homogeneous or stratified model atmospheres (Koester *et al.*, 2009; Liebert *et al.*, 1984; Pereira *et al.*, 2005). Additional examples include HS 0209+0832, with $T_{eff} = 36\,000$ K and a 2% helium abundance (Jordan *et al.*, 1993), and PG 1603+432, with $T_{eff} = 35\,000$ K and a 1% helium abundance (Vennes *et al.*, 2004). The existence of these systems adds to the mounting evidence that the DB gap is not real, as increasing numbers of He-rich stars are being discovered within and near its boundaries.

1.6.1 DBA White Dwarfs

The DBA stars have strong lines of He I and weaker Balmer lines, hydrogen-to-helium ratios (by number) in the range $N(\mathrm{H})/N(\mathrm{He}) \sim 10^{-5}$ to 10^{-3}, and, for the most part, an effective temperature below 20 000 K (Shipman *et al.*, 1987). Hence, they cluster at the low end of the DB temperature distribution. The DBA white dwarfs were previously thought to comprise roughly 20% of He-rich white dwarfs between 12 500 and 20 000 K (Shipman *et al.*, 1987). However, more recent surveys have dramatically changed this picture. The SPY project yielded a sample of 71 helium-rich degenerates, of which six were new DBA discoveries and 14 were DB stars reclassified as DBA due to the detection of hydrogen lines (Voss *et al.*, 2007). In all, 55% of their SPY samples were DBA stars. This is a factor of almost three times higher than the fraction of DBA stars first estimated by Shipman *et al.* (1987).

It remains unclear if the DBA stars have accreted their hydrogen or if the small H mass was originally primordial, diluted by convection, and then floated back to the surface as a result of convective mixing. This large fraction of DBA stars, coupled with the total hydrogen masses estimated for the DBA stars suggests the possibility that DB white dwarfs, as they cool, accrete interstellar hydrogen, thus raising their hydrogen mass to the point at which a DBA star appears. This channel for forming DBA white dwarfs was favored if the DB gap was real because DB white dwarfs could be masquerading as DA stars in the DB gap with only a thin hydrogen layer

($< 10^{-15} M_{wd}$). This layer could be mixed away and diluted, resulting in a DB star appearing at the cool end of the DB gap. It now appears that there is no DB gap. Hence, this constraint on the H-layer mass is no longer relevant. While interstellar accretion of hydrogen cannot be easily dismissed, it may still prove plausible that the hydrogen is accreted from volatile-rich debris or comets. The fact that there are DBAZ and DBZA stars with accreted calcium may support this scenario.

1.7 Hot DQ White Dwarfs

The discovery of hot DQ white dwarfs with carbon-dominated atmospheres (Dufour *et al.*, 2007, 2008, also see Chapter 3) in the SDSS Data Release 4 sample has raised new questions about white dwarf formation channels. These objects are distinctly different from the cooler, normal DQ stars, which have helium-dominated atmospheres and carbon abundances of $\log N(\mathrm{C})/N(\mathrm{He}) \sim 10^{-5}$ by number, with the highest carbon abundance measured for ordinary DQ stars being $\log N(\mathrm{C})/N(\mathrm{He}) \sim 10^{-3}$. The hot DQs all have effective temperatures between $\sim$ 18 000 K and 24 000 K. Their surface compositions are completely dominated by carbon, with no evidence of H or He I in their optical spectra. Their surface gravities are typically $\log g = 8$, with one object (SDSS J142625.70+575218.4) having a gravity near $\log g = 9$ (Dufour *et al.*, 2008). Their optical spectra contain numerous absorption lines of C II, which is the hallmark spectroscopic signature of the hot DQs. The strongest transitions of C II are at 4267, 4300, 4370, 4860, 6578, and 6583 Å.

Despite extensive searches of the vast SDSS sample, no carbon-dominated DQ stars have been found with T_{eff} higher than the hottest hot DQ white dwarf, at 24 000 K. Based upon this absence of hotter carbon-dominated DQ stars, it is quite possible that these objects appear helium-dominated at higher temperatures, but with very low mass helium layers that could be effectively mixed and diluted in the carbon-rich convection zone that forms and deepens due to carbon recombination as the hot DQ star cools (Dufour *et al.*, 2008). However, the helium layer would have to be thin in order to be hidden from spectroscopic detection as the hot DB transforms into a hot DQ star. Adding to the puzzle posed by the hot DQ stars, Dufour *et al.* (2009) point out an exceptionally high fraction of hot DQ stars with high magnetic fields ($\sim$ 40–60% among the hot DQ white dwarfs, compared with $\sim$ 10% for the sample of nearby white dwarfs of all types; Liebert *et al.*, 2003).

It seems likely that these objects could be cooled-down descendants of stripped carbon core objects like H 1504+65 (Nousek *et al.*, 1986). They could have experienced a late thermal pulse that eliminated most of the helium, a phenomenon similar to the one that is generally believed to explain the existence of other hydrogen deficient stars (Werner and Herwig, 2006). If the hot DQ white dwarfs are cooled versions of stripped carbon-oxygen cores like H 1504+001, then this would be an entirely new evolutionary channel.

1.8
Conclusion

It is clear that most of the progress achieved in understanding the competing physical processes in hot white dwarf envelopes, the spectral evolution of hot white dwarfs, and the identification of the white dwarf progenitor channels has arisen directly from an interactive combination of synthetic spectral abundances studies via space- and ground-based spectroscopy, with studies of nuclear astrophysics and thermal instabilities via AGB and post-AGB stellar evolutionary sequences including mass loss. A major triumph has been the successful prediction of surface abundances in hot white dwarfs from born-again thermal pulse models and AGB thermal pulse models. This has led to an agreement between the observed surface abundances from synthetic spectral analyses of high gravity post-AGB stars (PG 1159 and subluminous Wolf–Rayet planetary nebula nuclei) and the theoretical intershell abundances of their double shell-burning AGB progenitors (Werner and Herwig, 2006). That pre-white dwarf, post-AGB surface abundances shed light on nuclear astrophysical processes deep inside the intershell layers of the progenitor AGB star is all the more remarkable.

While episodic mass ejection and winds are directly observed in post-AGB stars, including planetary nebula nuclei, evolving on the plateau and knee portions of pre-white dwarf evolutionary tracks, it remains unknown whether theoretically-predicted weak wind mass loss and ion-selective winds exist along the white dwarf cooling tracks (Unglaub, 2008). Yet, along the white dwarf cooling tracks at $T_{\mathrm{eff}} < 50\,000$ K, where the most marked disagreement between observed and predicted surface abundances occur, weak winds may not exist. Complicating this scenario is the possible interplay of accretion with diffusion and radiative levitation at $T_{\mathrm{eff}} > 20\,000$ K. At $T_{\mathrm{eff}} > 50\,000$ K, the decreasing abundances of helium and metals are consistent with the combined effects of diffusion, radiative forces, and weak wind outflow. This wind outflow has not yet been detected and associated mass loss rates are unknown (Unglaub, 2008).

The future of research on hot white dwarfs will undoubtedly involve ever more sophisticated, multidimensional, quasistatic and hydrodynamic evolutionary calculations with more realistic mass loss prescriptions, and a greatly expanded database of atomic data and model atoms as well as the continued incorporation of diffusion and other envelope physical process into NLTE model atmosphere codes. These inevitable developments, along with the anticipated quantum leap in the number of white dwarfs with measured accurate parallaxes obtained by the Gaia Mission (e. g., Cacciari, 2009), will lead to an ever more detailed knowledge of white dwarf formation channels and the envelope physical processes governing and controlling spectral evolution down the white dwarf cooling sequence.

An added windfall is the availability of ultra-sophisticated model atmosphere codes (e. g., TLUSTY/SYNSPEC[2], Hubeny, 1988; Hubeny and Lanz, 1995; German Astrophysical Virtual Observatory[3]) to a widening circle of investigators, which

2) http://nova.astro.umd.edu (15 June 2011)
3) http://www.g-vo.org (15 June 2011)

should also serve to enhance the quantitative analyses shedding light on the formation and spectral evolution of hot degenerate stars.

Acknowledgments

It is my pleasure to thank Jay Holberg for useful discussions about hot DA stars and Patrick Dufour for discussions regarding hot DQ stars. I would also like to thank Klaus Werner for providing temperature data on hot non-DA stars in advance of publication. This work was supported by NSF grant AST1008845, and in part by NSF grant AST807892, both for Villanova University.

References

Allende Prieto, C., Hubeny, I., and Smith, J.A. (2009) *Mon. Not. R. Astron. Soc.*, **396**, 759.

Althaus, L.G., Serenelli, A.M., Panei, J.A., Córsico, A.H., García-Berro, E., and Scóccola, C.G. (2005) *Astron. Astrophys.*, **435**, 631.

Althaus, L.G., Panei, J.A., Miller Bertolami, M.M., García-Berro, E., Córsico, A.H., Romero, A.D., Kepler, S.O., and Rohrmann, R.D. (2009) *Astrophys. J.*, **704**, 1605.

Barstow, M.A. *et al.* (1993) *Mon. Not. R. Astron. Soc.*, **264**, 16.

Barstow, M. A., Good, S. A., Holberg, J. B., Burleigh, M. R., Bannister, N. P., Hubeny, I., and Napiwotzki, R. (2002) *Mon. Not. R. Astron. Soc.*, **330**, 425.

Barstow, M.A., Good, S.A., Burleigh, M.R., Hubeny, I., Holberg, J.B., and Levan, A.J. (2003) *Mon. Not. R. Astron. Soc.*, **344**, 562.

Bergeron, P., Wesemael, F., Beauchamp, A., Wood, M.A., Lamontagne, R., Fontaine, G., and Liebert, J. (1994) *Astrophys. J.*, **432**, 305.

Bragaglia, A., Renzini, A., and Bergeron, P. (1995) *Astrophys. J.*, **443**, 735.

Cacciari, C. (2009) *Mem. Soc. Astron. Ital.*, **80**, 97.

Chayer, P., Fontaine, G., and Wesemael, F. (1995) *Astrophys. J. (Suppl.)*, **99**, 189.

Demers, S., Beland, S., Kibblewhite, E.J., Irwin, M.J., and Nithakorn, D.S. (1986) *Astron. J.*, **92**, 878.

Dreizler, S. (1999) *Astron. Astrophys.*, **352**, 632.

Dreizler, S. and Heber, U. (1998) *Astron. Astrophys.*, **334**, 618.

Dreizler, S. and Werner, K. (1996) *Astron. Astrophys.*, **314**, 217.

Dreizler, S. and Werner, K. (1997) in *White Drafts. Astrophysics and Space Science Library*. Proceedings of the 10th European Workshop on White Dwarfs, vol. 214 (eds J. Isern, M. Hernanz, and E. Gracia-Berro), Kluwer Academic Publishers, Dordrecht, p. 213.

Dreizler, S., Werner, K., Jordan, S., and Hagen, H. (1994) *Astron. Astrophys.*, **286**, 463.

Dreizler, S., Werner, K., and Heber, U. (1995) in *White Dwarfs. Lecture Notes in Physics*, vol. 443 (eds D. Koester and K. Werner), Springer-Verlag, Berlin, p. 160.

Dufour, P., Liebert, J., Fontaine, G., and Behara, N. (2007) *Nature*, **450**, 522.

Dufour, P., Fontaine, G., Liebert, J., Schmidt, G.D., and Behara, N. (2008) *Astrophys. J.*, **683**, 978.

Dufour, P., Liebert, J., Swift, B., Fontaine, G., and Sukhbold, T. (2009) *J. Phys. Conf. Ser.*, **172**, 012012.

Dupuis, J., Vennes, S., Chayer, P., Hurwitz, M., and Bowyer, S. (1998) *Astrophys. J. Lett.*, **500**, L45

Dupuis, J., Vennes, S., and Chayer, P. (2002) *Astrophys. J.*, **580**, 1091.

Eisenstein, D.J. *et al.* (2006a) *Astrophys. J. (Suppl.)*, **167**, 40.

Eisenstein, D.J. *et al.* (2006b) *Astron. J.*, **132**, 676.

Finley, D.S., Jelinsky, P., Dupuis, J., and Koester, D. (1993) *Astrophys. J.*, **417**, 259.

Finley, D.S., Koester, D., and Basri, G. (1997) *Astrophys. J.*, **488**, 375.

Fontaine, G. and Michaud, G. (1979) *Astrophys. J.*, **231**, 826.

Fontaine, G. and Wesemael, F. (1987) *Second Conference on Faint Blue Stars*. Proceedings of IAU Colloq. 95 (eds A.G.D. Philip, D.S. Hayes, and J.W. Liebert), L. Davis Press Inc, Schenectady, NY, p. 319.

Gianninas, A., Bergeron, P., Dupuis, J., and Ruiz, M.T. (2010) *Astrophys. J.*, **720**, 581.

Good, S.A., Barstow, M.A., Holberg, J.B., Sing, D.K., Burleigh, M.R., and Dobbie, P.D. (2004) *Mon. Not. R. Astron. Soc.*, **355**, 1031.

Good, S.A., Barstow, M.A., Burleigh, M.R., Dobbie, P.D., Holberg, J.B., and Hubeny, I. (2005) *Mon. Not. R. Astron. Soc.*, **363**, 183.

Green, R.F., Schmidt, M., and Liebert, J. (1986) *Astrophys. J. (Suppl.)*, **61**, 305.

Greenstein, J.L. and Sargent, A.I. (1974) *Astrophys. J. (Suppl.)*, **28**, 157.

Hagen, H.-J., Groote, D., Engels, D., and Reimers, D. (1995) *Astron. Astrophys. (Suppl.)*, **111**, 195.

Han, Z. (1998) *Mon. Not. R. Astron. Soc.*, **296**, 1019.

Homeier, D. (2003) in *NATO Science Series II – Mathematics, Physics and Chemistry, White Dwarfs*, vol. 105 (eds D. de Martino, R. Silvotti, J.-E. Solheim, and R. Kalytis), Kluwer Academic Publisher, Dordrecht, p. 371.

Homeier, D., Koester, D., Hagen, H.-J., Jordan, S., Heber, U., Engels, D., Reimers, D., and Dreizler, S. (1998) *Astron. Astrophys.*, **338**, 563.

Hubeny, I. (1988) *Comput. Phys. Commun.*, **52**, 103.

Hubeny, I. and Lanz, T. (1995) *Astrophys. J.*, **439**, 875.

Hügelmeyer, S.D. and Dreizler, S. (2009) *J. Phys. Conf. Ser.*, **172**, 012048.

Hügelmeyer, S.D., Dreizler, S., Werner, K., Krzesiński, J., Nitta, A., and Kleinman, S.J. (2005) *Astron. Astrophys.*, **442**, 309.

Iben, I., Jr. (1984) *Astrophys. J.*, **277**, 333.

Iben, I., Jr., Kaler, J.B., Truran, J.W., and Renzini, A. (1983) *Astrophys. J.*, **264**, 605.

Jordan, S., Koester, D., Wulf-Mathies, C., and Brunner, H. (1987) *Astron. Astrophys.*, **185**, 253.

Jordan, S., Heber, U., Engels, D., and Koester, D. (1993) *Astron. Astrophys.*, **273**, L27.

Jordan, S., Wolff, B., Koester, D., and Napiwotzki, R. (1994) *Astron. Astrophys.*, **290**, 834.

Kahn, S.M., Wesemael, F., Liebert, J., Raymond, J.C., Steiner, J.E., and Shipman, H.L. (1984) *Astrophys. J.*, **278**, 255.

Kidder, K.M. (1991) Ph.D. Thesis, University of Arizona.

Koester, D. *et al.* (2001) *Astron. Astrophys.*, **378**, 556.

Koester, D., Voss, B., Napiwotzki, R., Christlieb, N., Homeier, D., Lisker, T., Reimers, D., and Heber, U. (2009) *Astron. Astrophys.*, **505**, 441.

Kruk, J.W., Kimble, R.A., Buss, R.H., Jr., Davidsen, A.F., Durrance, S.T., Finley, D.S., Holberg, J.B., and Kriss, G.A. (1997) *Astrophys. J.*, **482**, 546.

Krzesiński, J., Nitta, A., Kleinman, S.J., Harris, H.C., Liebert, J., Schmidt, G., Lamb, D.Q., and Brinkmann, J. (2004) *Astron. Astrophys.*, **417**, 1093.

Krzesiński, J., Kleinman, S.J., Nitta, A., Hügelmeyer, S., Dreizler, S., Liebert, J., and Harris, H. (2009) *Astron. Astrophys.*, **508**, 339.

Lamontange, R., Wesemael, F., Fontaine, G., Demers, S., Bergeron, P., Irwin, M.J., and Kunkel, W.E. (1997) in *White Dwarfs. Astrophysics and Space Science Library*, vol. 214 (eds J. Isern, M. Hernanz, and E. Gracia-Berro), Kluwer Academic Publishers, Dordrecht, p. 143.

Liebert, J., Wesemael, F., Sion, E.M., and Wegner, G. (1984) *Astrophys. J.*, **277**, 692.

Liebert, J., Wesemael, F., Hansen, C.J., Fontaine, G., Shipman, H.L., Sion, E.M., Winget, D.E., and Green, R.F. (1986) *Astrophys. J.*, **309**, 241.

Liebert, J., Bergeron, P., and Holberg, J.B. (2003) *Astron. J.*, **125**, 348.

Liebert, J., Bergeron, P., and Holberg, J.B. (2005) *Astrophys. J. (Suppl.)*, **156**, 47.

Limoges, M.-M. and Bergeron, P. (2010) *Astrophys. J.*, **714**, 1037.

Marsh, M.C. *et al.* (1997) *Mon. Not. R. Astron. Soc.*, **287**, 705.

Napiwotzki, R. (1999) *Astron. Astrophys.*, **350**, 101.

Napiwotzki, R. and Schoenberner, D. (1991) *Astron. Astrophys.*, **249**, L16.

Napiwotzki, R. *et al.* (2003) *The Messenger*, **112**, 25.

Nousek, J.A., Shipman, H.L., Holberg, J.B., Liebert, J., Pravdo, S.H., White, N.E., and Giommi, P. (1986) *Astrophys. J.*, **309**, 230.
Paczyński, B. (1970) *Acta Astron.*, **20**, 47.
Paerels, F.B.S. and Heise, J. (1989) *Astrophys. J.*, **339**, 1000.
Pereira, C., Bergeron, P., and Wesemael, F. (2005) *Astrophys. J.*, **623**, 1076.
Petre, R., Shipman, H.L., and Canizares, C.R. (1986) *Astrophys. J.*, **304**, 356.
Rauch, T., Heber, U., Hunger, K., Werner, K., and Neckel, T. (1991) *Astron. Astrophys.*, **241**, 457.
Rauch, T. and Werner, K. (2010) *The Impact of HST on European Astronomy*. Astrophysics and Space Science Proceedings (ed. F.D. Macchetto), Springer, Dordrecht, p. 11.
Sahnow, D.J. *et al.* (2000) *Proc. SPIE*, **4013**, 334.
Schoenberner, D. (1983) *Astrophys. J.*, **272**, 708.
Shipman, H.L., Green, R., and Liebert, J. (1987) *Astrophys. J.*, **315**, 239.
Sion, E.M. (1984) *Astrophys. J.*, **282**, 612.
Sion, E.M. (1986) *Publicat. ASP*, **98**, 821.
Sion, E.M., Greenstein, J.L., Landstreet, J.D., Liebert, J., Shipman, H.L., and Wegner, G.A. (1983) *Astrophys. J.*, **269**, 253.
Sion, E.M., Liebert, J., and Starrfield, S.G. (1985) *Astrophys. J.*, **292**, 471.
Sion, E.M., Liebert, J., Vauclair, G., and Wegner, G. (1989) in *White Dwarfs. Lecture Notes in Physics*, vol. 328 (ed. G. Wegner), Springer Verlag, Berlin, p. 354.
Unglaub, K. (2008) *Astron. Astrophys.*, **486**, 923.
Unglaub, K. and Bues, I. (1998) *Astron. Astrophys.*, **338**, 75.
Unglaub, K. and Bues, I. (2000) *Astron. Astrophys.*, **359**, 1042.
Vauclair, G., Vauclair, S., and Greenstein, J.L. (1979) *Astron. Astrophys.*, **80**, 79.
Vennes, S. (1999) *Astrophys. J.*, **525**, 995.
Vennes, S., Dupuis, J., Bowyer, S., Fontaine, G., Wiercigroch, A., Jelinsky, P., Wesemael, F., and Malina, R. (1994) *Astrophys. J. Lett.*, **421**, L35.
Vennes, S., Thejll, P.A., Wickramasinghe, D.T., and Bessell, M.S. (1996) *Astrophys. J.*, **467**, 782.
Vennes, S., Thejll, P.A., Galvan, R.G., and Dupuis, J. (1997) *Astrophys. J.*, **480**, 714.
Vennes, S., Dupuis, J., and Chayer, P. (2004) *Astrophys. J.*, **611**, 1091.
Voss, B. (2006) Ph.D. Thesis, Kiel University.
Voss, B., Koester, D., Napiwotzki, R., Christlieb, N., and Reimers, D. (2007) *Astron. Astrophys.*, **470**, p. 1079.
Wassermann, D., Werner, K., Rauch, T., and Kruk, J.W. (2010) Proceedings of the 17th European White Dwarf Workshop, AIP Conf. Proc. vol. 1273 (eds K. Werner and T. Rauch), AIP, Melville, New York, p. 105.
Webbink, R.F. (1984) *Astrophys. J.*, **277**, 355.
Werner, K. (1992) in *The Atmospheres of Early-Type Stars, Lecture Notes in Physics*, vol. 401 (eds U. Heber and C.S. Jeffery), Springer-Verlag, Berlin, p. 273.
Werner, K. (2010) *Recent Advances in Spectroscopy Theoretical, Astrophysical and Experimental Perspectives* (eds R.K. Chaudhuri, M.V. Mekkaden, A.V. Raveendran, and A. Satya Narayanan), Springer, Berlin, p. 199.
Werner, K. and Heber, U. (1991) *Astron. Astrophys.*, **247**, 476.
Werner, K., Heber, U., and Hunger, K. (1991) *Astron. Astrophys.*, **244**, 437.
Werner, K., Hamann, W.-R., Heber, U., Napiwotzki, R., Rauch, T., and Wessolowski, U. (1992) *Astron. Astrophys.*, **259**, L69.
Werner, K., Dreizler, S., Heber, U., Rauch, T., Wisotzki, L., and Hagen, H.-J. (1995) *Astron. Astrophys.*, **293**, L75.
Werner, K., Dreizler, S., Heber, U., Rauch, T., Fleming, T.A., Sion, E.M., and Vauclair, G. (1996) *Astron. Astrophys.*, **307**, 860.
Werner, K., Rauch, T., Napiwotzki, R., Christlieb, N., Reimers, D., and Karl, C.A. (2004) *Astron. Astrophys.*, **424**, 657.
Werner, K. and Herwig, F. (2006) *Publ. Astron. Soc. Pac.*, **118**, 183.
Werner, K., Rauch, T., and Kruk, J.W. (2007) *Astron. Astrophys.*, **474**, 591.
Werner, K., Rauch, T., and Kruk, J.W. (2008) *Astron. Astrophys.*, **492**, L43.
Werner, K., Rauch, T., and Kruk, J.W. (2010) *Astrophys. J. Lett.*, **719**, L32.
Wisotzki, L., Koehler, T., Groote, D., and Reimers, D. (1996) *Astron. Astrophys. (Suppl.)*, **115**, 227.
Wolff, B., Jordan, S., and Koester, D. (1996) *Astron. Astrophys.*, **307**, 149.
York, D.G. *et al.* (2000) *Astron. J.*, **120**, 1579.

2
Cool White Dwarfs

Mukremin Kilic

2.1
White Dwarf Cosmochronology

White dwarfs start their lives with surface temperatures of $\sim$ 100 000 K. They are effectively without nuclear energy sources; therefore, they radiate energy without replenishment for the rest of their lives. Normal stars have a safety mechanism that prevents secular changes in the star. If the star produces more energy through nuclear burning than it can radiate, the resulting increase in energy leads to an expansion and to a decrease in temperature. Since nuclear burning is highly sensitive to temperature, after a small decrease in temperature, thermal equilibrium will be re-established. The same mechanism works in the other case as well: if the star is radiating more energy than it produces, then a slight contraction increases the temperature, which in turn increases the rate of energy production, and the thermal equilibrium is re-established.

White dwarfs have no such mechanism to keep themselves in thermal equilibrium. The pressure of a degenerate gas depends on density and is almost independent of the temperature. This means that a white dwarf cannot keep itself in thermal equilibrium by contracting. The electron degeneracy pressure prevents the contraction. A normal star heats up as it loses energy, whereas a white dwarf cools down – white dwarfs spend all of their lives simply cooling.

The interiors of white dwarfs are highly degenerate, and since degenerate electrons are good conductors, the interiors are nearly isothermal. The core temperature is approximately the same as the temperature at the core/envelope boundary. Thus, a white dwarf can be treated as having an isothermal core that contains most of the mass of the star and a thin nondegenerate outer layer that acts like an insulating blanket and controls the rate at which the energy is radiated into space. The rate of radiation is controlled by the radiative opacity at the boundary between these two layers. Radiative transfer of energy in the core of the star is negligible compared to thermal conduction. Mestel (1952) created a simple analytical model for white dwarf evolution and showed that to first order, there is a simple relation

White Dwarf Atmospheres and Circumstellar Environments. First Edition. Edited by D. W. Hoard.

between the age, $\tau_{\rm cool}$, and luminosity, $L_{\rm wd}$, of a white dwarf,

$$\log(\tau_{\rm cool}) \propto -\frac{5}{7}\log\left(\frac{L_{\rm wd}}{L_\odot}\right) . \tag{2.1}$$

The beauty of this equation is that the age of a white dwarf can be deduced simply from its luminosity. This technique, white dwarf cosmochronology, can be used to infer the ages of the white dwarf parent populations, including open and globular clusters, and the thin disk, thick disk, and halo of the Galaxy. The Mestel age-luminosity relation is in good agreement with the predictions of detailed numerical models (Iben and Tutukov, 1984) and is applicable to white dwarfs with $-1 \geq \log(L_{\rm wd}/L_\odot) \geq -3$ (Wood, 1990). However, several important physical effects produce deviations from the classical Mestel cooling theory. For example, neutrino cooling, surface convection, and crystallization can increase or decrease the cooling rate.

Shining with residual thermal energy, white dwarfs cool down fastest at the beginning of their lives, and the rate of cooling subsequently slows down with time. The same cooling trend applies to any object with thermal energy; for example, a cup of boiling water. However, at the extreme temperatures expected in a young white dwarf star, plasmon neutrinos dominate the energy loss from the star. The energy loss due to neutrinos is an order of magnitude higher than the loss from photons. Plasmon neutrinos drive the evolution of hot white dwarfs. The effects of neutrinos diminish with decreasing temperature of the star and are predicted to be insignificant below $\sim$ 20 000 K (Winget *et al.*, 2004).

As the surface temperature of a white dwarf drops below 15 000 K, a convection zone (associated with partial ionization of the dominant element at the surface, hydrogen or helium) develops. This is due to the large temperature gradient between the stellar core and the surface layers. White dwarfs are predicted to have core temperatures around $\sim 5\times10^6$ K at surface effective temperatures of 8000 K. When the base of the convection zone reaches into the interior, it increases the energy transfer rate through the outer layers. This process, convective coupling, significantly changes the cooling rates of cool white dwarfs (Fontaine *et al.*, 2001).

Another important physical effect that takes place in the interiors of cool white dwarfs is crystallization. As the white dwarfs cool, the Coulomb interactions between ions become significant in the interiors. Kirzhnits (1960), Abrikosov (1961), and Salpeter (1961) showed that the ions should crystallize into a lattice. Crystallization of the interiors of white dwarfs releases latent heat and supplies the stars with an additional energy source (Van Horn, 1968). This extra energy delays the cooling of white dwarfs. Studying the white dwarf cooling sequence of the halo globular cluster NGC 6397, Winget *et al.* (2009) confirm the release of latent heat during crystallization.

A typical white dwarf with $M = 0.6M_\odot$ is expected to crystallize at $T_{\rm eff} \approx$ 6000 K (Wood, 1992). Crystallization begins at higher surface temperatures ($T_{\rm eff} \sim$ 10 000 K) in more massive white dwarfs. More than 40 years after Salpeter's prediction of crystallization in white dwarf interiors (Salpeter, 1961), the first empirical test was achieved by Metcalfe *et al.* (2004). Fitting the observed pulsation periods of the massive DAV white dwarf BPM 37093, they concluded that 90% of the star is crystallized (in an independent analysis, Brassard and Fontaine, 2005, conclude that this white dwarf is only 32 to 82% crystallized). However, the effect of crystallization on white dwarf cooling is not yet empirically constrained, and it is one of the main uncertainties for the age estimates of the coolest white dwarfs.

Theoretical investigators have also proposed delays in white dwarf cooling due to chemical fractionation (phase separation) of carbon and oxygen at the time of interior crystallization, resulting in the release of significant gravitational energy (Chabrier *et al.*, 1993; Fontaine *et al.*, 2001; Stevenson, 1980; Wood, 1995). That is, carbon and oxygen might not crystallize together, but undergo phase separation upon crystallization. White dwarfs crystallize from the center outward. When crystallization starts, oxygen is transported inward and carbon outward. Since carbon is slightly lighter than oxygen, redistribution of the ions releases additional gravitational energy that further delays the white dwarf cooling (Winget *et al.*, 2009).

The relative distribution of carbon and oxygen in the core is important for determining the magnitude of the additional energy source provided by phase separation. Unfortunately, due to the uncertainties in the $C^{12}(\alpha,\gamma)O^{16}$ reaction rate, the interior chemical profiles are uncertain (Salaris *et al.*, 1997). Nevertheless, Montgomery *et al.* (1999) demonstrate that the largest possible increase in white dwarf ages due to phase separation is $\sim$ 1.5 Gyr, with a most likely value of approximately 0.6 Gyr. Hence, the unknown initial chemical profiles do not prevent the use of cool white dwarfs as accurate chronometers.

After crystallization, the heat capacity in the crystallized regions will be due to lattice vibrations in the form of phonons, and this will result in lower heat capacity. The heat capacity will drop off in proportion to T^3 and the white dwarf will cool rapidly. This relatively rapid cooling phase is referred to as Debye cooling.

Figure 2.1 shows cooling curves for 0.2–1.3 $M_\odot$ white dwarfs from Fontaine *et al.* (2001). The quantity dt_{cool}/dM_{bol} is the inverse of the cooling rate. The cooling proceeds from left to right (from higher to lower luminosity). The cooling rate is initially fast due to neutrino cooling and it is also relatively fast (especially for the most massive white dwarfs) at very low luminosities due to Debye cooling. The first open circle along a track indicates the onset of crystallization at the center. The more massive models crystallize at higher effective temperatures and luminosities. After the onset of crystallization, dt_{cool}/dM_{bol} increases (or the cooling rate decreases) due to the release of latent heat. The second open circle on each track marks the point where 98% of the star is crystallized. At this point, most of the latent heat has been released. The onset of convective coupling is indicated by a filled circle on each curve. Until that point, the thermal reservoir of the core remains insulated by the envelope. When the base of the convection zone reaches the interior, there is initially excess thermal energy. This excess energy slows the cooling process (an increase in dt_{cool}/dM_{bol}) for a while and produces a bump in the cooling tracks. However, once this excess energy is radiated away, convection speeds up the cooling process because the thermal reservoir of the interior is not as well insulated after convective coupling (Fontaine *et al.*, 2001).

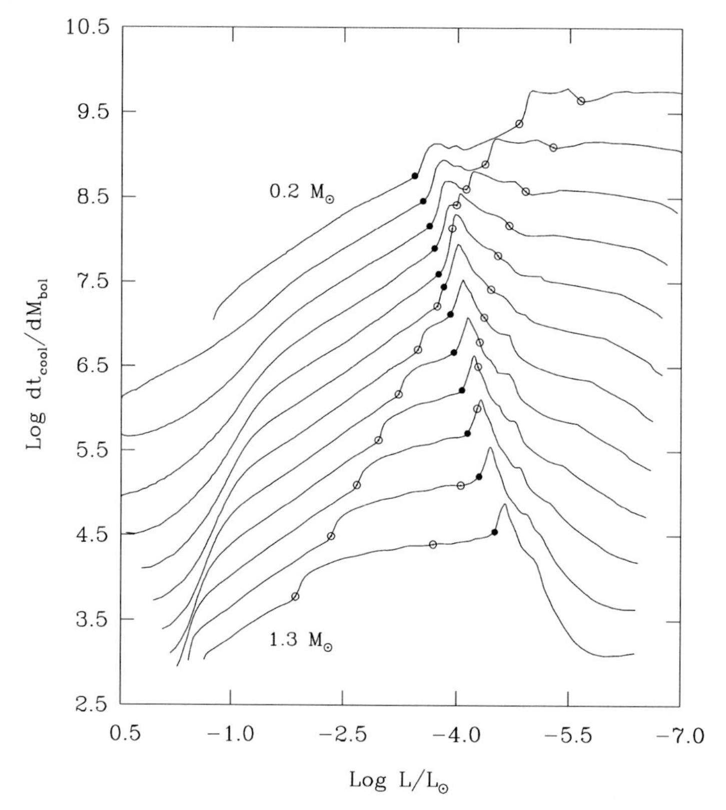

Figure 2.1 Evolutionary models for 0.2–1.3 $M_{\odot}$ white dwarfs showing the derivative of the cooling curve as a function of the luminosity. Only the $0.2 M_{\odot}$ curve is on the correct vertical scale; the other curves have been arbitrarily shifted downward for visualization purposes. The first open circle (at higher luminosity) on each curve corresponds to the onset of crystallization at the center of the evolving model, and the second one indicates the epoch when 98% of the mass of the star has solidified. The filled circle on each curve indicates the onset of convective coupling. From Fontaine *et al.* (2001), reproduced by permission of the University of Chicago Press, © 2001. The Astronomical Society of the Pacific.

Even though the evolution of white dwarfs is described as a simple cooling process, Figure 2.1 demonstrates that the cooling rate of a white dwarf is a strong function of its mass and surface temperature. Input physics in the evolutionary models, including crystallization, phase separation, and convective coupling have major effects on the derived ages. The total uncertainty from the input physics is on the order of 1–2 Gyr for the oldest white dwarfs in the Galactic disk and halo. Cool white dwarfs remain powerful tools for near-field cosmology and they provide 10% or better accuracy for the age estimates of the oldest white dwarfs in the solar neighborhood.

In addition to the evolutionary calculations, white dwarf model atmospheres are essential for cosmochronology. These atmosphere models provide surface boundary conditions for the cooling calculations and also provide emergent fluxes and colors to be compared with the observed spectra and colors. Observations of nearby cool white dwarfs at a variety of wavelengths offer a unique opportunity to study the relatively dense atmospheres of these objects.

2.2 Cool White Dwarf Atmospheres

The previous section focused on the use of white dwarfs as cosmic chronometers. White dwarfs are typically observed in the optical and near-infrared, and it is then assumed that the amount of flux they emit at all wavelengths is known by applying bolometric corrections. Using the estimated total fluxes, cooling ages for white dwarfs are determined. However, the accuracy of the bolometric corrections depends on the state of understanding of cool white dwarf atmospheres, unless empirical constraints are acquired at *all* wavelengths. In this context, recent discoveries (e. g., Gates *et al.*, 2004; Harris *et al.*, 2008) of a dozen ultracool ($T_{\text{eff}} \leq 4000$ K) white dwarfs are puzzling. Current theoretical models cannot be fit to the optical and near-infrared spectral energy distributions of these stars. This section will focus on the efforts to understand the cool white dwarf atmospheres.

Cool white dwarfs have atmospheres dominated by hydrogen or helium. Both hydrogen and helium are neutral below 5000 K; white dwarfs cooler than this temperature have featureless optical spectra unless they have trace amounts of metals. The primary opacity sources in cool hydrogen-rich ($T_{\text{eff}} \leq 5500$ K) white dwarf atmospheres are believed to be collision induced absorption due to molecular hydrogen in the infrared (Bergeron *et al.*, 1995; Hansen, 1998; Saumon and Jacobson, 1999) and Lyman-α in the ultraviolet (Koester and Wolff, 2000; Kowalski and Saumon, 2006).

The primary opacity source in pure helium atmosphere white dwarfs is He^- free–free absorption. This opacity is considerably weaker than those of hydrogen. Thus, the atmospheric pressure and density are significantly larger than in the hydrogen-rich white dwarfs. As a consequence of this, the ideal gas approximation cannot be used for pure helium atmosphere white dwarfs (Bergeron *et al.*, 1995). The spectral energy distributions of these stars are similar to those of blackbodies. Hence, differentiating between hydrogen- and helium-rich atmosphere solutions and, therefore, determining the temperatures and bolometric luminosities of cool white dwarfs requires a detailed analysis of their optical and infrared spectral energy distributions (Bergeron *et al.*, 1997, 2001).

2.2.1 Collision Induced Absorption

Under normal circumstances, molecular hydrogen (H_2) does not absorb or emit dipole radiation since there is no change of dipole moment during rotation and vibration. However, in dense environments, H_2-H_2 or H_2-He interactions may induce a temporary dipole moment and collision induced absorption (CIA) becomes possible (see Borysow *et al.*, 1997, and references therein). CIA opacity is strongly wavelength dependent and is expected to produce broad absorption features in the near-infrared. This flux deficiency can be seen in late type stars, brown dwarfs, and cool white dwarfs. Hydrogen-rich white dwarfs are predicted to become redder as they cool until the effects of CIA become significant below 5500 K.

The CIA opacity strongly depends on the temperature, surface gravity, and composition (hydrogen to helium ratio) of the star. The opacity becomes stronger with decreasing temperature and increasing surface gravity (i. e., denser atmospheres).

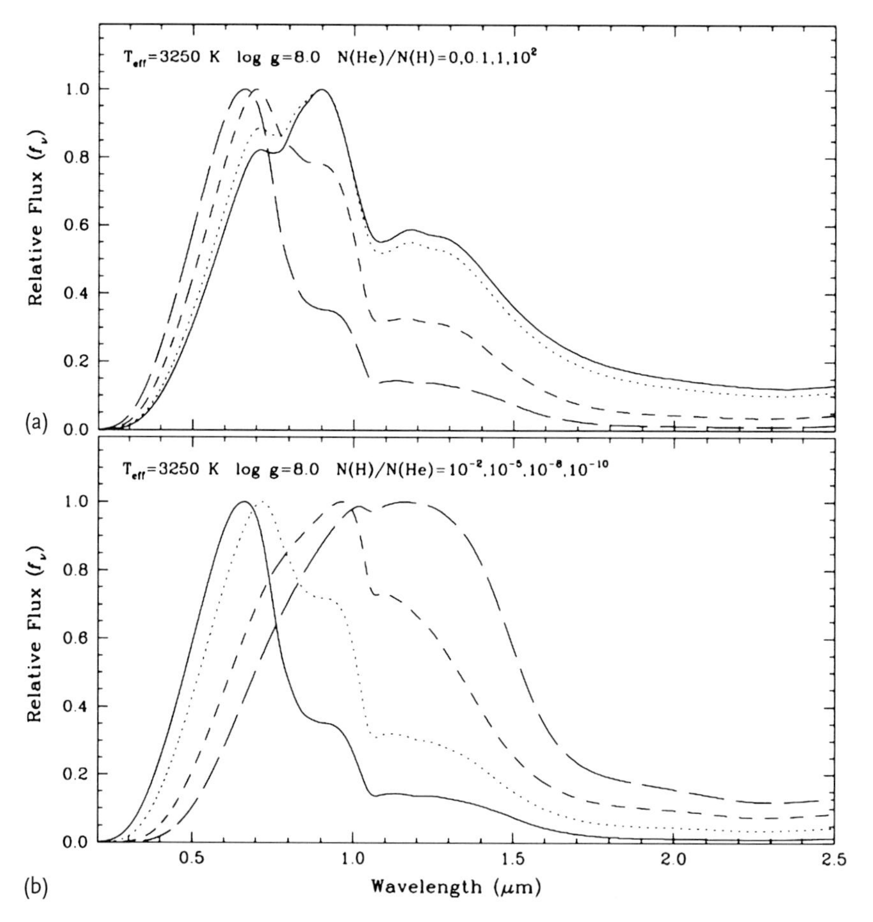

Figure 2.2 The effect of CIA on the emergent flux of a $T_{\text{eff}} = 3250$ K white dwarf (P.-E. Tremblay, private communication; also see Figure 5 of Bergeron and Leggett, 2002). Panel (a) compares models with $\log g = 8.0$ from a pure hydrogen composition to a value of $\text{H/He} = 10^{-2}$, for which the infrared flux deficiency is the strongest. In (b), the hydrogen abundance is further decreased from a value of $\text{H/He} = 10^{-2}$ to 10^{-10}. In both panels, the plotted line proceeds through solid, dotted, short-dashed, and long-dashed as the H/He ratio is successively incremented to lower values. From Kilic *et al.* (2010b), reproduced with the permission of the AAS.

However, the dependence on the ratio of hydrogen to helium is more complicated. Figure 2.2 shows an illustrative sequence of models at constant T_{eff} and $\log g$, but with different hydrogen/helium abundances. Since helium-rich atmospheres are denser, the CIA opacity becomes important at slightly higher temperatures in mixed H/He atmospheres compared to pure hydrogen atmosphere white dwarfs. Depending on the composition of the atmosphere, whether it is hydrogen or helium dominated, the strength of the CIA opacity changes. The maximum CIA opacity occurs in an atmosphere with H/He $\sim 10^{-2}$ (P.-E. Tremblay, private communication). For very cool temperatures, as shown in the figure, broad absorption features at 0.8 μm and 1.1 μm are expected.

2.2.2
The Missing Opacity Source in the Blue: Lyman-α Absorption

A detailed model atmosphere analysis of cool white dwarfs in the solar neighborhood by Bergeron *et al.* (1997, 2001) shows that there are discrepancies between the observations and the model atmosphere predictions in the ultraviolet. These authors find evidence for an unrecognized opacity source in the ultraviolet and interpret it in terms of a pseudocontinuum opacity originating from the Lyman edge.

Theoretical work by Kowalski and Saumon (2006) indicate that the missing opacity source in the blue is the far red wing of the pressure-broadened Lyman-α line. This absorption mechanism is the result of perturbations of hydrogen atoms by their interactions with H and H_2. These perturbations lower the Lyman-α transition energy by more than 5 eV, enabling bound-bound absorption from the ground state of a hydrogen atom. This occurs in close-range, rare collisions that perturb

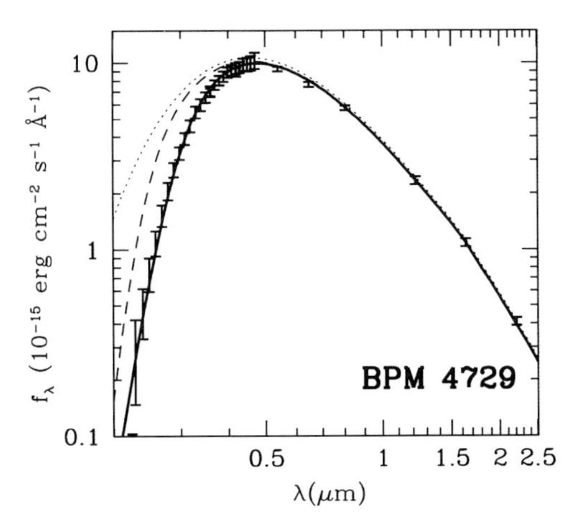

Figure 2.3 Spectral energy distribution of the DA white dwarf BPM 4729. The solid and dotted lines represent the pure hydrogen models (T_{eff} = 5820 K) with and without the opacity from the red wing of the Lyman-α line, respectively. The dashed line represents the spectrum obtained when only H–H collisions are included in the Lyman-α opacity calculations. From Kowalski and Saumon (2006), reproduced with the permission of the AAS.

the bound states of the hydrogen atom. Kowalski and Saumon (2006) include this opacity source in their model atmosphere calculations, which reproduce the spectral energy distributions and colors of cool white dwarfs fairly well.

Figure 2.3 displays the spectral energy distribution of a $T_{\text{eff}} = 5820$ K DA white dwarf, BPM 4729, compared to the model predictions with and without the opacity from the red wing of the Lyman-α line. The models with Lyman-α absorption match the spectral energy distribution of this cool DA white dwarf extremely well at ultraviolet and blue wavelengths, indicating that the previously unidentified opacity source in cool white dwarf atmospheres is, indeed, the red wing of the Lyman-α line.

2.2.3
Model Atmospheres Versus Observations

The state-of-the-art white dwarf model atmospheres match the observations extremely well for the majority of the known cool white dwarfs. Figure 2.4 shows the spectral energy distributions of nine cool white dwarfs with $T_{\text{eff}} \leq 6000$ K. The models and the observations agree over the entire wavelength range from the ultraviolet to the mid-infrared. This gives us confidence that white dwarf model atmo-

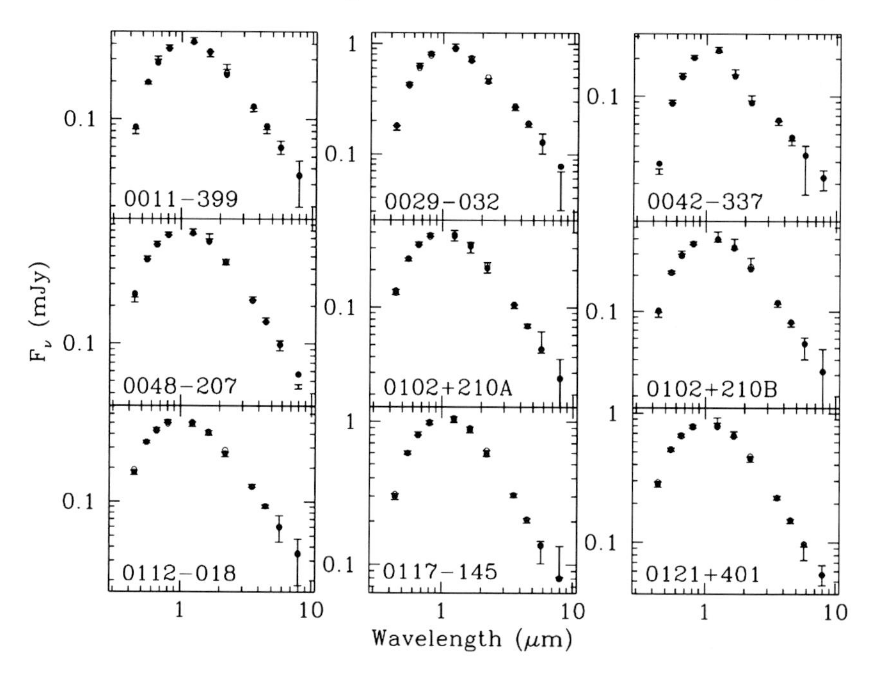

Figure 2.4 Spectral energy distributions of nine cool white dwarfs ($T_{\text{eff}} \leq 6000$ K) observed with the Spitzer Space Telescope. The observed fluxes are shown as error bars, whereas the expected flux distributions from pure hydrogen and mixed H/He atmosphere models are shown as open and filled circles, respectively. From Kilic *et al.* (2009), reproduced with the permission of the AAS.

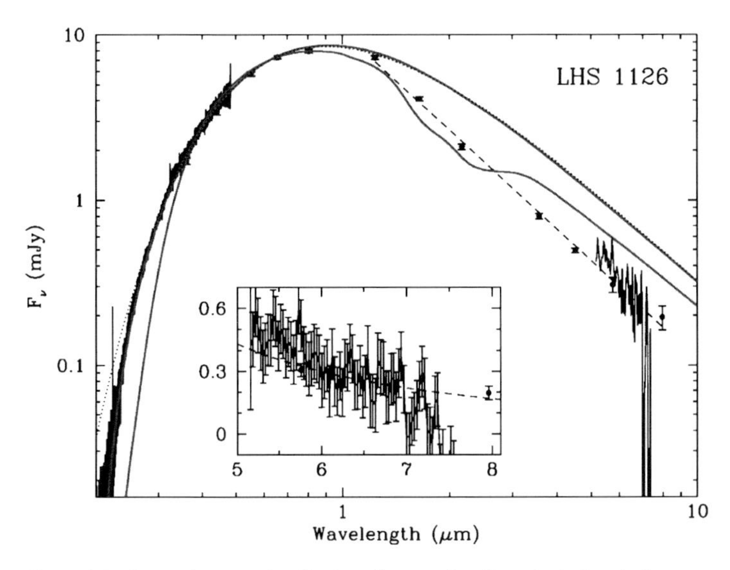

Figure 2.5 Spectral energy distribution of LHS 1126, along with a 5400 K pure He atmosphere model (dotted line). The upper and lower solid lines show the model atmospheres (including the CIA opacity) with log (H/He) $= -5.51$ and -1.86, respectively. The dashed line shows a least square power law fit to the infrared photometry data only. The infrared portion of the spectral energy distribution is best fit with a power law index of -1.99. The inset shows the 5–8 μm spectral region in detail. From Kilic *et al.* (2008), reproduced with the permission of the AAS.

spheres are accurate enough to derive reliable effective temperatures for cool white dwarfs. However, there are several stars that show strong infrared flux absorption and for which the models fail.

Figure 2.5 displays the spectral energy distribution for one of these anomalous white dwarfs, LHS 1126. This star shows shifted Swan bands in the optical that are attributed to molecular absorption from carbon. LHS 1126 shows significant mid-infrared flux absorption and a featureless spectrum in the 5–8 μm range. Even though this flux absorption is attributed to the CIA opacity due to molecular hydrogen, the shape of the flux deficit compared to the models cannot be explained with current CIA opacity calculations. The infrared portion of the LHS 1126 spectral energy distribution is best-fit with a power law index of -1.99, identical to a Rayleigh–Jeans spectrum. This argues that the observed infrared absorption may be due to an unrecognized gray-like opacity source in the infrared. Alternatively, the current CIA opacity calculations might be wrong under certain conditions; for example, mixed H/He atmospheres. This issue will be revisited later in this chapter.

The detailed model atmosphere calculations and theoretical understanding of the evolutionary histories of white dwarfs demonstrate that cool white dwarfs can be used as accurate age indicators for different stellar populations in the Galaxy. The full potential of white dwarf cosmochronology can be utilized with a large sample of cool white dwarfs. Such a sample can be constructed based on a photometric or astrometric search.

2.3
Identification of Large Samples of Cool White Dwarfs

2.3.1
Photometric Selection

A magnitude-limited, kinematically-unbiased sample of white dwarfs can be obtained through a photometric survey. To do so, a unique color signature is necessary to photometrically identify a white dwarf among the many other types of field stars. The magnitude limit is also a critical factor in the search for cool white dwarfs; if the survey cannot provide sufficiently high signal to noise ratio data for $M_V \sim 16$, then it cannot recognize cool low luminosity white dwarfs.

Broad-band photometric surveys can be used to find hot white dwarfs due to their blue colors. Eisenstein *et al.* (2006) found 9317 new white dwarfs with $T_{\text{eff}} \geq 7000$ K in the Sloan Digital Sky Survey (SDSS) Data Release 4 (DR4). However, broad-band filter photometry has a limited capacity to distinguish metal poor subdwarfs from cool white dwarfs. In the absence of significant line blanketing, both white dwarfs and subdwarfs have broad-band colors that closely approximate those of a blackbody. However, by comparing the flux through a magnesium absorption line-centered filter (e. g., DDO51), Claver (1995) suggested that cool white dwarfs could be distinguished from other field stars of similar temperature. This is because the majority of cool white dwarfs have essentially featureless spectra around 5150 Å, whereas subdwarfs and main sequence stars show significant absorption from the Mg b triplet and/or MgH.

Even though, a priori, the use of a narrow-band filter centered on the Mg absorption feature seems like a promising technique, Kilic *et al.* (2004) demonstrated that this method is not as effective as expected at separating white dwarfs from subdwarfs. White dwarfs with temperatures in the range 5000–7000 K are still photometrically indistinguishable from observed field stars. Figure 2.6 shows the equivalent width of the Mg/MgH feature in main sequence stars. Mg absorption becomes strong enough to affect the photometry in K0 ($T_{\text{eff}} \sim 5000$ K) and later type stars. Due to the spread in colors and weak Mg absorption in the F–G type stars, white dwarfs with $T_{\text{eff}} = 5000$–7000 K have colors similar to F–G stars.

White dwarfs with temperatures in the range 3500–5000 K are also not cleanly separated from subdwarfs in color–color diagrams using the DDO51 filter. Until recently, cool white dwarfs were thought to have spectral energy distributions similar to blackbodies. In fact, this is why Claver (1995) suggested that a narrow-band filter centered on the MgH feature would place cool white dwarfs above the observed field star sequence; the DDO51 filter would separate blackbodies from subdwarfs. Although subdwarfs have strong MgH absorption in this temperature range and they deviate from blackbodies, observed cool white dwarfs deviate from blackbodies, as well (see the discussion above on cool white dwarf atmospheres).

The most prominent features in the optical spectra of subdwarf stars are Mg/MgH and CaH + TiO. In addition to the DDO51 filter, Claver (1995) suggested the use of an intermediate-band filter centered on the CaH + TiO band at

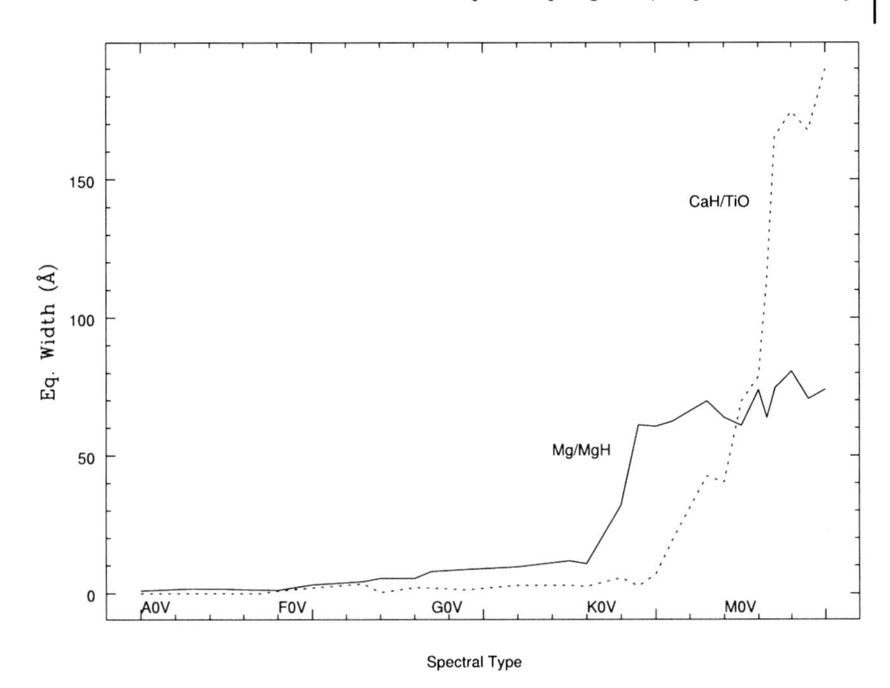

Figure 2.6 Equivalent width of the Mg/MgH and CaH + TiO features. The Mg/MgH feature becomes strong enough to affect the photometry in K0 and later type stars, whereas the CaH + TiO feature at ∼ 6850 Å dominates for M0 and later type stars. From Kilic *et al.* (2004), reproduced with the permission of the AAS.

∼ 6850 Å to identify white dwarfs. CaH + TiO absorption becomes strong in M0 ($T_{\text{eff}} \sim 3800$ K) and later type stars (Figure 2.6). White dwarfs in this temperature range show depressed infrared colors due to CIA if they have pure-H or mixed H/He atmospheres, and they can be identified by using the DDO51 filter if they have pure-He atmospheres. The CIA exhibited by ultra cool white dwarfs is extremely broad-band and monotonically varies throughout the red–infrared region, whereas the CaH/TiO band is very narrowly confined in wavelength. Thus, the CaH + TiO filter, if ratioed with another nearby pseudocontinuum filter, could show a much stronger dependency on temperature and metallicity in main sequence and subdwarf stars than it does in ultracool white dwarfs. Therefore, the CaH + TiO filter photometry may be useful for the identification of cool hydrogen-rich or mixed atmosphere white dwarfs, though broad-band photometry surveys are also successful in finding ultracool white dwarfs (e. g., Gates *et al.*, 2004; Harris *et al.*, 2008).

2.3.2
Proper Motion Selection

Current photometric surveys are not very efficient in finding cool white dwarfs. The reduced proper motion technique (Luyten, 1918) offers an alternative approach and efficient means to identify cool white dwarfs, as well as halo white dwarfs, by their underluminosity in comparison to main sequence stars with similar colors and their high space motions, respectively. The reduced proper motion, defined as $H = m + 5 \log \mu + 5$, where m is the apparent magnitude and μ is the proper motion in arcseconds per year, has long been used as a proxy for the absolute magnitude of a star, for a sample with similar kinematics. A clean selection of white dwarfs from the much larger sample of Population I main sequence stars and Population II subdwarfs is possible using a reduced proper motion diagram.

Within the last decade, the SDSS emerged as the main resource for Galactic stellar population studies thanks to its accurate photometry and astrometry. Munn *et al.* (2004) use the USNO-B positions and the SDSS astrometry to derive proper motions with $\sim 3.5\,\mathrm{mas\,yr^{-1}}$ accuracy with 90% completeness down to $g = 19.7$ mag.

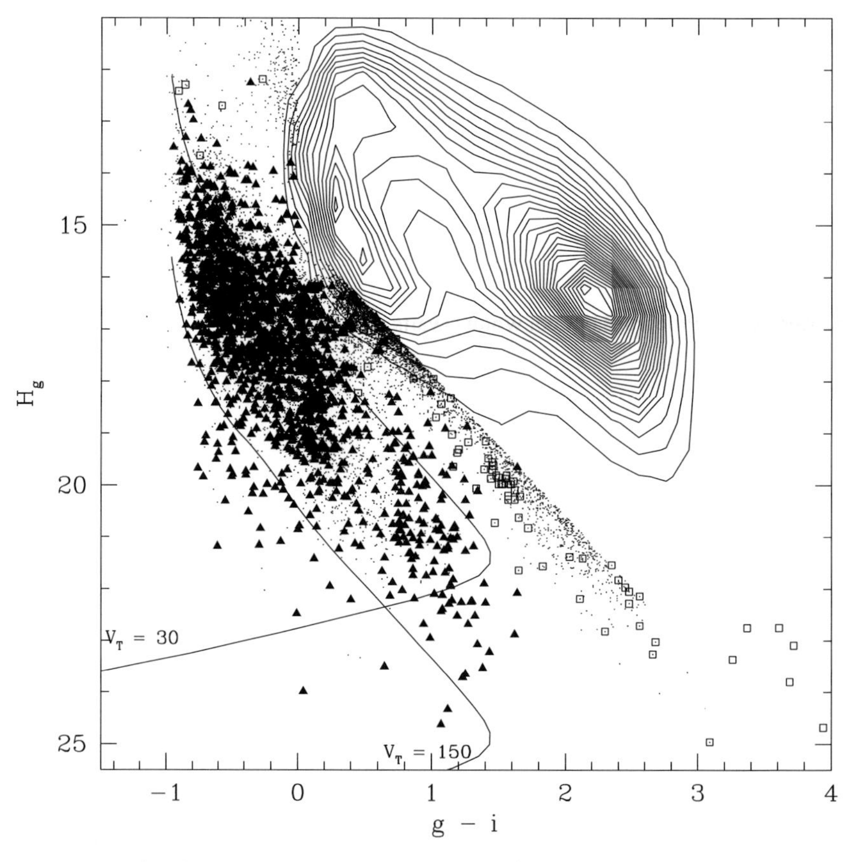

Figure 2.7 Reduced proper motion diagram for stars in the SDSS DR3. Individual stars are only plotted in the region of interest for white dwarfs, the remaining stars are represented by the contours. Spectroscopically confirmed white dwarfs and subdwarfs are shown as triangles and squares, respectively. White dwarf cooling curves for different tangential velocities are shown as solid lines. The $V_{\mathrm{tan}} = 30\,\mathrm{km\,s^{-1}}$ curve marks the expected location of disk white dwarfs, whereas the $V_{\mathrm{tan}} = 150\,\mathrm{km\,s^{-1}}$ curve represents the halo white dwarfs. From Kilic *et al.* (2010b), reproduced with the permission of the AAS.

Follow-up spectroscopy is usually necessary to confirm that the astrometrically selected targets are white dwarfs. High signal-to-noise ratio spectra with a few angstroms resolution is often needed to reject the high velocity Population II metal-weak "sdK" stars, which show few, weak, narrow features until they become cool enough to exhibit MgH (5200 Å) and CaH molecular bands.

Kilic *et al.* (2006, 2010b) performed such a spectroscopic survey of hundreds of cool white dwarf candidates selected using a reduced proper motion diagram from the SDSS. Figure 2.7 shows the reduced proper motion diagram for the SDSS Data Release 3 (DR3) area including spectroscopically confirmed white dwarfs and subdwarfs. The cool white dwarf selection works well for $V_{tan} \geq 30\,\mathrm{km\,s^{-1}}$. There is a roughly 1.5% contamination rate due to incorrectly measured proper motions. Only one of the 75 cool white dwarf candidates with $V_{tan} \geq 30\,\mathrm{km\,s^{-1}}$ and $M_{bol} > 14.6$ mag is actually a subdwarf, which corresponds to a contamination rate of 1.3%. The reduced proper motion diagram can thus be used to define a statistically complete sample of white dwarfs, including the coolest white dwarfs, which are difficult to efficiently select using other techniques. Now that a large sample of cool white dwarfs can be selected, their observational and group properties can be studied.

2.4 Observational Properties of Cool White Dwarfs

2.4.1 Color–Color Diagrams

Figure 2.8 shows the optical and infrared color–color diagrams for the cool white dwarfs studied by Bergeron *et al.* (2001) and Kilic *et al.* (2010b). The observed ranges of $g - r$ and $r - i$ colors agree well with the pure hydrogen or helium atmosphere model predictions except for the white dwarfs with significant i- or z-band absorption. The $g - r$ versus $r - i$ diagram provides an efficient way to identify white dwarfs that show strong absorption in the i-band. The near-infrared colors of the $T_{eff} = 4000$–$10\,000$ K white dwarfs also agree with either pure hydrogen or pure helium atmosphere models. However, there are more than a dozen white dwarfs with strong infrared absorption ("infrared-faint", labeled by name in the figure) that deviate from these models. The observed sequence for infrared-faint white dwarfs is significantly different than the pure hydrogen model sequence indicating that these white dwarfs most likely do not have pure hydrogen atmospheres. The similarities between the colors for the mixed H/He atmosphere models and the infrared-faint stars suggest that these stars have mixed H/He atmospheres.

Figure 2.8 demonstrates that the optical and near-infrared colors of cool white dwarfs agree well with the model predictions for pure hydrogen, pure helium, or mixed H/He atmospheres. Of course, a detailed model atmosphere analysis of the photometric and spectroscopic data is required in order to derive accurate temperatures, compositions, and ages for individual objects.

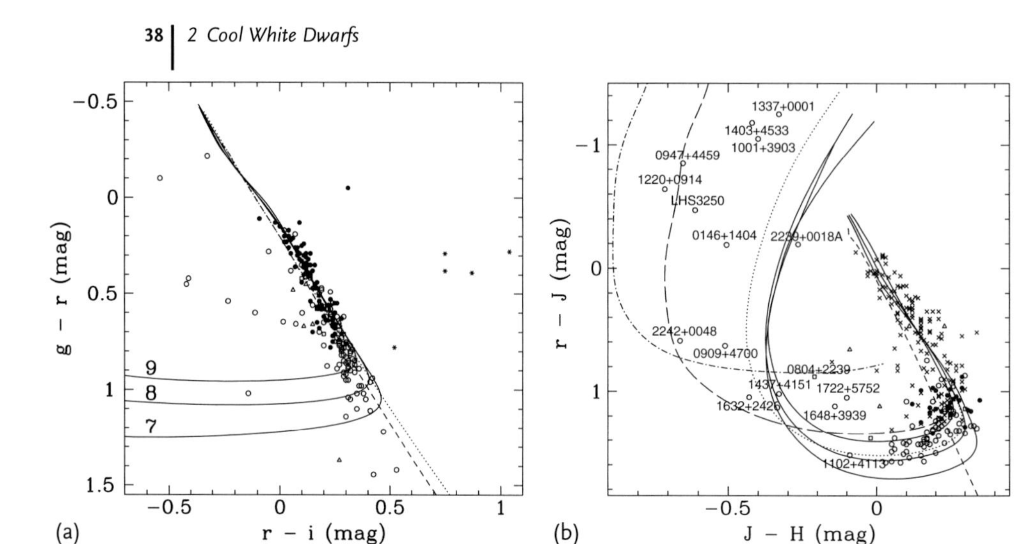

Figure 2.8 Color–color diagrams for DA (filled circles), DC (open circles), DQ (open triangles), DZ (open squares), and white dwarf + d*M* (asterisks) in the SDSS. The cool white dwarf sample of Bergeron *et al.* (2001) is shown as crosses. The solid lines show the predicted colors for pure hydrogen atmosphere white dwarfs with $T_{\rm eff} = 2000$–30 000 K and $\log g = 7$, 8, and 9. The dashed line shows a pure helium atmosphere white dwarf sequence with $T_{\rm eff} = 3000$–30 000 K and $\log g = 8$. The dotted, long-dashed, and dashed-dotted lines show the predicted color sequences for mixed atmosphere models ($T_{\rm eff} = 2000$–6000 K and $\log g = 8$) with H/He = 10, 1, and 0.01, respectively. From Kilic *et al.* (2010b), reproduced with the permission of the AAS.

2.4.2
Pure Hydrogen and Pure Helium Atmosphere White Dwarfs

Figure 2.9 shows a representative sample of fits for DA white dwarfs with $T_{\rm eff} =$ 5000–6000 K. All six stars in this figure show Hα absorption, and their optical and near-infrared spectral energy distributions are matched fairly well by pure hydrogen atmosphere models. The spectroscopic observations of Hα are not used directly in the fitting procedure, but they serve as an internal check of the photometric solutions. The theoretical line profiles are calculated using the parameters obtained from the spectral energy distribution fits. This figure shows that the predicted line profiles are in good agreement with the pure hydrogen atmosphere model solutions derived from the photometric observations. The excellent match between the theoretical Hα line profiles and the observations rules out significant amounts of helium in the atmospheres of these white dwarfs.

Figure 2.10 shows sample fits for helium-rich DC white dwarfs. The spectroscopic fits are not shown here since all of the stars are featureless near the Hα region. Several of the white dwarfs shown in the figure are warm enough to exhibit Hα if they were hydrogen-rich. The lack of Hα absorption reveals a helium-rich composition, and the pure helium models provide excellent fits to the spectral energy distributions of these objects. There are many stars with $T_{\rm eff} \geq 4500$ K that are best explained as pure helium atmosphere objects, but there are none below this tem-

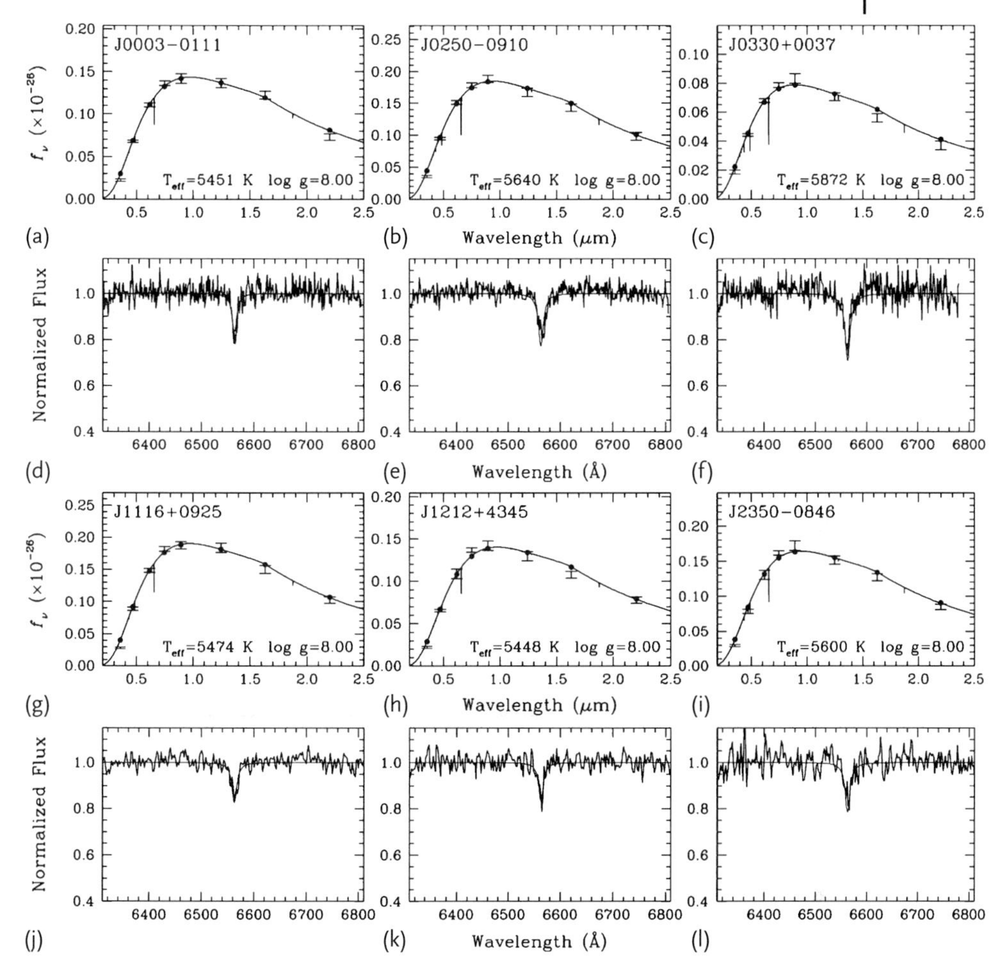

Figure 2.9 Fits to the spectral energy distributions of six DA white dwarfs with pure hydrogen models. Here and in the following figures, the $ugriz$ and JHK photometric observations are represented by error bars, while the model monochromatic fluxes are shown as solid lines. The u-band photometry is excluded from the fits since the Lyman-α opacity is missing in these models. The filled circles represent the average of the model over the filter bandpasses. The lower panels show the normalized spectra together with the synthetic line profiles for the parameters obtained from the spectral energy distribution fits. From Kilic *et al.* (2010b), reproduced with the permission of the AAS.

perature. Given the observed infrared colors of cool white dwarfs, perhaps this is not surprising. The $r - J$ versus $J - H$ color–color diagram (see Figure 2.8) shows that the coolest white dwarfs display absorption in the infrared, indicating that they have hydrogen in their atmospheres.

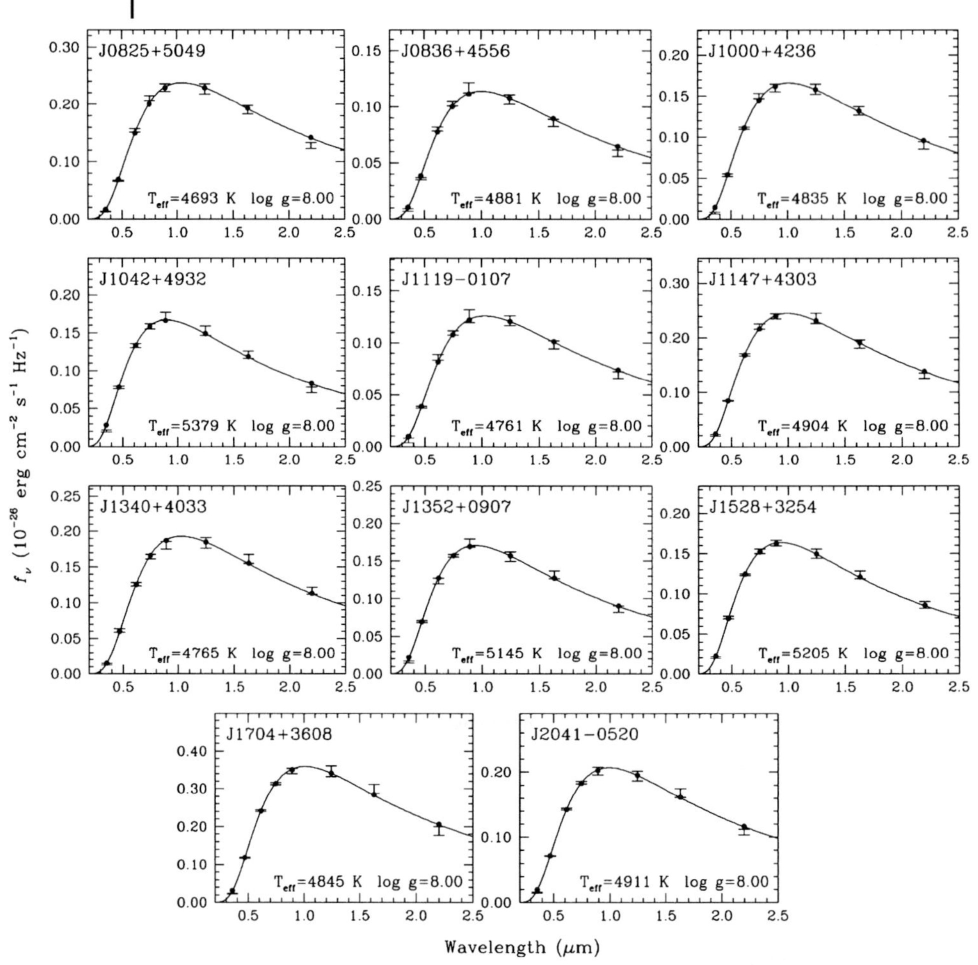

Figure 2.10 Fits to the spectral energy distributions of eleven DC white dwarfs with pure helium models. All objects have featureless spectra near the Hα region. From Kilic *et al.* (2010b), reproduced with the permission of the AAS.

2.4.3
Mixed H/He Atmosphere White Dwarfs

Bergeron *et al.* (2001) do not find a large population of cool white dwarfs with mixed hydrogen and helium atmospheres. Such stars would show up as outliers in the optical and infrared color–color diagrams due to the H_2-He CIA, which produces strong flux deficits in the infrared. The SDSS cool white dwarf sample has half a dozen new stars with significant absorption in the infrared. Pure hydrogen and pure helium models fail to reproduce the spectral energy distributions for these stars. Figure 2.11 presents mixed H/He atmosphere model fits to eight DC white dwarfs. The mixed H/He atmosphere models with $\log(\text{H/He}) = -5.9$ to -3.4

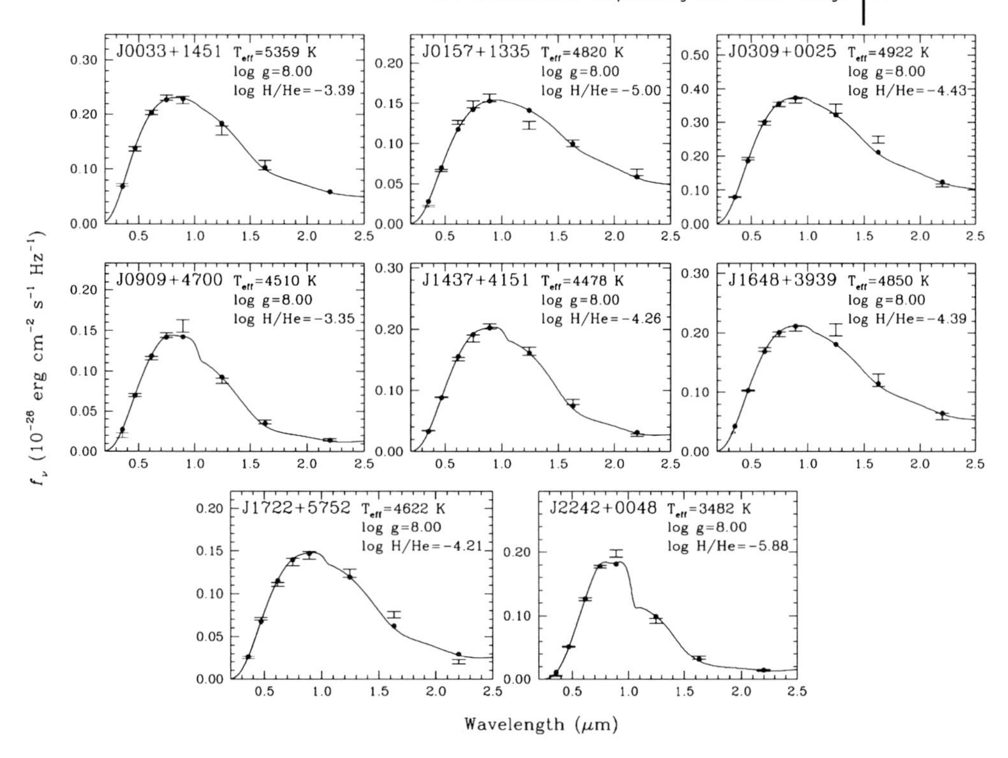

Figure 2.11 Fits to the spectral energy distributions of eight DC white dwarfs with mixed H/He atmosphere models. From Kilic *et al.* (2010b), reproduced with the permission of the AAS.

fit the observed spectral energy distributions relatively well for these stars. Six of these targets have temperatures in the range 4500–5000 K, in which many helium-rich DC white dwarfs are found. An important implication of these temperature assignments is that not all white dwarfs that show infrared flux deficits are ultracool (i. e., have temperatures below 4000 K).

2.4.4 Ultracool (or Infrared-Faint) White Dwarfs

White dwarfs cooler than about 4000 K are classified as ultracool. Starting with Hambly *et al.* (1997) and Harris *et al.* (1999), the SDSS and various other proper motion surveys have discovered ultracool white dwarfs. LHS 3250 is the best studied ultracool white dwarf with significant absorption in the optical and infrared. Bergeron and Leggett (2002) performed detailed model atmosphere analysis of this star based on optical and near-infrared photometry and a parallax measurement. While none of their fits reproduce the LHS 3250 spectral energy distribution perfectly (see Figure 2.12), they ruled out pure hydrogen composition based on the nondetection of the strong CIA feature near 0.8 μm. Instead, the observations are better fit with helium-dominated atmosphere models with small amounts of

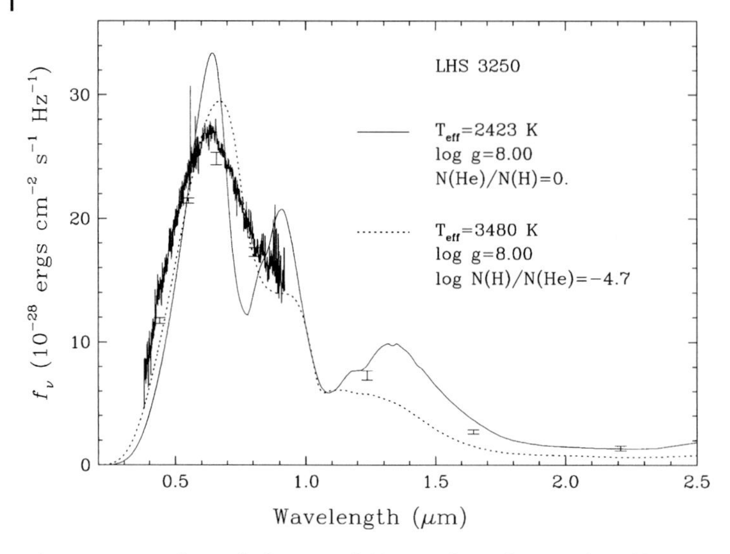

Figure 2.12 Best-fit pure hydrogen (solid line) and mixed H/He (dotted line) atmosphere model fits to the spectral energy distribution of LHS 3250. From Bergeron and Leggett (2002), reproduced with the permission of the AAS.

hydrogen. Bergeron and Leggett (2002) find best-fit values of $T_{\rm eff} = 3480$ K and $\log({\rm H/He}) = -4.7$ for this star.

Studying the spectral energy distributions of a dozen ultracool white dwarf stars in the SDSS, Kilic *et al.* (2010b) found that none of these stars show the strong CIA feature at 0.8 μm, ruling out pure hydrogen composition models. Pure helium models also fail to reproduce the spectral energy distributions as these stars all show significant absorption in the optical and infrared. The best-fit models have mixed H/He compositions with temperatures as low as 2300 K. Like LHS 3250, the peaks of the spectral energy distributions of the ultracool white dwarfs are predicted to be too sharp compared to the observations. This is probably an indication that the CIA opacities are wrong at such low temperatures. Despite the fact that current model atmospheres do not produce perfect fits to the observed photometry, extreme hydrogen-rich compositions for these stars can be ruled out based on the current CIA opacity calculations.

2.5 Spectral Evolution of Cool White Dwarfs

Understanding the overall spectral evolution of cool white dwarfs is important for white dwarf cosmochronology since the cooling time depends on the insulating effects of the outer layers. There are strong indicators that spectral evolution takes place at different temperature ranges. For example, the DB gap between 45 000 and 30 000 K, the increasing number of helium-rich objects below 10 000 K (Tremblay and Bergeron, 2007), and the non-DA gap between 5000–6000 K (Bergeron *et al.*, 1997) indicate that a sizable fraction of white dwarfs change their spectral types during their lifetimes. Even though it is impossible to know the exact history of individual objects, statistical studies of large samples of cool white dwarfs can be used to understand the overall spectral evolution of cool white dwarfs.

Based on a detailed model atmosphere analysis of 150 white dwarfs with $T_{\rm eff} \leq$ 12 000 K, Bergeron *et al.* (2001) find the frequency of pure hydrogen and pure helium atmosphere white dwarfs to be 64 and 33%, respectively. They find helium-rich atmosphere white dwarfs down to about 4500 K and hydrogen-rich white dwarfs down to 4000 K. The coolest and oldest white dwarfs are likely to accrete from the interstellar medium in their $\sim$ 10 Gyr lifetimes. The lack of pure helium white dwarfs below 4500 K supports this scenario. Bergeron *et al.* (2001) also find a non-DA gap (or a deficiency in number) between about 5000 and 6000 K. They find non-DA stars above and below this temperature range, but they find only three non-DA stars in the gap. In addition, they do not find a large population of mixed H/He atmosphere white dwarfs. In contrast, the SDSS cool white dwarf sample of Kilic *et al.* (2010b) is restricted to stars cooler than about 6600 K and it contains 48% pure hydrogen, 35% pure helium, and 17% mixed H/He atmosphere white dwarfs. The SDSS sample fills in the non-DA gap somewhat. However, there is still a gap between 5600 and 6200 K in both samples.

Based on the Bergeron *et al.* (1995) models, a significant fraction of white dwarfs in the temperature range 4500–5000 K are He-rich. Since Hα is invisible at these temperatures, the choice of composition depends on the quality of the fits to the spectral energy distributions. The best H-rich model fit is sometimes not too different from the He-rich model fit. It is possible that small shifts in the optical and near-infrared photometric calibration may explain the overabundance of He-rich objects in this temperature range. The noninclusion of the Lyman-α opacity in the models or problems with the CIA calculations may cause incorrect assignment of the atmospheric types to white dwarfs in this temperature range. Further work is required to understand if the observed overabundance of He-rich atmosphere white dwarfs at this temperature range is real.

Overall, the spectral evolutionary history of cool white dwarfs is complex (Bergeron *et al.*, 2001). The model atmosphere analysis by Kowalski and Saumon (2006) presents a completely different picture, in which white dwarfs below 6000 K are hydrogen-rich. They come to this conclusion by excluding the DQ and DZ white dwarfs from their sample and also using a different set of pure helium atmosphere models that have colors essentially the same as blackbodies. Since the cool white dwarf spectral energy distributions are not blackbodies, they assign hydrogen-rich composition for most cool white dwarfs and they propose a simple evolutionary scenario in which white dwarfs accrete hydrogen from the interstellar medium and turn into hydrogen-rich white dwarfs even if they start with pure helium atmospheres. The Bergeron *et al.* (1995) pure helium atmosphere models have colors slightly different than simple blackbodies and more similar to the observed colors of cool DC white dwarfs. Resolving the discrepancy between these two different interpretations would require a thorough study of the differences between these models. DZ white dwarfs are the only cool ($T_{\rm eff} < 5000$ K) white dwarfs with atomic absorption lines. A detailed model atmosphere analysis of cool DZ white dwarfs (e. g., Dufour *et al.*, 2007) would be a crucial test for identifying problems with both sets of models.

2.6 Ages for Individual White Dwarfs

The ultimate goal of white dwarf cosmochronology is to derive accurate ages for individual white dwarfs. This goal can be achieved if accurate optical and near-infrared photometry and trigonometric parallax measurements are available. Without a parallax measurement, the mass is unknown. Since the white dwarf cooling ages strongly depend on the mass, the resultant ages are uncertain as well.

Figure 2.13 displays model fits to the spectral energy distributions of eight cool white dwarfs with SDSS photometry (Kilic *et al.*, 2010b). Omitting the *u*- and *g*-band photometry from the fits (due to the missing Lyman-α opacity in these models), the observations are best explained with pure hydrogen atmosphere mod-

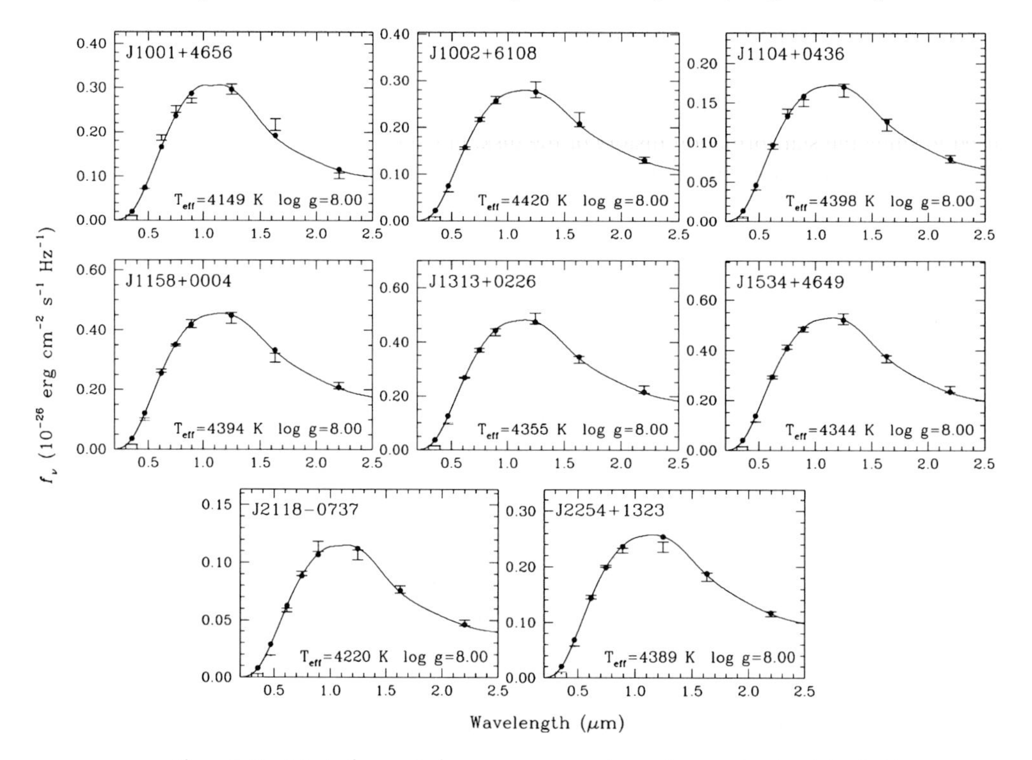

Figure 2.13 Fits to the spectral energy distributions of eight white dwarfs with $T_{\text{eff}} = 4150$–4420 K. All objects have featureless spectra near the Hα region, and the observations are best explained with pure hydrogen atmosphere models. From Kilic *et al.* (2010b), reproduced with the permission of the AAS.

els with $T_{\text{eff}} = 4150$–4420 K, corresponding to cooling ages of 7.8–8.6 Gyr for $\log g = 8$ white dwarfs. However, if these white dwarfs are slightly undermassive (i. e., $M = 0.5 M_\odot$), then they would be 6.5–7.4 Gyr old. Similarly, if they are slightly more massive (i. e., $M = 0.7 M_\odot$), then they would be 9.3–9.9 Gyr old. Hence, the unknown mass introduces uncertainties of several Gyr in the cooling age estimates.

Bergeron *et al.* (1997, 2001) avoid this problem by studying a large sample of cool white dwarfs with accurate optical/infrared photometry and parallax observations. Figure 2.14 compares the masses and surface temperatures of their sample of white dwarfs against cooling isochrones. These isochrones do not apply to low-mass white dwarfs, which have helium cores. The shape of the isochrones is informative; low-mass white dwarfs evolve faster due to their larger radii. However, crystallization starts earlier in more massive white dwarfs and they begin to cool off rapidly, creating the parabola-shaped isochrones. Crystallization gradually starts in lower mass white dwarfs, changing the shape of the isochrones for older systems. Again, these isochrones demonstrate the importance of mass measurements in assigning ages to cool white dwarfs.

The coolest white dwarfs in Figure 2.14 have disk kinematics and total (main-sequence lifetime + white dwarf cooling) ages of ≈ 9 Gyr, indicating that the Galactic disk has to be ≥ 9 Gyr old to produce such white dwarfs. The existence of young and old white dwarfs implies that the Galactic disk has been forming stars for the past 9 Gyr. The exact number distribution of cool white dwarfs can be used to study the star formation history of the disk, but such studies usually rely on the white dwarf luminosity function.

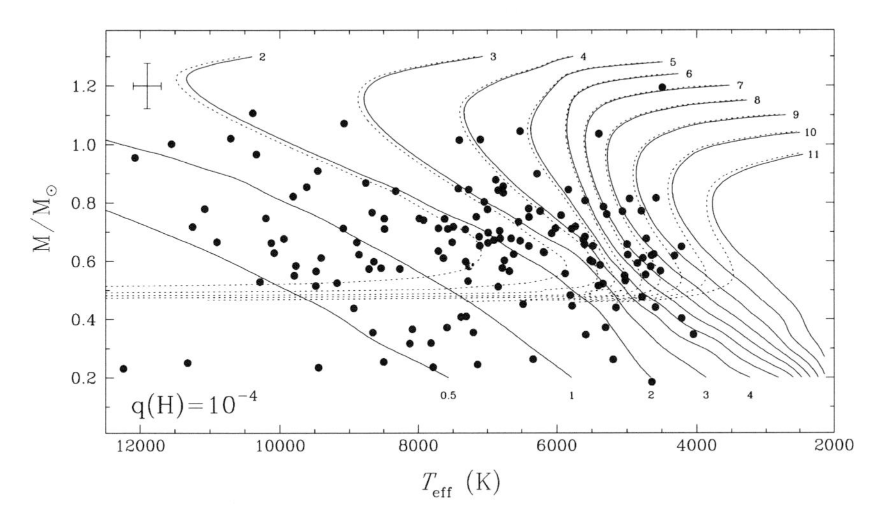

Figure 2.14 Masses of white dwarfs in the trigonometric parallax sample of Bergeron *et al.* (2001) as a function of temperature along with the isochrones (solid lines) with C/O core compositions, and helium and hydrogen layer mass fractions of 10^{-2} and 10^{-4}, respectively. The isochrones are labeled in units of 1 Gyr. The dotted lines show isochrones including the main sequence lifetime. A typical error bar is shown in the top left corner. From Bergeron *et al.* (2001), reproduced with the permission of the AAS.

2.7 The White Dwarf Luminosity Function

White dwarfs are the remnants of the earliest and all subsequent generations of star formation and, thus, they are tracers of the evolution of the Galaxy. Since the white dwarf cooling rate slows as their temperature drops, the oldest white dwarfs remain visible with current technology. Because the cooling rate slows, any census finds more and more white dwarfs at lower and lower temperatures (and luminosities) until, quite abruptly, no more are found. The result of such a census is called the white dwarf luminosity function.

Attempts to exploit white dwarfs as chronometers showed that the white dwarf luminosity function is a map of the history of star formation in the disk, and that there is a shortfall of low luminosity white dwarfs – the inevitable consequence of the finite age of the disk (Winget *et al.*, 1987). The most commonly used luminosity function for cool white dwarfs (Liebert *et al.*, 1988) is based on a sample of only 43 stars selected on the basis of large proper motion from the Luyten Half Second Proper Motion Survey. An investigation of the cool end of the white dwarf luminosity function that is focused on obtaining accurate ages, and also disentangling theoretical uncertainties in the cooling process, would greatly benefit from a much larger sample of cool white dwarfs.

Based on a reduced proper motion diagram for the SDSS DR3 footprint, Harris *et al.* (2006) derived cool white dwarf samples by taking all stars below and blueward of the white dwarf model curves for $V_{\rm tan} = 20$, 30, and $40\,{\rm km\,s^{-1}}$. As discussed in Section 2.3.2, the cool white dwarf selection works well for $V_{\rm tan} \geq 30\,{\rm km\,s^{-1}}$. Only one of the 75 cool white dwarf candidates in the Harris *et al.* (2006) $V_{\rm tan} \geq 30\,{\rm km\,s^{-1}}$ and $M_{\rm bol} > 14.6$ mag sample is actually a subdwarf.

Harris *et al.* (2006) used the Bergeron *et al.* (1995) models to fit all five SDSS magnitudes to determine temperature, distance, bolometric magnitude, and tangential velocity for each star. The choice of hydrogen or helium atmosphere models has little effect on the estimated $M_{\rm bol}$ for relatively warm white dwarfs. However, the colors are significantly different for pure hydrogen and pure helium atmosphere models for $T_{\rm eff} \leq 5300$ K white dwarfs due to CIA opacity. Lacking infrared data, Harris *et al.* (2006) make a weighted H/He assignment for each star based on the fraction of each type from the studies in the literature. The luminosity function is derived by using the $1/V_{\rm max}$ method (summing the inverse volume of space in which each star potentially would have been included within the sample limits and calculating each luminosity bin separately). The new white dwarf luminosity function is shown in Figure 2.15.

The new luminosity function is remarkably smooth and featureless. The only noticeable feature in the range $8 < M_{\rm bol} < 15$ mag is the small plateau near $M_{\rm bol} = 10.5$ mag. The ZZ Ceti instability strip is at slightly fainter $M_{\rm bol}$, so it is probably not related to this feature. No feature is predicted (Fontaine *et al.*, 2001) from a pause in white dwarf cooling and dimming. If the feature is real, it could reflect a nonuniform star formation rate in the Galactic disk. The cooling time for a normal-mass white dwarf to reach $M_{\rm bol} \sim 10.5$ is 0.3 Gyr, and the main-sequence lifetime

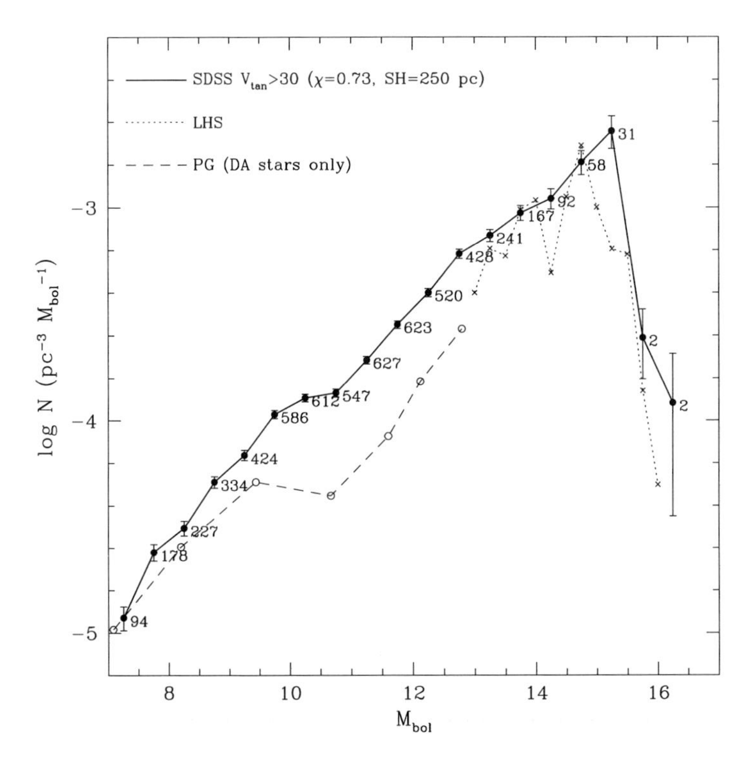

Figure 2.15 White dwarf luminosity function from the SDSS DR3. The number of stars used for each data point is indicated. Two results from the literature are shown for comparison: the dotted line at the faint end is taken from Leggett *et al.* (1998), based on the LHS Catalogue; the dashed line at the bright end is taken from Liebert *et al.* (2005), based on analysis of the DA white dwarfs in the PG Survey. From Harris *et al.* (2006), reproduced with the permission of the AAS.

of a likely progenitor is $\sim$ 2.5 Gyr, suggesting that a drop in the Galactic star formation rate about 3 Gyr ago (after a burst or a long-duration higher rate of star formation) might be the cause of the plateau seen in Figure 2.15.

The result at the bright end is in acceptable agreement with results from the hydrogen-rich (DA) stars from the Palomar–Green Survey (Liebert *et al.*, 2005). Accounting for non-DA stars will raise the densities from the PG Survey by a small amount. The result at the faint end is in excellent agreement with that from the local sample selected from the LHS Survey (Leggett *et al.*, 1998). The abrupt drop in the luminosity function, signaling the absence of older, cooler white dwarfs, occurs at $M_{\text{bol}} = 15.40$ mag. For an $M = 0.6 M_{\odot}$, pure H atmosphere white dwarf, this corresponds to an age of 7.8 Gyr (Bergeron *et al.*, 1995). However, the exact shape of the drop and the exact luminosity at which it occurs both depend on the unknown H/He type of white dwarfs in this sample. Also evident in this plot is a rise in the luminosity function at $M_{\text{bol}} = 15.1$–15.2 mag. This rise is consistent with the predicted onset of convective coupling between the convective hydrogen atmosphere and the degenerate core (see Figure 2.1), causing temporary additional release of internal energy and delayed cooling. Observationally, the rise in this figure, like the following drop, is sensitive to the assumptions on H/He types and to the luminosity of the coolest stars. Additional near-infrared photometry of the coolest white dwarfs in this sample shows that they have $T_{\rm eff} \geq 4150$ K and white dwarf cooling ages of ≤ 8.6 Gyr (see Section 2.6). Parallax observations are still needed for accurate mass and age assignments for these targets. However, both individual age assignments and the drop in the luminosity function point to a Galactic disk age of about 8 Gyr. The current best estimate on the age of the Galactic disk is 8 ± 1.5 Gyr (Leggett *et al.*, 1998).

The integral of the Harris *et al.* luminosity function shown in Figure 2.15 gives a space density of white dwarfs in the solar neighborhood of $(4.6 \pm 0.5) \times 10^{-3}$ pc^{-3}. The assumptions about the scale height and the fraction of He-atmosphere stars affect this result, though not drastically. This space density is consistent with the density measured from the 20 pc sample of Sion *et al.* (2009), $(4.9 \pm 0.5) \times 10^{-3}$ pc^{-3}.

2.8 Halo White Dwarfs

The quest to identify halo white dwarfs has been hampered by the lack of proper motion surveys that go deep enough to detect the cool halo white dwarfs. The initial claims for a significant population of halo white dwarfs in the field (Oppenheimer *et al.*, 2001) and in the Hubble Space Telescope (HST) Deep Field (Ibata *et al.*, 2000; Méndez and Minniti, 2000) were later rejected by detailed model atmosphere analysis (e. g., Bergeron *et al.*, 2005) and additional proper motion measurements. To date, the coolest known probable halo white dwarfs are WD 0346+246 and SDSS J1102 + 4113, both with $T_{\rm eff} \approx 3800$ K. In addition, some of the ultracool white dwarfs detected in the SDSS may belong to the thick disk or halo, but current atmosphere models have problems reproducing their intriguing spectral energy distributions. Therefore, their temperatures and ages remain uncertain.

A substantial investment of HST time on two globular clusters, M4 and NGC 6397, revealed clean white dwarf cooling sequences. Figure 2.16 shows a color-magnitude diagram of the point sources in the region that encloses the white dwarf population of the globular cluster NGC 6397. The white dwarf cooling sequence of this cluster extends down to several hundred degrees cooler than $T_{\rm eff} = 4000$ K. The coolest white dwarfs in this cluster are about 650 ± 230 K cooler than the coolest white dwarfs in the disk (Kowalski, 2007). Hansen *et al.* (2004, 2007) use these data to derive cooling ages of ≈ 12 Gyr for the two clusters. These studies demonstrate that the Galactic halo is older than the disk by ≥ 2 Gyr.

Even though the white dwarfs in globular clusters provide reliable age estimates, the long exposure times required to reach the white dwarf terminus limit these studies to the nearest few clusters. In addition, only two-filter (V and I) photometry was used to model the absolute magnitude and color distribution of the oldest white dwarfs to derive ages. The far closer and brighter white dwarfs of the local

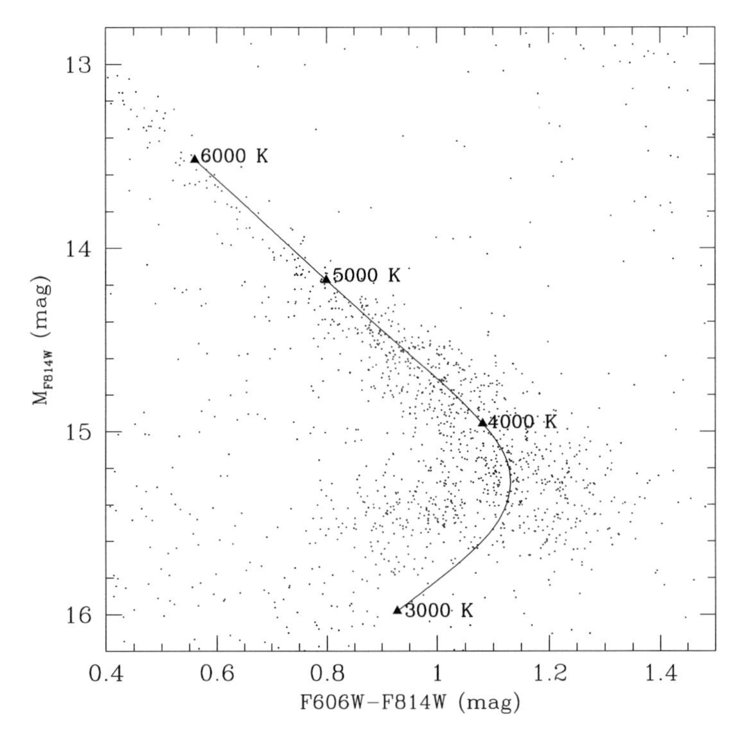

Figure 2.16 Color–magnitude diagram of the white dwarfs in the globular cluster NGC 6397 (Hansen *et al.*, 2007). The white dwarf sequence is dereddened by $E(F606W - F814W) = 0.16$ mag and vertically shifted by $\mu = 12.0$ mag. The solid line shows the colors for $0.53\,M_\odot$ white dwarfs with $T_{\text{eff}} = 3000$–6000 K. From Kilic *et al.* (2010a), reproduced with the permission of the AAS.

halo field are an enticing alternative, as well as complementary targets, with the additional potential to constrain the age range of the Galactic halo.

The SDSS cool white dwarf sample of Harris *et al.* (2006) suffered from the magnitude limit of the Palomar Observatory Sky Survey plates. Nevertheless, Harris *et al.* (2006) constructed a luminosity function for halo white dwarfs assuming that all high-velocity candidates belong to the halo. This luminosity function is only an upper limit for the halo since the actual composition of the sample is not well understood. The shape of the luminosity function (rising toward lower luminosities) and the integrated space density ($4 \times 10^{-5}\,\text{pc}^{-3}$) are both consistent with models of an old, single-burst population.

Deep proper motion surveys provide exciting opportunities for discovering many more halo white dwarfs in the field. Such a proper motion survey (with a limiting magnitude of $r = 21$ mag) done at the Bok and USNO telescopes reveal hundreds of potential halo white dwarf candidates (Kilic *et al.*, 2010a). Figure 2.17 shows the spectral energy distributions of three such candidates. The temperatures for these objects range from 3730 to 4110 K, corresponding to white dwarf cooling ages of 8.7–9.7 Gyr and total ages of 10–11 Gyr for average mass white dwarfs. Based on

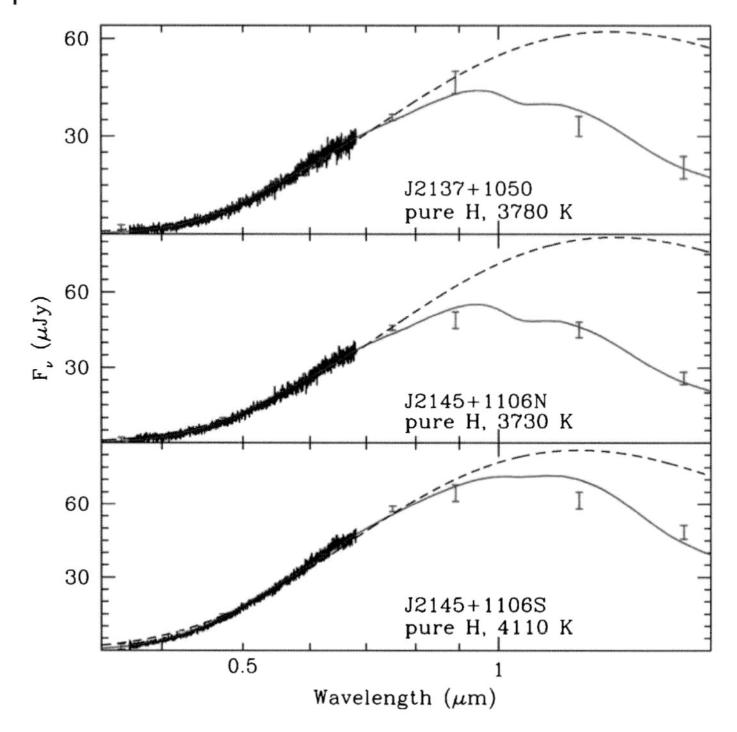

Figure 2.17 Observed spectra, SDSS photometry, and near-infrared photometry of 3 halo white dwarf candidates compared to the best-fit pure hydrogen atmosphere model spectra (solid lines, assuming $\log g = 8$) and blackbody spectral energy distributions with the same temperatures (dashed lines). From Kilic *et al.* (2010a), reproduced with the permission of the AAS.

the white dwarf cooling sequences of the globular clusters M4 and NGC 6397, and the several field halo white dwarf candidates discussed above, it can already be seen that there is a gap of 1–2 Gyr between the star formation in the halo and the star formation in the disk at the solar annulus. Future deep, wide-field proper motion surveys should find many old halo white dwarfs that can be used to constrain the age of the Galactic halo more accurately.

2.9
Conclusions and Future Prospects

White dwarfs are relatively simple objects with well understood evolutionary histories. Devoid of any nuclear energy, they simply cool with time. This makes them excellent chronometers. White dwarf cosmochronology provides an independent and accurate age dating method for different Galactic populations, including the solar neighborhood, open and globular clusters, and the Galactic thin disk, thick disk, and halo.

White dwarf atmospheres are relatively dense and complicated to model. The effects of CIA due to molecular hydrogen and Lyman-α absorption are evident in the spectra of hydrogen-rich white dwarfs. However, the state-of-the-art model atmospheres match the observations fairly well, even for white dwarfs as cool as 3800 K (see Figure 2.17). The models fail to explain about a dozen infrared-faint white dwarfs with puzzling spectral energy distributions, but the remainder of the known cool white dwarfs in the SDSS and the literature are explained fairly well by the current model atmosphere calculations. Hence, accurate temperatures can be measured for all but the infrared-faint white dwarfs.

Open problems remain in the understanding of the evolution of cool white dwarfs, including the delays introduced by crystallization, phase separation, and convective coupling as well as the uncertainties on individual ages due to the unknown evolutionary history of each object. However, the theoretical uncertainties due to the unknown core composition, helium layer mass, crystallization, and phase separation are on the order of 1–2 Gyr for 10 Gyr old white dwarfs (M. Montgomery, private communication). These uncertainties can be lowered with a better understanding of the interior physics. Accurate luminosity functions derived from larger samples of white dwarfs will be helpful in lowering these uncertainties.

The future of white dwarf cosmochronology is bright. Survey data provided by facilities such as the Panoramic Survey Telescope and Rapid Response System (Pan-STARRS), the Large Synoptic Survey Telescope (LSST), and the Gaia satellite mission will discover significant numbers of disk, thick disk, and halo white dwarfs. In addition, the LSST and Gaia will deliver distances and absolute magnitudes for all of these white dwarfs, enabling a detailed study of the properties of these objects. Based on the Liebert *et al.* (1988) white dwarf luminosity function for the Galactic thin disk and a single-burst, 12 Gyr-old population with 10 and 0.4% local normalization for the thick disk and halo, there should be 3200 thick disk and 140 halo white dwarfs per 1000 square degrees (for a Galactic latitude of 45°) down to a limiting magnitude of $V = 21.5$ mag. For a limiting magnitude of $V = 24$ mag and assuming 50% sky coverage, the Pan-STARRS and LSST will image $\sim$ 1.3 million thick disk and $\sim$ 80 000 halo white dwarfs. These surveys will be invaluable resources for disk and halo white dwarf studies.

References

Abrikosov, A.A. (1960) *Zh. Eksp. Teor. Fiz.*, **39**, 1798.

Bergeron, P. and Leggett, S.K. (2002) *Astrophys. J.*, **580**, 1070.

Bergeron, P., Saumon, D., and Wesemael, F. (1995) *Astrophys. J.*, **443**, 764.

Bergeron, P., Ruiz, M.T., and Leggett, S.K. (1997) *Astrophys. J. (Suppl.)*, **108**, 339.

Bergeron, P., Leggett, S.K., and Ruiz, M.T. (2001) *Astrophys. J. (Suppl.)*, **133**, 413.

Bergeron, P., Ruiz, M.T., Hamuy, M., Leggett, S.K., Currie, M.J., Lajoie, C.-P., and Dufour, P. (2005) *Astrophys. J.*, **625**, 838.

Brassard, P., and Fontaine, G. (2005) *Astrophys. J.*, **622**, 572.

Borysow, A. *et al.* (1997) *Astron. Astophys.*, **324**, 185.

Chabrier, G., Ségretain, L., Hernanz, M., Isern, J., and Mochkovitch, R. (1993) *White*

Dwarfs: Advances in Observation and Theory. NATO ASIC Proc. 403, p. 115.
Claver, C.F. (1995) Ph.D. Thesis, University of Texas.
Dufour, P. *et al.* (2007) *Astrophys. J.*, **663**, 1291.
Eisenstein, D.J. *et al.* (2006) *Astrophys. J. (Suppl.)*, **167**, 40.
Fontaine, G., Brassard, P., and Bergeron, P. (2001) *Publicat. ASP*, **113**, 409.
Gates, E. *et al.* (2004) *Astrophys. J. Lett.*, **612**, 129.
Hambly, N.C., Smartt, S.J., and Hodgkin, S.T. (1997) *Astrophys. J. Lett.*, **489**, L157.
Hansen, B.M.S. (1998) *Nature*, **394**, 860.
Hansen, B.M.S. *et al.* (2004) *Astrophys. J. (Suppl.)*, **155**, 551.
Hansen, B.M.S. *et al.* (2007) *Astrophys. J.*, **671**, 380.
Harris, H.C., Dahn, C.C., Vrba, F.J., Henden, A.A., Liebert, J., Schmidt, G.D., and Reid, I.N. (1999) *Astrophys. J.*, **524**, 1000.
Harris, H.C. *et al.* (2006) *Astron. J.*, **131**, 571.
Harris, H.C. *et al.* (2008) *Astrophys. J.*, **679**, 697.
Ibata, R., Irwin, M., Bienaymé, O., Scholz, R., and Guibert, J. (2000) *Astrophys. J. Lett.*, **532**, L41.
Iben, I. Jr., and Tutukov, A.V. (1984) *Astrophys. J.*, **282**, 615.
Kilic, M., Winget, D.E., von Hippel, T., and Claver, C.F. (2004) *Astron. J.*, **128**, 1825.
Kilic, M. *et al.* (2006) *Astron. J.*, **131**, 582.
Kilic, M., Kowalski, P.M., Mullally, F., Reach, W.T., and von Hippel, T. (2008) *Astrophys. J.*, **678**, 1298.
Kilic, M., Kowalski, P.M., Reach, W.T., and von Hippel, T. (2009) *Astrophys. J.*, **696**, 2094.
Kilic, M. *et al.* (2010a), *Astrophys. J. Lett.*, **715**, L21.
Kilic, M. *et al.* (2010b), *Astrophys. J. (Suppl.)*, **190**, 77.
Kirzhnits, D.A. (1960) *Sov. Phys. – JETP*, **11**, 365.
Koester, D. and Wolff, B. (2000) *Astron. Astrophys.*, **357**, 587.
Kowalski, P.M. (2007) *Astron. Astrophys.*, **2007**, 474., 491
Kowalski, P.M. and Saumon, D. (2006) *Astrophys. J. Lett.*, **651**, L137.
Leggett, S.K., Ruiz, M.T., and Bergeron, P. (1998) *Astrophys. J.*, **497**, 294.
Liebert, J., Dahn, C.C., and Monet, D.G. (1988) *Astrophys. J.*, **332**, 891.
Liebert, J., Bergeron, P., and Holberg, J.B. (2005) *Astrophys. J. (Suppl.)*, **156**, 47.
Luyten, W.J. (1918) *Lick Obs. Bull.*, **10**, 135.
Méndez, R.A., and Minniti, D. (2000) *Astrophys. J.*, **529**, 911.
Mestel, L. (1952) *Mon. Not. R. Astron. Soc.*, **112**, 583.
Metcalfe, T.S., Montgomery, M.H., and Kanaan, A. (2004) *Astrophys. J. Lett.*, **605**, 133.
Montgomery, M.H., Klumpe, E.W., Winget, D.E., and Wood, M.A. (1999) *Astrophys. J.*, **525**, 482.
Munn, J.A. *et al.* (2004) *Astron. J.*, **127**, 3034.
Oppenheimer, B.R., Hambly, N.C., Digby, A.P., Hodgkin, S.T., and Saumon, D. (2001) *Science*, **292**, 698.
Salaris, M., Dominguez, I., Garcia-Berro, E., Hernanz, M., Isern, J., and Mochkovitch, R. (1997) *Astrophys. J.*, **486**, 413.
Salpeter, E.E. (1961) *Astrophys. J.*, **134**, 669.
Saumon, D. and Jacobson, S.B. (1999) *Astrophys. J.*, **511**, 107.
Sion, E.M., Holberg, J.B., Oswalt, T.D., McCook, G.P., and Wasatonic, R. (2009) *Astron. J.*, **138**, 1681.
Stevenson, D.J. (1980) *J. Phys.*, **41**, 2.
Tremblay, P.-E. and Bergeron, P. (2007) *Astrophys. J.*, **657**, 1013.
Van Horn, H.M. (1968) *Astrophys. J.*, **151**, 227.
Winget, D.E., Hansen, C.J., Liebert, J., Van Horn, H.M., Fontaine, G., Nather, R.E., Kepler, S.O., and Lamb, D.Q. (1987) *Astrophys. J.*, **315**, L77.
Winget, D.E., Kepler, S.O., Campos, F., Montgomery, M.H., Girardi, L., Bergeron, P., and Williams, K. (2009) *Astrophys. J. Lett.*, **693**, L6.
Winget, D.E. *et al.* (2004) *Astrophys. J. Lett.*, **602**, 109.
Wood, M.A. (1990) Ph.D. Thesis, Texas University, Austin.
Wood, M.A. (1992) *Astrophys. J.*, **386**, 539.
Wood, M.A. (1995) in *Lecture Notes in Physics*, vol. 443 (eds D. Koester and K. Werner), Springer-Verlag, Berlin, p. 41.

3
Stars with Unusual Compositions: Carbon and Oxygen in Cool White Dwarfs

Patrick Dufour

3.1
Introduction

White dwarfs represent the end products of the evolution of all stars on the main sequence that had initial masses below $\sim 8 M_{\odot}$. After some important mass loss episodes in the red giant phases, followed by the end of thermonuclear activity in the interiors of such stars, they ultimately shrink to Earth-size objects with masses of about 0.4 to $1.2 M_{\odot}$. The vast majority of them are believed to be composed mainly (i. e., more than 99% by mass) of carbon and oxygen, the products of hydrogen and helium nuclear burning. Intuitively, then, one might assume that it would not be surprising to observe a significant amount of carbon and oxygen in white dwarf photospheres. However, nuclear burning and mass loss episodes do not consume 100% of the hydrogen and helium initially present in each star at birth. With a surface gravity of the order of $\log g \sim 8$, gravitational settling in white dwarf stars is quite efficient and the light elements leftover from previous evolutionary phases rapidly float to the surface, while heavier elements sink out of sight. Since there is ultimately more than enough hydrogen and helium to form an optically thick photosphere, it is not possible to directly observe the white dwarf core material. Thus, the majority of white dwarfs are found to have a surface composition that is completely pure in hydrogen or helium.[1]

Nevertheless, carbon and oxygen features are still found in the spectra of several classes of hydrogen-deficient objects, namely, the PG 1159 stars, and the DQ, DBQ, and Hot DQ white dwarfs. In the case of the hot PG 1159 stars ($T_{\text{eff}} \gtrsim 75\,000$ K), the presence of these elements is somewhat easier to understand since the gravitational separation of the elements is simply not completed yet. A thorough review of the

1) Recent studies have revealed that many cool white dwarfs also have Ca II H and K lines when observed at sufficiently high resolution (see Zuckerman *et al.*, 2010, and references therein). The presence of heavy elements in these objects is now believed to be the result of the accretion of nearby planetesimals or asteroids. Since these objects represent a class of their own, and are detailed elsewhere in this volume, this chapter contains no further discussion of the impure atmospheres found in DAZ, DZ, and DBZ white dwarfs.

White Dwarf Atmospheres and Circumstellar Environments. First Edition. Edited by D. W. Hoard.

observed properties of the extremely hot, hydrogen-deficient post-asymptotic giant branch (post-AGB) stars has already been written by Werner and Herwig (2006), and they are not discussed further here.

In cooler stars, for which the process of radiative levitation can be considered negligible and gravitational settling has had sufficient time to transform PG 1159 stars into helium-dominated objects, other physical mechanisms must be called upon to explain the presence of observable amounts of carbon (and sometimes oxygen). The following chapter is a broad review of the observational signatures, physical properties, and evolution of DQ, DBQ, and Hot DQ white dwarfs, and also presents an overview of the main challenges that future investigations of these types of object should try to address. Although these spectral types together represent only a small fraction of the total number of white dwarfs, they nevertheless provide extremely valuable information about the evolution of stars following the AGB phase as well as the spectral evolution of white dwarfs in general.

3.2 DQ White Dwarfs

3.2.1 Historical Introduction and General Properties

The first mention of the existence of this class of white dwarf can be traced back to Greenstein and Matthews (1957). Very strong, broad, and unidentified features observed at the same wavelength in the spectra of two stars, Luyten 879-14 and Wolf 219, were believed to be correlated with high pressure and possibly also an abundance anomaly. These "λ4670" stars, as they were referred to in the early literature, were too cool for the features to be identified with He II λ4686 or C and N ions near $\lambda\lambda$4640–4660. It was noted that similar bands also appear in "cooler stars than those in which the Balmer lines are strong" and it was proposed that a molecular compound, maybe C_2, could be responsible for the bands. This hunch was later found to be correct and, soon thereafter, model atmospheres including C_2 opacity were used for a quantitative analysis of these white dwarfs (see below). The designation λ4670 was eventually abandoned and replaced by the "C_2" type as the origin of the bands became clear. Finally, as some members of this class also showed atomic C I lines (e. g., G47-18; see Figure 3.1) and other white dwarfs only showed carbon lines in the ultraviolet (see below), they collectively became known as DQ white dwarfs (McCook and Sion, 1987).

Figure 3.1 shows typical spectra of well known members of the DQ spectral class. The principal characteristic of these objects is the presence of a series of broad bands originating from the Swan electronic system of the C_2 molecule. Each band is actually the result of the superposition of thousands of rotational transitions for a given initial and final vibrational quantum number ν. Bands with the same difference in vibrational quantum number, $\Delta\nu$ (0-0, 1-1, 2-2, or 1-0, 2-1, 3-2, etc.; see Figure 3.1) blend to form broad asymmetric features with sharp band heads

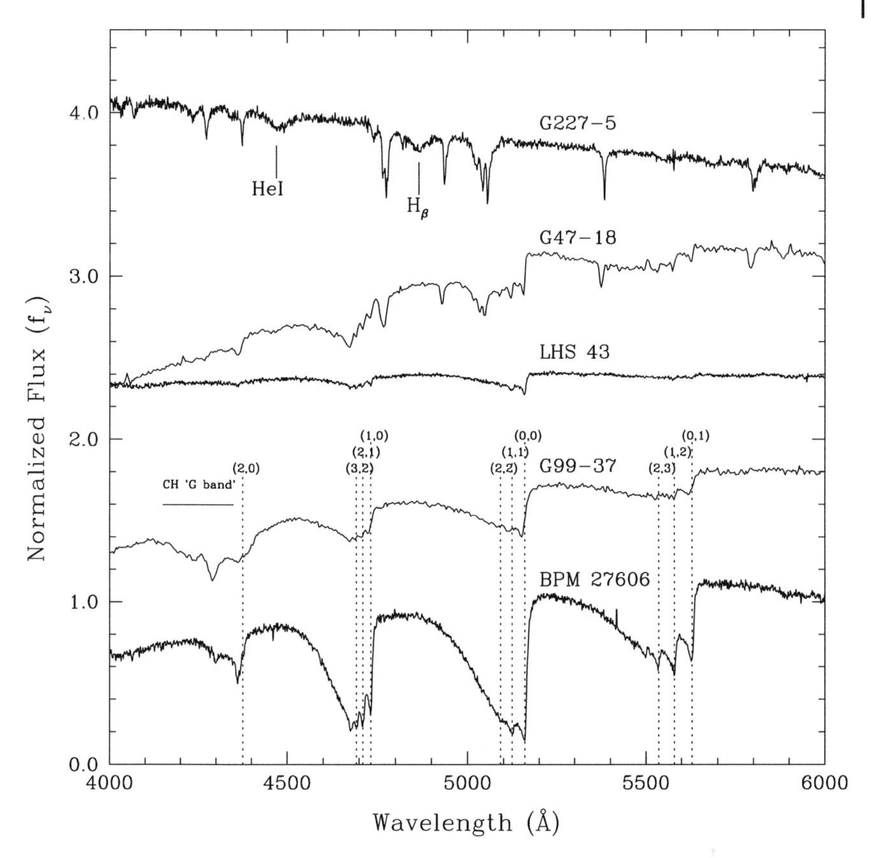

Figure 3.1 Typical DQ white dwarf spectra. The spectra are normalized to unity at 5300 Å and separated from each other by an arbitrary offset for clarity. These objects are characterized by strong asymmetrical molecular bands originating from the C_2 Swan system. The top two objects (G227-5 and G47-18) also show atomic C I lines. Other rare features for DQ stars, such as Hβ, He I, and the CH G band, are indicated in the figure. The location of the band heads for various initial and final vibrational quantum numbers are indicated by vertical dotted lines.

at 5160, 4730, and 5635 Å for $\Delta\nu = 0, +1$, and -1, respectively (see the first few chapters of Herzberg, 1950, for a description of the physics behind the formation of these molecular bands; although the remainder of that book is somewhat dated, it still contains one of the best-written introductions to this subject.). Because the high gas pressure in the atmosphere of cool DQ white dwarfs causes important line broadening, and the individual rotational lines are numerous, lines overlap and the band structure may be regarded as continuous. This is a very fortunate situation for theorists since this permits the use of the so called "Just Overlapping Line Approximation", a formalism that is much more efficient, time-wise, than a full line-by-line computation (Zeidler-K.T. and Koester, 1982). Atomic carbon lines are also observed in the optical for DQ stars with $T_{\rm eff} \geq 9500$ K (e. g., see G227-5 and G47-

18 in Figure 3.1) as well as a weak He I λ4471 line for the hottest members of this class. Lines from heavy elements other than carbon are rarely seen in these objects.

From the general absence of Balmer lines, it was rapidly understood that these white dwarfs had a hydrogen deficient atmosphere. The first analyses based on model atmospheres including C_2 opacity (Grenfell, 1972, 1974) also revealed that the dominant element in the atmosphere was helium, not carbon, in spite of the absence of helium lines and the presence of strong carbon features in the DQ spectra they investigated (a similar conclusion, based on molecular formation calculations, was also reached by Bues, 1973). This is simply because stars below $\sim$ 11 000 K are too cool to populate the electronic levels giving rise to He I transitions (only the hottest DQ white dwarfs, such as G227-5 and G35-26, show weak helium lines). Furthermore, the high pressure together with the low continuum opacity in helium-rich photospheres (relative to that of hydrogen-rich white dwarfs of the same effective temperature) allow even minute amounts of contaminant to appear quite strongly in these cool white dwarf spectra.

By the end of the 1970s and the beginning of the 1980s, the sample of "C_2" stars had grown rapidly, as many previously classified DC stars (white dwarfs with featureless continuum spectra) turned out to be weak DQ white dwarfs when better observations became available (e. g., Wegner, 1983a). At the same time, observations with the International Ultraviolet Explorer (IUE) revealed the presence of many atomic C I lines in the ultraviolet spectra of not only "C_2" white dwarfs, but also in many stars previously classified as DC from optical spectroscopy alone (see Figure 3.2, as well as Weidemann and Koester, 1995, for a complete list of references for these ultraviolet observations). Note that these objects are also classified as DQ type since, according to the classification scheme of McCook and Sion (1987), DQ white dwarfs are objects that show carbon features, either atomic or molecular, in any part of the electromagnetic spectrum (DB stars with carbon lines are an exception and are classified as DBQ; see Section 3.3). With the exception of Procyon B, which was observed with the Hubble Space Telescope (HST; Provencal and Shipman, 1999; Provencal *et al.*, 2002), observations of DQ white dwarfs at ultraviolet wavelengths have only revealed carbon lines and no other elements (see Figure 3.2).

The beginning of the 1980s also saw the appearance of the first comprehensive analyses of DQ white dwarfs as a group (Koester *et al.*, 1982a; Wegner and Yackovich, 1984). In Koester *et al.* (1982a), a sample of DQ and DC white dwarfs was thoroughly analyzed based on observations in the optical as well as in the ultraviolet (from IUE). Using trigonometric parallaxes, they derived radii (and, hence, masses via the mass–radius relationship) for 17 DC and DQ white dwarfs. The mean mass of their sample, $\langle M \rangle = 0.55 M_\odot$, was found to be comparable to that of DA white dwarfs (more recently, Dufour *et al.*, 2005, analyzed 16 DQ white dwarfs with trigonometric parallaxes and found no significant difference in the mean mass compared to that of DB white dwarfs, the most likely progenitors of DQ white dwarfs; see below). In Wegner and Yackovich (1984; also see Yackovich 1982), model atmosphere analyses indicated that these stars were helium-rich objects with traces of carbon ($-3 \leq \log \mathrm{C/He} \leq -7$) and effective temperatures in

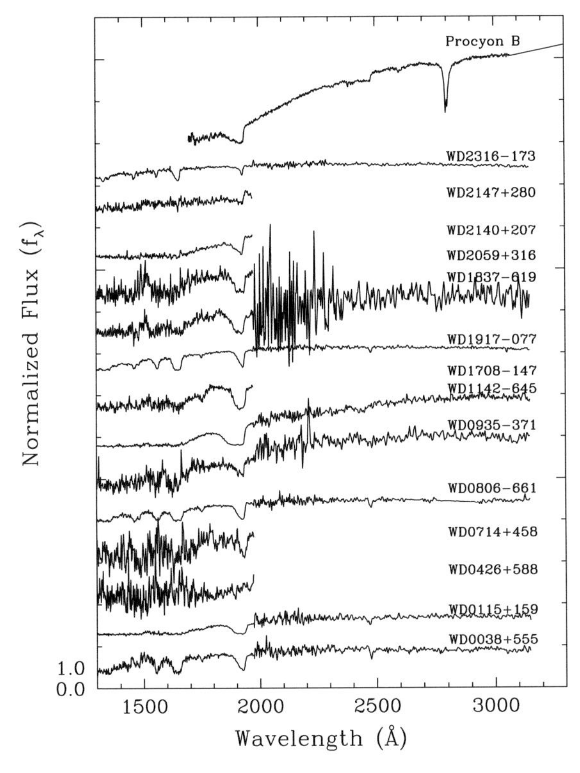

Figure 3.2 Archival spectroscopic observations of DQ white dwarfs in the ultraviolet. All spectra were obtained with IUE except for Procyon B, which was observed with HST. The spectra are normalized to unity and separated from each other by an arbitrary offset for clarity.

the ∼ 6000–11 000 K range. A clear correlation between the abundance of carbon and the effective temperature is also revealed by their analyses, with the hottest objects having the largest amount of atmospheric pollution. That the hottest objects are more polluted is not, a priori, obvious from a simple glance at the optical spectra since the carbon features in the coolest objects tend to be stronger than in their hotter counterparts. For example, the modern carbon abundance for LHS 43 (= L145-141) is more than one order of magnitude higher than that of G99-37 even though the C_2 Swan bands of the latter are almost three times deeper (see Figure 3.1). This is simply a reflection of the relative importance of the different sources of opacity at different effective temperatures. Detailed model atmosphere analyses are, thus, crucial for the interpretation of these objects.

The sample of known DQ white dwarfs has significantly increased (hundreds of new specimens) in the last few years, thanks to the Sloan Digital Sky Survey (SDSS; see Eisenstein *et al.*, 2006, and references therein). Excluding those numerous recent discoveries from the SDSS, Table 3.1 lists parameters for all of the DQ white dwarfs that have been analyzed in the literature to date (refer to Dufour *et al.*, 2005; Gänsicke *et al.*, 2010; Koester and Knist, 2006, for a complete list of 82 new DQ white dwarfs from the SDSS). In cases where an object has been analyzed

Table 3.1 List of known DQ white dwarfs (excluding SDSS) analyzed in the literature.

Name	WD Name	T_{eff} (K)	log C/He	References
G218-8	0038+555	10 900	−6.0	13, 16
G268-40	0042−238	10 500	−2.7	3
G33-49	0115+159	9050	−4.33	3, 14, 16, 19
G35-26	0203+207	12 000	−1.5	11
W219	0341+182	6510	−6.41	3, 8, 14, 16, 19
Stein 2051B	0426+588	6800	−6.7	7
L879-14	0435−088	6300	−6.41	3, 8, 14, 16, 19
G99-37	0548−001	6070	−6.82	1, 2, 14, 16, 19
G87-29	0706+377	6590	−6.4	8, 14, 16, 19
Procyon B	0736+053	7740	−5.5	15, 17
GD 84	0714+458	9000	−7.0	13
L97-3	0806−661	10 250	−5.55	2, 3, 14, 23
G47-18	0856+331	9920	−2.88	1, 2, 3, 8, 14, 16, 19
LP 487-021	0913+104	9000	−4.7	10
LDS 275A	0935−371	9380	−4.21	5, 14, 19
G195-42	0946+534	8100	−5.33	8, 14, 16, 19
LP 93-21	1042+593	8500	−2.9	2
LP 612-033	1115−029	9270	−4.25	16, 19
L145-141	1142−645	7900	−5.14	2, 3, 16, 19
LEHPM 2-4051	1149−272	6188	−7.2	22
ESO 267-110	1157−462	7190	−5.66	14, 19
1222+102	1222+102	11 100	−2.6	21
LHS 5222	1235+422	6250	−5.0	20
GSC2U J131147.2+292348	1309+296	5120	−5.8	18
GD 352	1550+626	8000	−5.0	8
G257-38	1641+732	7000	−3.7	2
L845-70	1708−147	9250	−7.0	13, 14
G227-5	1727+560	12 500	−2.5	9
G184-12	1831+197	7110	−5.87	8, 16, 19
BPM 11668	1837−619	8500	−5.0	6
LDS 678B	1917−077	10 200	−6.7	12
G187-15	2059+316	9360	−3.99	5, 8, 16, 19
G126-27	2140+207	8200	−5.28	2, 3, 16, 19
G188-27	2147+280	11 000	−7.0	13
BPM 27606	2154−512	7000	−4.2	2, 3, 8
EG 156	2254+076	12 230	–	14
G157-034	2311−068	7440	−5.67	16, 19
L791-40	2317−173	10 800	−6.5	4, 13
G171-27	2352+401	7710	−5.19	8, 14, 16, 19

1 Grenfell, (1974); 2 Bues, (1979); 3 Koester *et al.*, (1982a); 4 Koester *et al.*, (1982b); 5 Wegner, (1983c); 6 Wegner, (1983b); 7 Wegner and Yackovich, (1983); 8 Wegner and Yackovich, (1984); 9 Wegner and Koester, (1985); 10 Wegner *et al.*, (1985); 11 Thejll *et al.*, (1990); 12 Oswalt *et al.*, (1991); 13 Weidemann and Koester, (1995); 14 Bergeron *et al.*, (1997); 15 Provencal *et al.*, (1997); 16 Bergeron *et al.*, (2001); 17 Provencal *et al.*, (2002); 18 Carollo *et al.*, (2003); 19 Dufour *et al.*, (2005); 20 Kawka and Vennes, (2006); 21 de Martino *et al.*, (2007); 22 Subasavage *et al.*, (2007); 23 Subasavage *et al.*, (2009).

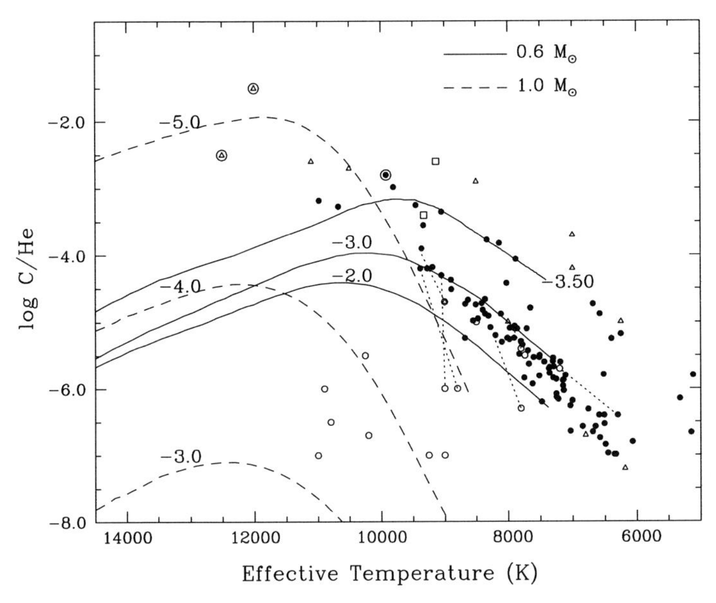

Figure 3.3 Measured carbon abundance as a function of T_{eff} found in the literature. Filled circles were determined as part of the author's work, while the triangles (optical) and open circles (ultraviolet) are from other sources (see Table 3.1). The two squares are the oxygen-rich DQ stars recently uncovered by Gänsicke *et al.* (2010). Note that, to ensure homogeneity in the atmospheric parameter determinations presented here, all the stars from Koester and Knist (2006) have been refitted using the same grid as in Dufour *et al.* (2005). The results are, within the uncertainties, essentially the same. Atmospheric parameters for five stars that have been analyzed from both ultraviolet and optical observations are also connected by dotted lines (see text). The three massive (determined from trigonometric parallax measurements) DQ white dwarfs G47-18, G35-26, and G227-5 are surrounded by larger open circles. Predictions of evolutionary models with 0.6 and $1.0 M_\odot$ are also shown, with the value of $\log M_{\text{He}}/M_\star$ for the various curves indicated in the figure. Note that the evolutionary sequences stop near 7500 K only because of the absence of OPAL opacity data at lower temperatures.

more than once, the table gives the most recent published atmospheric parameters. This census reveals that 39 DQ stars (121 if the SDSS sample is included) have been analyzed in the last $\sim$ 40 years using various model atmosphere codes and techniques as well as data from different regions of the electromagnetic spectrum. Figure 3.3 shows the carbon abundance as a function of effective temperature for all of these DQ white dwarfs (including the SDSS sample). What has been learned from these analyses, as well as some of the most outstanding puzzles that have emerged, are discussed below.

3.2.2 Formation Mechanism

What is the origin of the relatively large amount of carbon found in cool DQ white dwarfs? Several theoretical explanations were put forth in the late 1970s (see Vauclair and Fontaine, 1979, and references therein), though none turned out to be entirely satisfactory. Since gravitational settling timescales and cooling ages were known with sufficient accuracy to rule out a primordial origin, other mechanisms had to be considered. One of the early hypotheses considered was the accretion of carbon-rich material from the interstellar medium. A serious blow to the accretion hypothesis was the absence of hydrogen and other metallic lines in the optical spectra of DQ stars. For instance, limits on the abundance of calcium and magnesium are particularly constraining due to the strong resonance line of Ca II (H and K lines) in the optical and Mg II ($\lambda\lambda$2795.5, 2802.7) in the ultraviolet. Prior to the SDSS, three DQ white dwarfs were known to have traces of hydrogen (G227-5, G35-26, and G99-37; the last of these being deduced from the presence of a molecular band of CH) and only one with metal lines (Procyon B, with $\log \mathrm{Fe/He} = -10.7$, $\log \mathrm{Mg/He} = -10.4$, and $\log \mathrm{Ca/He} = -12$, Provencal *et al.*, 2002; also see Figure 3.2 for a comparison of its ultraviolet spectrum to those of other DQ white dwarfs.). The absence of calcium (optical) or magnesium (ultraviolet) in most DQ white dwarfs essentially eliminated the accretion hypothesis as a viable explanation. Moreover, the derived upper limits on the total accretion rates of hydrogen were too low to account for the carbon abundance pattern found in DQ stars.

Since the cool DQ white dwarfs have helium-dominated atmospheres, it was only natural to assume that their progenitors were the hotter helium-rich DB white dwarfs. Early proposals for the very existence of helium atmosphere white dwarfs envisioned that convective mixing could transform a hydrogen-rich DA into a helium-rich DB white dwarf during the cooling lifetime of the star (Shipman, 1972; Strittmatter and Wickramasinghe, 1971). While this idea has been abandoned as a plausible explanation for the presence of helium-rich white dwarfs in general,[2] it was nevertheless an inspiration to Vauclair and Fontaine (1979) for trying a similar mechanism to explain the presence of carbon in cool helium-rich white dwarfs. In the scenario they explored, convective mixing episodes taking place between the helium envelopes and the underlying carbon layer of a helium-rich white dwarf could indeed produce helium-dominated envelopes with traces of carbon. However, this only worked for some fine-tuned circumstances in which the initial helium content was quite close to the maximum possible mass of the helium convection zone, a situation Vauclair and Fontaine (1979) believed was quite unlikely. The main conclusion they reached was that any lower value of the helium content would lead to an almost pure carbon surface composition, something that was not observed empirically at the time.[3] At the time, the presence of carbon in white dwarf spectra had yet to be explained in a satisfactory manner.

The idea that the helium convective zone penetrates the deeper regions where carbon is abundant was too attractive to be completely dismissed. Carbon had to come from the core, but a way to get around the problem of producing an almost

2) Note that it is believed that DA stars do mix to form DB stars in a few cases, and convective mixing is also the favored mechanism to explain the increase in the non-DA/DA ratio at lower effective temperature; see Tremblay and Bergeron (2008) and references therein.

3) It is amusing to note that a very similar idea has been invoked $\sim$ 30 years later to explain the carbon-dominated atmospheres of Hot DQ white dwarfs; see Section 3.4.

pure carbon atmosphere when mixing occurs had to be found. This was essentially achieved qualitatively with the original proposition of Koester *et al.* (1982a), who speculated that the bottom of the helium convection zone could meet an extended transition zone of mixed carbon and helium with a structure governed by diffusive equilibrium. Within this theoretical framework, when the deep helium convection zone meets this carbon diffusion tail, only a small amount of carbon is dredged up and mixed in the photosphere, producing carbon abundances roughly compatible with the observations. More quantitative formulations of this scenario were later developed by Fontaine *et al.* (1984) and Pelletier *et al.* (1986). To this day, there is considerable confidence that the dredge-up mechanism is the correct explanation for the carbon pollution observed in DQ white dwarfs.

3.2.3 Relation between Carbon Abundance and Temperature: An Overview

What has been learned from the analysis of hundreds of cool helium-rich white dwarfs with traces of carbon? According to the convective dredge-up scenario, the carbon pollution should reach a maximum near $\sim$ 12 000 K (i. e., when convection is close to maximum), and then slowly decrease at lower effective temperatures. This prediction is in agreement with the observed trend in the carbon abundance as a function of effective temperature that emerged from early optical data (Wegner and Yackovich, 1984). The idea of the trend was somewhat challenged when ultraviolet measurements were considered (Weidemann and Koester, 1995, also see the unfilled circles in Figure 3.3), but with the recent addition of the numerous SDSS objects, the existence of a well defined sequence in carbon pollution becomes quite clear (discrepant points are discussed below). Interestingly, since the carbon pollution depends on the distance between the rich carbon core and the base of the helium convective zone, the analysis of a large sample of DQ white dwarfs offers a unique opportunity to empirically measure the thickness of white dwarf helium envelopes, a poorly known quantity. Early attempts to exploit this potential by Pelletier *et al.* (1986) revealed that the mass of the helium layer was about $\log q(\mathrm{He}) = \log M_{\mathrm{He}}/M_\star \sim -3.75$, implying relatively thin white dwarf helium layers by post-AGB evolutionary standards. More recent work by Dufour *et al.* (2005), using a large homogeneous sample of DQ spectra taken mostly from the SDSS, and a new generation of evolutionary models (Fontaine and Brassard, 2005, note that these models incorporated significant improvements over those of Pelletier *et al.* 1986, in particular the use of the latest OPAL opacities), finds that the bulk of DQ stars forms a well defined sequence with $\log q(\mathrm{He}) \sim -2.5$, a value more in line with the predictions of post-AGB models (see the $0.6 M_\odot$ dredge-up model predictions in Figure 3.3).

Thus, it appears that the carbon abundance in most DQ white dwarfs can successfully be explained by standard evolutionary models of $0.6 M_\odot$ white dwarfs with about the expected thickness of the helium envelope. However, Figure 3.3 also shows that many objects are located far above and below the "main sequence" defined by the bulk of the DQ sample (to avoid confusion in terminology, this feature is henceforth referred to as the "DQ main sequence"). Noteworthy is the presence of several white dwarfs with very high carbon abundances, forming an almost parallel sequence one dex above the DQ main sequence (Dufour *et al.*, 2005; Koester and Knist, 2006). The presence of a few stars with a very small carbon pollution is also notable. Most of these have featureless optical spectra and show traces of carbon only in the ultraviolet.

According to the standard dredge-up theory, these unusual carbon abundances can be interpreted as a manifestation of the presence of white dwarfs with different stellar masses and/or a helium envelope thinner or thicker than the average. One of the goals of white dwarf evolution theory is to understand how this diversity of carbon pollution arises and its connection to prior evolutionary phases. If the explanation resides in the thickness of the helium layers, then the problem is transferred to anterior evolutionary phases and a comprehensive theory will be needed to explain how such a diversity in the distribution of helium layer thicknesses is produced. On the other hand, variations in the mass distribution can be verified empirically by examining how the few DQ white dwarfs with mass determinations[4] are distributed in the $\log \mathrm{C/He}$ versus T_{eff} diagram.

Can variations in DQ white dwarf masses be the sole explanation for the presence of the "discrepant" carbon abundances? Based on the fact that the most polluted white dwarf in their sample, G47-18, was significantly more massive ($M = 1.05\,M_\odot$) than the other DQ white dwarfs in the DQ main sequence, Dufour *et al.* (2005) proposed that the discrepant highly polluted DQ stars were probably all massive white dwarfs (the main idea being that a more massive AGB core is expected to leave a smaller helium content, which favors the dredge-up of larger amounts of carbon at DQ temperatures). The massive star hypothesis was further supported by trigonometric parallax measurements of two other highly polluted DQ white dwarfs, G227-5 and G35-26, which confirmed that these objects were also massive ($M_{\mathrm{wd}} \sim 1.0\,M_\odot$). With three out of three highly polluted white dwarfs with trigonometric parallax having high mass (see Figure 3.3), the massive white dwarf hypothesis to explain the "second sequence" appeared very promising, but evolutionary sequences at $1.0\,M_\odot$, recently presented by Fontaine and Brassard (2005), cast some doubt on this interpretation. The result of these calculations, which are reproduced in Figure 3.3, show that the higher carbon pollution found in the three massive DQ stars discussed above can indeed be explained by the evolution of high mass white dwarfs with very thin helium envelopes, as expected for the evolution of more massive stars.

4) Unlike the case of hydrogen-rich DA or helium-rich DB white dwarfs, spectroscopic and photometric analysis of DQ white dwarfs cannot provide an unambiguous determination of the surface gravity (the atmospheric parameters cannot be determined uniquely because a change in $\log g$ can be compensated for by a corresponding change in T_{eff} and $\log \mathrm{C/He}$). Thus, fits are usually done by assuming the canonical $\log g = 8$, unless a trigonometric parallax measurement is available – if the distance from Earth, D, is known, then the solid angle $4\pi(R_{\mathrm{wd}}/D)^2$, obtained by fitting the spectral energy distribution, can be used to obtain the radius of the white dwarf, R_{wd}, which, in turn, can be converted into the mass using the mass–radius relationship.

While high mass evolutionary models provide a logical interpretation for the massive DQ white dwarfs, they completely fail to explain the high carbon abundance of the coolest white dwarfs (again, see Figure 3.3). The slopes of the $1.0 M_\odot$ curves are much steeper than those for $0.6 M_\odot$, and by the time such a massive white dwarf has cooled down to $\sim$ 8000 K, no Swan bands should be detectable, contrary to what is observed. Furthermore, the other two cool white dwarfs on the second sequence that have parallax measurements, BPM 27606 (Koester *et al.*, 1982a) and GSC2U J131147.2+292348 (H. C. Harris, private communication) are both found to have "normal" white dwarf masses near $0.65 M_\odot$, indicating that some of the unusually polluted DQ stars are not necessarily massive. These cool, highly polluted white dwarfs simply appear to be normal mass white dwarfs with thinner helium envelopes. The explanation, within stellar evolution theory, for how these thin helium envelope white dwarfs form, and why there seems to be a dichotomy (as opposed to a continuous distribution between $\log q = -3.5$ and -3.0) in the helium layer thicknesses remains elusive.

A similar conclusion can be reached from the white dwarfs with low carbon abundances since many of these stars appear to also have "normal" (i. e., near $0.6 M_\odot$) masses. Here, again, it might be necessary to invoke differences in the helium envelope thickness to account for these objects. It should be noted, however, that the atmospheric parameters obtained from analysis of ultraviolet data are somewhat more uncertain. The main reason is that many C I atomic lines are found to be asymmetric (in particular, see the C I λ1930 line in all of the DQ stars shown in Figure 3.2) and shifted with respect to laboratory wavelengths. The classical treatment, using van der Waals broadening within the impact approximation, clearly fails for these lines and statistical theory must be used. Attempts by Koester *et al.* (1982a) gave some good results, but the fits are not entirely satisfactory by modern criteria. Some concerns also emerge when the atmospheric parameters are examined for the few stars that have both ultraviolet and Swan band analyses. For example, the atmospheric parameters for five DQ white dwarfs analyzed by Dufour *et al.* (2005) are found to be quite different from those derived from ultraviolet observations (the two sets of $T_{\rm eff}$–log C/He for these five objects are connected with dotted lines in Figure 3.3).

While it might be tempting to blame the differences in the analysis results on systematic differences between the model atmosphere codes that were utilized, additional, as yet unpublished, fits to these ultraviolet data (made by this author) indicate that such is not the case. Although these new models do not yet incorporate asymmetrical profiles for the atomic C I lines, the achieved fits are sufficiently good[5] to conclude that something more fundamental is happening (probably a problem with one of the sources of opacity, atomic/molecular data, or line broadening theories used in the model atmosphere calculations). Nevertheless, these are only "minor" concerns that do not overshadow the fact that there is indeed a population of white dwarfs with a much smaller carbon pollution than is predicted by the standard dredge-up theory. It is possible that the atmospheric parameters

5) For example, the fit to the IUE data is qualitatively similar, in most cases, to that presented in Figure 6 of Subasavage *et al.* (2009) for L97-3 (= WD 0806–661).

presented in Figure 3.3 will shift in one or the other direction when the problems noted above are resolved, but the overall picture is unlikely to change much. Again, the explanation for the presence of these white dwarfs with a low carbon pollution remains elusive.

3.2.4
DQ White Dwarfs with Oxygen

The presence of large amounts of oxygen in the atmosphere of DQ white dwarfs is not expected within the standard diffusion/dredge-up model. Indeed, in the PG 1159→DO→DB→DQ evolutionary scheme, because gravitational diffusion depends on the mass of the elements, only a very small amount of oxygen is expected to be dredged-up along with carbon once lower effective temperatures are reached.[6] This was also empirically supported by various studies (e. g., Kilic *et al.*, 2003; Koester *et al.*, 1982a) since no cool white dwarfs, prior to the SDSS, had ever been found to show any oxygen features in the optical. Only upper limits of the order of C : O = 10 : 1 had been obtained from the absence of the fourth positive system of CO (1700–2000 Å) in ultraviolet observations of cool white dwarfs (Koester *et al.*, 1982a). For hotter stars, C : O $\sim$ 1 : 1 could not be ruled out. However, since there was no observational or theoretical justification for the inclusion of such a large amount of oxygen, this element received practically no attention in subsequent studies.

This situation will no doubt change with the recent discovery in the SDSS of four DQ white dwarfs with oxygen lines. The first two objects, SDSS J090157.92+575135.9[7] and SDSS J142342.64+572949.3, were originally identified by Liebert *et al.* (2003b). They appear to be normal DQ white dwarfs with temperatures of about 12 000 K, but also show atomic O I lines near 6156–6158 and 7772–7775 Å. However, no detailed model atmosphere analysis of these two stars has yet been published. More recently, two new DQ white dwarfs with oxygen lines, SDSS J092208.19+292810.9 and SDSS J110239.69+205439.4, have been thoroughly analyzed by Gänsicke *et al.* (2010). The oxygen abundances found for these two objects are extremely large (about 1%), particularly for SDSS 1102, whose remarkable spectrum shows mainly strong O I lines with only a few weak C I lines (see Figure 1 of Gänsicke *et al.*, 2010). Even more interesting, the O/C ratios for these two oxygen-rich (but still helium-dominated) DQ white dwarfs are found to be $>$ 1 (log O/C $\sim$ 0.6 and 1.4–1.8, respectively), an extremely surprising result that still has no satisfactory explanation within standard evolutionary scenarios. Most certainly, objects with such a large amount of oxygen are the result of the evolution of massive white dwarfs with an oxygen/neon core surrounded by a

6) Note that this assumes a typical PG 1159 composition, by mass fraction, of He/C/O = 0.42/0.43/0.15 at the start of the cooling track. Evolutionary calculations exploring other compositions are highly anticipated.

7) Throughout this chapter, the full name of an SDSS white dwarf is given the first time it is mentioned. Thereafter, only the first four digits of the right ascension portion of its name is used. For example, SDSS J090157.92+575135.9 = SDSS 0901.

carbon/oxygen envelope. Unfortunately, no detailed evolutionary models exploring that hypothesis and explaining the strange surface composition of oxygen-rich DQ stars have been published yet.

To conclude, it is notable that these two DQ white dwarfs, presented as open squares in Figure 3.3,[8] also have carbon abundances that place them above the DQ main sequence in the $\log C/He$ versus $T_{\rm eff}$ diagram (in the so-called "second sequence"). While it is certainly premature at this time to suggest any connection between all of these stars, the idea that they might actually follow a different evolutionary path, that of heavier core white dwarfs, certainly deserves more scrutiny.

3.2.5 The Peculiar ("C_2H"?) DQ White Dwarfs

A group of cool white dwarf stars that exhibit strong unidentified molecular absorption bands similar to the Swan bands are classified as "peculiar DQ" stars (or DQpec; see Figure 3.4 for typical examples). While the first discovered objects of this class (Bergeron *et al.*, 1994; Greenstein and Liebert, 1990) all showed a series of bands that were shifted to the blue by about 100 Å (with respect to the position of the C_2 Swan band), it now appears, thanks to the many new discoveries from the SDSS, that there is a continuum of displacements between the expected wavelength of the C_2 band and the most shifted of the DQpec features. The displaced features also appear to be more "rounded" than normal C_2 band heads, which usually show sharp edges followed by slow degradation toward the blue. The exact opacity source responsible for these bands has not yet been identified although the designation "C_2H" star has tentatively been attributed to them by Schmidt *et al.* (1995). There are several hypotheses, each with their own set of problems, that have been proposed to explain these strange objects. These hypotheses are reviewed and re-examined in the remainder of this section, in light of new information provided by recent white dwarf studies. In the end, the exact nature of these objects remains mysterious, although the belief of this author is that they are the product of C_2 molecules "quenched" by very high atmospheric pressures (this suggestion nicely complements the recent investigation by Hall and Maxwell, 2008, who reached a similar conclusion).

One of the first interpretations by Liebert and Dahn (1983) for the unknown bands observed in LHS 1126 was that they were simply C_2 bands that had been shifted blueward due to the extreme gas pressure in the line-forming region. They also suggested that C_2, under these "crowded" atmospheric conditions, might also be responsible for the infrared flux deficiency detected for this object. This interpretation was later refuted by Bergeron *et al.* (1994), who interpreted the peculiar spectral energy distribution of LHS 1126, especially the large infrared deficit, as the result of collision-induced absorption (CIA) by molecular hydrogen due to collisions with helium (see Chapter 2 by M. Kilic for a discussion of this mechanism). Within the context of mixed hydrogen and helium atmospheres (as opposed to a

8) Since the effective temperature derived from photometry is in disagreement with that determined from spectroscopy, an average of the values quoted in Gänsicke *et al.* (2010) is used.

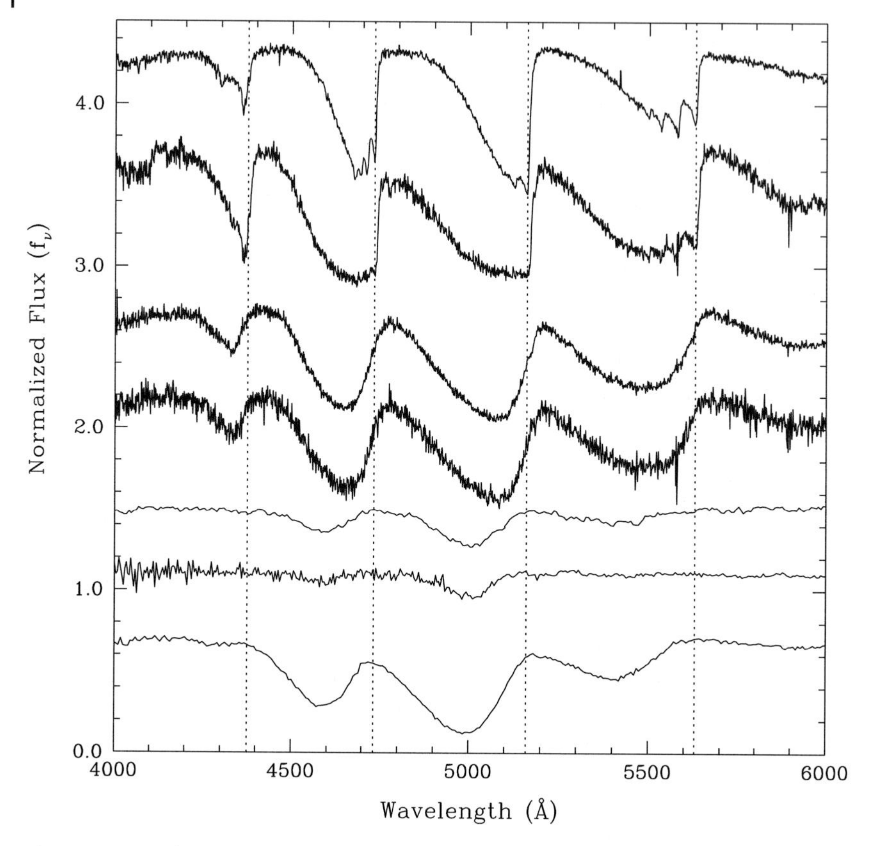

Figure 3.4 Typical DQpec (bottom five) and DQ (top two) spectra showing a gradual progression of the shifted features. Note that the second white dwarf from the top seems to be on the verge of becoming a DQpec. The spectra are normalized to unity and are separated from each other by an arbitrary offset for clarity.

helium-rich atmosphere with traces of carbon), the pressure in the continuum-forming region is much lower (in fact, even lower than that of some normal DQ white dwarfs), and the pressure-shift interpretation is no longer viable. A similar conclusion was also reached for three other peculiar DQ stars which all showed infrared deficits and an abundance of hydrogen near $\log \mathrm{H/He} \sim -2$ (see Bergeron *et al.*, 1997, for a detailed photometric analysis of these stars).

Although the spectral energy distribution seemed to be well explained by mixed H/He models, the strong bands were still lacking a plausible explanation. A shift due to a strong magnetic field was considered, but spectropolarimetric measurements rapidly ruled out that possibility (Schmidt *et al.*, 1995). At the time, no "normal" DQ white dwarfs were found with temperatures below $\sim$ 6500 K (note that this cannot simply be explained by cooling timescale, since the Galactic disk is old enough for DQ stars to have cooled to much lower effective temperatures). Meanwhile, the estimated effective temperatures derived from the spectral energy distributions of DQpec stars were all found to be cooler than this apparent DQ cutoff

(Bergeron *et al.*, 1997). Thus, it was strongly suspected that the two spectral types were somehow connected. Furthermore, given that the DQpec stars all appear to have a large hydrogen abundance (derived from the infrared flux deficiency), it was only natural to suspect that the formation of a molecular compound involving carbon and hydrogen might be responsible for a "replacement" of the C_2 bands by the shifted features at lower effective temperature. Based on a detailed chemical equilibrium analysis of H/He/C mixtures under the estimated physical conditions encountered in DQpec, Schmidt *et al.* (1995) found that the C_2H molecule was preferentially formed at the photosphere of these objects.

However, although this interpretation is very attractive, it has a few problems. The first, and not the least, is that experimental and theoretical investigations of the C_2H molecule do not find shifted bands near the observed position of the DQpec bands (see the recent review by Hall and Maxwell, 2008, and references therein). Second, the parameter space explored by Schmidt *et al.* (1995) was perhaps not entirely appropriate for such cool DQ white dwarfs after all. In particular, they examined abundances of carbon that were much higher ($\log \text{C/He} = -2$ to -5) than what is currently found for the coolest DQ white dwarfs ($\log \text{C/He} \sim -7$; see Figure 3.3). The remainder of this section presents a re-examination of the molecular formation of various C_nH_n compounds using state-of-the-art DQ model atmospheres.[9] The approach used here differs from that of Schmidt *et al.* (1995) in that the atmospheric pressure is not fixed to a constant value; instead, it utilizes real DQ model atmospheres with chemical compositions that are, in light of recent DQ investigations (Dufour *et al.*, 2005; Koester and Knist, 2006), more realistic for such stars. It should first be noted that, with the exception of three stars with temperatures near 5000 K (discussed further below), no DQ white dwarfs are found below 6000 K, and that the carbon abundance of DQ stars at that effective temperature is $\log \text{C/He} \sim -7.0$ (see the end of the DQ main sequence in Figure 3.3). Thus, it seems only logical to assume that if DQ white dwarfs transform into DQpec near 6000 K, as proposed by Schmidt *et al.* (1995), the latter would have about the same carbon abundance. Assuming for the moment that this hypothesis is correct, and that a significant quantity of hydrogen somehow appears at the photosphere of such stars (also see Bergeron *et al.*, 1997), then could the sudden disappearance of DQ stars below $\sim$ 6000 K be explained as the result of the formation of a molecular species involving hydrogen and carbon?

White dwarf model atmospheres with $\log g = 8.0$, $T_{\text{eff}} = 6000$ K, $\log \text{C/He} = -7.0$, and various hydrogen abundances were calculated to address this question. According to these synthetic spectrum calculations, the C_2 bands indeed disappear as the hydrogen abundance approaches $\log \text{H/He} \sim -2.0$, with molecular species such as CH and C_2H_2 being preferentially formed. However, even for a hydrogen abundance as low as $\log \text{H/He} \sim -3.0$, the CH G band should easily be detected in a DQ star with such parameters. With the exceptions of G99-37 (see Figure 3.1) and BPM 27606 (Vornanen *et al.*, 2010), the CH G band has never been detected in

9) Note that the molecular equilibrium code used here has been explicitly verified to reproduce the results presented in Figure 4 of Schmidt *et al.* (1995).

any DQ white dwarf, which imposes a stringent upper limit on the hydrogen abundance of cool DQ stars near the observed cutoff. Empirical evidence thus seems to indicate that a $C_2 \rightarrow C_nH_n$ transition is highly improbable as an explanation for the absence of DQ stars below $T_{\rm eff} \sim 6000$ K. Finally, cooler models with $T_{\rm eff} = 5500$ K, $\log {\rm C/He} = -7.5$ (the approximate carbon abundance expected from an extrapolation of the DQ sequence to 5500 K), and $\log {\rm H/He} = -2.0$ (about what is needed to explain the infrared deficit with CIA) were explored, and reveal that CH, CH_2, CH_3, CH_4, C_2H_2, and C_2H are the most abundant molecules under such conditions. None of these molecules is known to have transitions that match those of the DQpec white dwarfs (see Hall and Maxwell, 2008, for a review of the various candidates).

To conclude, the sudden appearance of a large quantity of hydrogen (due to an as yet unknown mechanism) is a plausible explanation for the absence of objects with Swan bands near 6000 K. However, in this scenario, many objects with the CH G band would be expected near the alleged DQ→DQpec transition. This, combined with the fact that no C_nH_n molecules are known to show shifted (relative to the Swan bands) absorption bands, seems to indicate that a satisfactory interpretation for the DQpec white dwarfs has not yet been formulated. Finally, it is possible that the shifted features originate from a species that has not been considered, or that the inconsistencies between ultraviolet and optical analyses noted above turn out to significantly affect the arguments presented here. However, until a suitable candidate is found from laboratory observations (or theoretical calculations), other avenues should be explored.

A New Look at the High Pressure Arguments

Current observations of DQ white dwarfs seem to indicate that the hydrogen abundances near the cutoff are relatively low ($\log {\rm H/He} < -4.0$). Without a significant contribution from hydrogen, carbon becomes the main free electron donor in the atmospheric plasma of these stars. As the carbon abundance decreases with lower effective temperature, the dominant He^- free–free opacity, which is proportional to the number of free electrons, also decreases and the atmosphere becomes more and more transparent. Consequently, as the photosphere becomes deeper, the gas pressure in the line-forming region reaches extremely high values. According to the DQ model atmospheres explored here, the pressure at an optical depth of $\tau_{\rm R} = 1$ for DQ white dwarfs near the cutoff is such that the mean distance between the particles is not much larger than the size of the C_2 molecule.

Under such extreme conditions, it is certainly reasonable to question the survivability of the C_2 molecule as well as its absorption properties. Unfortunately, pressure dissociation of molecules and high pressure effects in helium fluids are not included in the models. Nevertheless, with relatively simple calculations, it is possible to assess the influence of the high atmospheric pressure on the C_2 molecules. One way to take into account, to first order, the effect of the interactions with surrounding particles is the occupation probability formalism of Hummer and Mihalas (1988). Within this formalism, the perturbation of excited levels by neutral particles is treated by considering that atoms are hard spheres with a characteristic radius associated with each excited level. A level is considered destroyed if its characteristic radius is larger than the mean interparticle distance (the excluded volume effect). For given physical conditions, an electron in level i has a probability w_i of being bound to the atom and a probability $1 - w_i$ of being ionized. The effect of the interactions are taken into account by including the occupation probability of each level in the calculation of a modified partition function and Boltzmann equation (see Hummer and Mihalas, 1988, for details of how to evaluate the occupation probability w_i). This procedure has been employed successfully for the modeling of the higher Balmer lines in DA white dwarfs (Bergeron *et al.*, 1991) and He I lines in DB white dwarfs (Beauchamp, 1995), and is now the standard approach to treat lines perturbed due to the high densities encountered in white dwarf atmospheres.

In order to evaluate possible high pressure effects on molecules, similar ideas can be used to evaluate the partition function of C_2 under the atmospheric conditions encountered in cool DQ white dwarfs. Current calculations[10] show that by the time the gas pressure has reached $\sim 10^{10}\,\mathrm{dyn\,cm^{-2}}$, the partition function is practically reduced to zero, indicating that molecules may not survive in such environments.

More interestingly, the accompanying synthetic spectrum calculations including occupation probability for C_2 indicate that for carbon abundances similar to those found on the DQ main sequence, the Swan bands smoothly disappear as the effective temperature approaches 6000 K. Hence, the interpretation of the DQ cutoff as a natural consequence of the higher atmospheric pressure with cooling appears promising. Moreover, the three cool DQ stars below the cutoff, near 5000 K (see Figure 3.3), can also be explained within this theoretical hypothesis. Indeed, these stars are found to have carbon abundances that are much higher, and the corresponding gas pressure at the photosphere is lower (see Figure 3.5). These calculations indicate that for a carbon abundance that is one dex higher than on the DQ main sequence, the Swan bands can survive until $\sim$ 5000 K.

Of course, the calculations presented here cannot be used as a definite confirmation that molecular dissociation actually occurs. High pressure effects in the equation of state and opacities are more complex (see Bergeron *et al.*, 1994, and Kowalski, Saumon, and Mazevet 2005 for an exploration of such effects) and this simple exploratory toy model can only be considered, at best, qualitative in nature. Nevertheless, faced with the lack of a satisfactory explanation for the DQpec stars, it is probably not too far-fetched to imagine that high pressure might also be responsible for the displaced bands. Under these conditions, one could speculate that the absorption properties of the C_2 molecules might be affected in such a way that they resemble those observed in DQpec stars. A more daring speculation is that a high density carbon/helium mixture could also be responsible, as originally proposed by Liebert and Dahn (1983) for the infrared absorption observed in these stars. If true, this would give a rather coherent view of the evolutionary path followed by these objects. Unfortunately, calculations confirming this hypothesis are not yet

10) Details of the assumptions made for the calculation of the partition function and the size of the C_2 molecule were first presented in the last chapter of Dufour (2007).

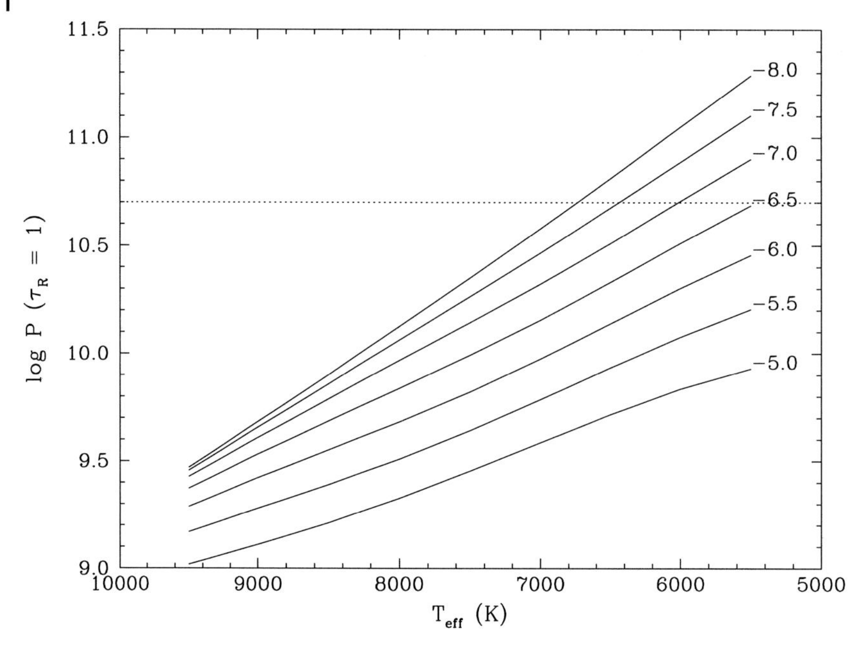

Figure 3.5 Atmospheric pressure at $\tau_R = 1$ as a function of effective temperature and carbon abundance. The new models discussed here indicate that for atmospheric parameters near $T_{eff} \sim 6000$ K and $\log C/He = -7.0$ (the approximate location of the DQ cutoff), the pressure is about the same ($\sim 10^{10.7}$ dyn cm^2; dotted line in the figure) as for $T_{eff} \sim 5000$ K and $\log C/He = -6.0$, parameters similar to that of the three cool stars below the DQ cutoff.

available. Until more sophisticated equations of state and non-ideal treatment for the opacity of helium and C_2 are included in model atmosphere calculations, all this is highly speculative and the nature of the DQpec stars remains mysterious.[11)]

3.3
Carbon and Oxygen in DBQ White Dwarfs

White dwarfs found with traces of carbon in the ultraviolet and an optical spectrum dominated by He I lines are classified as DBQ. The detection of carbon in such stars has been made possible only recently thanks to the highly sensitive instruments on board HST and the Far-Ultraviolet Spectroscopic Explorer (FUSE); only upper limits had been derived from low resolution observations with IUE in the 1980s (see Wegner and Nelan, 1987). This spectral class currently comprises eight members (see Table 3.2), but it is likely that most (if not all) DB white dwarfs have their atmospheres polluted at some level since carbon or oxygen was detected in

11) As this work was being finalized, the author learned that new calculations by Kowalski (2010) show that, indeed, high pressure effects on the C_2 molecule can explain the shifted features.

Table 3.2 Carbon abundance of DB white dwarfs observed in the ultraviolet.

Name	WD name	T_{eff} (K)	log C/He	Other elements	References
GD 408	0002+729	14 660	−8.8	Si, Fe	3
G270-124	0110−068	20 820	−7.8	Si, S, Fe	3
PG 0112+104	0112+104	31 300	−6.15	–	1, 4
BPM 17088	0308−565	22 600	−5.5	–	2
GD 190	1542+182	22 000	−5.5	–	1, 2
GD 358	1645+325	24 700	−6.0	–	1
GD 378	1822+410	16 600	−6.9	O, Si, S, Fe	3
EC 20058-5234	2005−525	28 000	−6.0	–	2

1 Provencal *et al.*, (2000); 2 Petitclerc *et al.*, (2005); 3 Desharnais *et al.*, (2008); and 4 Dufour *et al.*, (2010).

every object but one that has been observed with sufficient sensitivity (Desharnais *et al.*, 2008; Petitclerc *et al.*, 2005; Provencal *et al.*, 2000).

While the presence of carbon in helium-rich white dwarfs of lower effective temperature can easily be explained by the dredge-up scenario (see previous section), the origin of carbon in DB stars is more challenging. As discussed at length in Petitclerc *et al.* (2005), none of the traditional mechanisms normally called upon are entirely adequate in the DB temperature range. Indeed, past investigations showed that radiative support is essentially negligible for stars below 30 000 K, while the convective dredge-up of carbon from the core, which reaches a maximum near 12 000 K, decreases significantly for effective temperature above 13 000 K. The latter mechanism predicts carbon abundances between log C/He ~ -8 and -5 for effective temperatures of $\sim$ 19 000 and 14 000 K, respectively. However, empirical determinations only offer mixed results (Desharnais *et al.*, 2008, also see their Figure 12 for a summary of carbon abundances in helium-rich white dwarfs): GD 378 has a carbon abundance that is consistent with dredge-up predictions while GD 408 has a measured carbon abundance several orders of magnitude below what is expected. Note, however, that the presence of carbon in GD 408 can alternatively be explained by the accretion of small asteroids with chondritic compositions, since Si and Fe are also detected in its photosphere. Nevertheless, the low carbon abundance of GD 408 certainly challenges the theory on the hot side of the dredge-up model.

Also interesting is the detection of oxygen in two DB stars with temperatures near 17 000 K (GD 61 and GD 378; Desharnais *et al.*, 2008). Unfortunately, the published literature concerning the dredge-up of oxygen along with carbon is rather limited. Based on simple estimates, Desharnais *et al.* (2008) concluded that the observed oxygen in these two objects is not inconsistent with the expectation from dredge-up theory. It must be noted, however, that both of these stars also exhibit traces of Si and Fe, probably indicating that some, if not all, of the detected oxygen originates from an accreted asteroid. Thus, it is desirable to increase the sample of known DB stars with oxygen (as well as that of DQ white dwarfs with oxygen;

see Section 3.2.4), preferably uncontaminated by other heavy elements since O/C pattern ratios can potentially be linked to the oxygen core mass fraction, which is related to the astrophysically important, and not well constrained, $^{12}C(\alpha, \gamma)^{16}O$ reaction rate.

For effective temperatures above 19 000 K, the presence of carbon is somewhat more troublesome. Indeed, the only remaining viable scenario to explain the presence of carbon in that temperature regime is that it is left over from previous evolutionary phases. However, by the time a PG 1159 star cools to 30 000 K, the C/He ratio should be less than 10^{-15} if gravitational diffusion is left unimpeded. Faced with a dearth of explanations, Fontaine and Brassard (2005) proposed a model in which the gravitational settling is slowed by the presence of a small residual wind. According to their calculations, the abundance of carbon observed in hot DB white dwarfs can be accounted for by various mass loss rates in the range of (1.0–5.0) $\times 10^{-13} M_\odot \text{ yr}^{-1}$. This small wind would presumably die as the star cools to about 20 000 K, at which point convective dredge-up would gradually take over.

However, a close inspection of the carbon abundance pattern found in the hottest DB white dwarfs reveals that the wind model, even though it is the best hypothesis so far, is not devoid of problems. Indeed, the carbon abundances for the DB white dwarfs above 21 000 K are all found in the range log C/He ~ -5.5 to -6.0 (see Table 3.2), almost forming a plateau over a range spanning 10 000 K (see Figure 12 of Desharnais *et al.*, 2008). If a weak residual wind is responsible for slowing down gravitational diffusion, and this wind is slowly fading away as the star cools off (Fontaine and Brassard, 2005), then a range of carbon abundances going roughly from log C/He ~ -3.0 to -10.0 is expected, contrary to what is observed. The wind strength must instead be artificially increased toward lower effective temperatures in order to obtain the correct carbon abundance near 22 000 K. One possibility to explain the apparent plateau is that the current sample of stars have different wind strengths and just happen to have similar carbon abundances. Such a situation would be unlucky and unlikely, but with a sample of only five hot DB white dwarfs above 21 000 K that have a carbon abundance determination, it is too soon to draw an accurate picture of the evolution of objects in this temperature regime where both wind and gravitational settling operate simultaneously. As usual, more observations and theoretical study are needed.

3.4 Hot DQ White Dwarfs

3.4.1 Historical Introduction and General Properties

White dwarfs that show atomic carbon features in their optical spectra and that are situated in the DB temperature range (12 000 K and higher) are commonly called "Hot DQ" stars. Unfortunately, this terminology is confusing since the Hot DQ class, when defined this way, comprises stars with atomic C I features, which are simply the hottest versions of the DQ stars discussed in Section 3.2 as well as the fascinating carbon atmosphere white dwarfs that show mainly ionized C II lines which are the subject of this section[12]. In the following, the Hot DQ stars are considered to be only those white dwarfs with atomic C II lines in the optical while the original DQ designation is retained for the hottest DQ white dwarfs with neutral C I lines in their optical spectra (i. e., stars with spectra similar to those of the well known G227-5 and G35-26; see Figure 3.1 and Table 3.1).

Optical spectra of Hot DQ white dwarfs were first presented by Liebert *et al.* (2003b) in a paper that discussed many newly found white dwarfs from the SDSS with atomic oxygen and/or carbon lines. While many of the stars in their list of carbon white dwarfs were simply hot (12 000–14 000 K) DQ stars with neutral carbon lines consistent with maximal dredge-up pollution, some were found to be extremely hot (by DQ standards). These hot stars showed mainly ionized carbon lines and no helium lines (see Figure 4 of Liebert *et al.*, 2003b for typical SDSS spectra of these objects; Figures 6 and 7 in this work show higher signal-to-noise ratio observations of all Hot DQ white dwarfs known at this time). They were first believed to be simply hotter versions of the helium rich (but highly carbon polluted) white dwarfs G227-5 and G35-26, the hottest DQ stars known at the time. Indeed, simple estimates of the effective temperature, based on pure helium model atmosphere fits of the SDSS $g - i$ colors, placed these few rare objects in the DB temperature range (20 000–30 000 K). Since the spectra of pure helium atmospheres in that temperature range are expected to show strong He I lines, it was argued at the time that the higher opacity of the carbon contaminants could mask the dominant helium constituent[13]. However, it was recognized that estimates based on pure helium models were likely to be in substantial error (for example, a pure helium model color fit for the well known star G227-5 was about 5000 K hotter than the value derived by Wegner and Koester, 1985), and that they could be much cooler. Nevertheless, the presence of C II lines certainly indicated that these objects were much hotter than "normal" DQ white dwarfs with atomic C I lines, although rigorous stellar atmosphere modeling including all the appropriate opacities was needed to determined their effective temperatures with more precision.

This is exactly the task undertaken by this author about four years after the first publication of Hot DQ star spectra by Liebert *et al.* (2003b). The new model atmosphere code developed for the DQ white dwarfs in Dufour (2007) was perfectly appropriate (after a few updates for some opacity sources) for modeling these hot carbon rich white dwarfs. While fitting the spectra of the many warm objects similar to G227-5 and G35-26 was relatively simple, the hot C II line white dwarfs could not be reproduced with helium-rich models with traces of carbon. The prob-

12) To add even more confusion to this classification scheme, the DB stars that show ionized carbon lines in the ultraviolet are simply called DBQ white dwarfs, not "Hot DQ".

13) This is a good illustration of how a dominant paradigm, in this case the dredge-up of carbon in a helium-rich atmosphere, can sometimes unduly influence thought processes and prevent even good researchers from thinking "outside the box". In hindsight, the obvious alternative was that these objects had practically no helium after all. To be fair, however, at the time, nobody had ever calculated what a carbon atmosphere spectrum would look like.

lem was that every combination of $\log g$, $T_{\rm eff}$, and $\log \mathrm{C/He} < 0$ predicted strong He I lines, contrary to what is observed in the Hot DQ spectra. The hypothesis that the helium lines were simply masked by the high carbon opacity, as proposed by Liebert *et al.* (2003b), was thus shown to be incorrect and an alternative had to be found to explain the absence of helium lines. A stunning realization followed, that is, a carbon dominated atmosphere not only solved the problem of the helium lines but also provided quite a nice fit to the observed C II lines (Dufour *et al.*, 2007a). Preliminary analysis indicated effective temperatures of about 18 000 to 24 000 K with little or no trace of hydrogen or helium (Dufour *et al.*, 2008a). This was totally unexpected as no spectral evolution theory predicted the existence of objects with such an extreme composition. As noted in Section 3.3, within the standard diffusion/dredge-up/wind theory, the predicted carbon abundances in that temperature range are, at most, of the order of $\log \mathrm{C/He} \sim -5$. This is several orders of magnitude lower than the lower limit determination for Hot DQ stars (about $\log \mathrm{C/He} > +0$ to $+2$ depending on the effective temperature and surface gravity considered). Clearly, DQ white dwarfs with carbon as the dominant constituent of the atmosphere were "something different" (Dufour *et al.*, 2008a).

The catalogue of spectroscopically confirmed white dwarfs from the 4th data release of the SDSS (Eisenstein *et al.*, 2006) uncovered nine such stars with carbon dominated atmospheres, all of which have been analyzed in detail in Dufour *et al.* (2008a). Atmospheric parameters presented in that paper should, however, only be considered preliminary as they are based on the first generation of model atmospheres for this type of object (considerable improvement has recently been

Table 3.3 List of the 14 known hot DQ white dwarfs.

Name	Plate	MJD	Fiber	Notes
SDSS J000555.90−100213.3	650	52143	37	magnetic
SDSS J010647.92+151327.8	422	51878	422	oxygen, magnetic?, He I λ4471?, Hβ
SDSS J023637.42−073429.5	455	51909	403	oxygen?, magnetic
SDSS J081839.23+010227.5	2077	53846	575	oxygen
SDSS J110406.68+203528.6	2488	54149	149	oxygen?, magnetic?
SDSS J115305.54+005646.2	284	51943	533	oxygen, Hβ
SDSS J120027.73+225212.9	2643	54208	469	oxygen?, magnetic?
SDSS J133710.19−002643.7	299	51671	305	pulsating, oxygen, Hβ
SDSS J140222.25+381848.8	1642	53115	449	magnetic, He I λ4471
SDSS J142625.70+575218.4	789	52342	197	pulsating, magnetic, He I λ4471
SDSS J161531.71+454322.4	814	52443	577	oxygen?
SDSS J220029.09−074121.5	717	52468	462	pulsating, oxygen?, magnetic
SDSS J234843.30−094245.2	648	52559	585	pulsating, oxygen?
LAWDS NGC 2168 28	–	–	–	in cluster M35, magnetic?

incorporated into the models and will be published by this author in due course) and utilized rather noisy SDSS spectra (considerable progress has been made on the observational front, as well – see below). Subsequent SDSS data releases have recently revealed the existence of four new Hot DQ white dwarfs (S. Kleinman, private communication). Finally, one Hot DQ white dwarf has been discovered in the open star cluster M35 (NGC 2168) by Williams *et al.* (2006), for a total of 14 objects known so far. These 14 Hot DQ white dwarfs, as well as their main characteristics (to be discussed below), are listed in Table 3.3.

In order to obtain atmospheric parameters that are as reliable as those obtained from DA and DB star fitting, much higher signal-to-noise (S/N) ratio observations than those provided by the SDSS are needed. Following the discovery of the extreme composition of Hot DQ white dwarfs, a spectroscopic follow-up program was established to obtain high S/N ratio observations of all the known Hot DQ white dwarfs (see Dufour *et al.*, 2009b, for an early progress report). To date, all but two stars have been re-observed using the 6.5 m MMT telescope and the 10 m Keck-I telescope. The best spectroscopic observations in the optical currently avail-

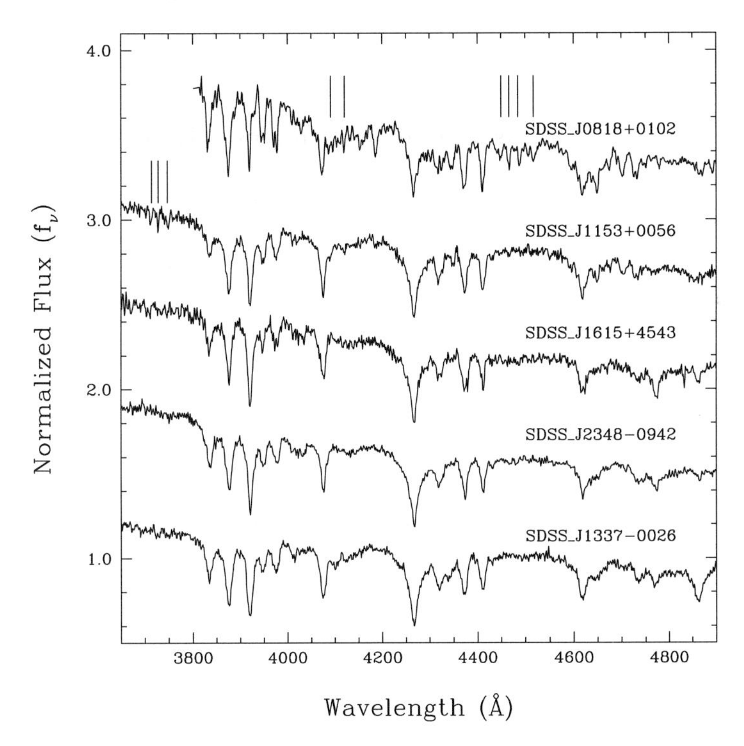

Figure 3.6 Optical spectroscopic observations for the nonmagnetic Hot DQ white dwarfs. All spectra were obtained with the 6.5 m MMT telescope, except for that of SDSS J081839.23+010227.5, which is an SDSS spectrum (a three-point average window smoothing has been applied in the display of this object). The spectra are normalized to unity at 4500 Å and offset from each other for clarity. Tick marks note the position of oxygen lines.

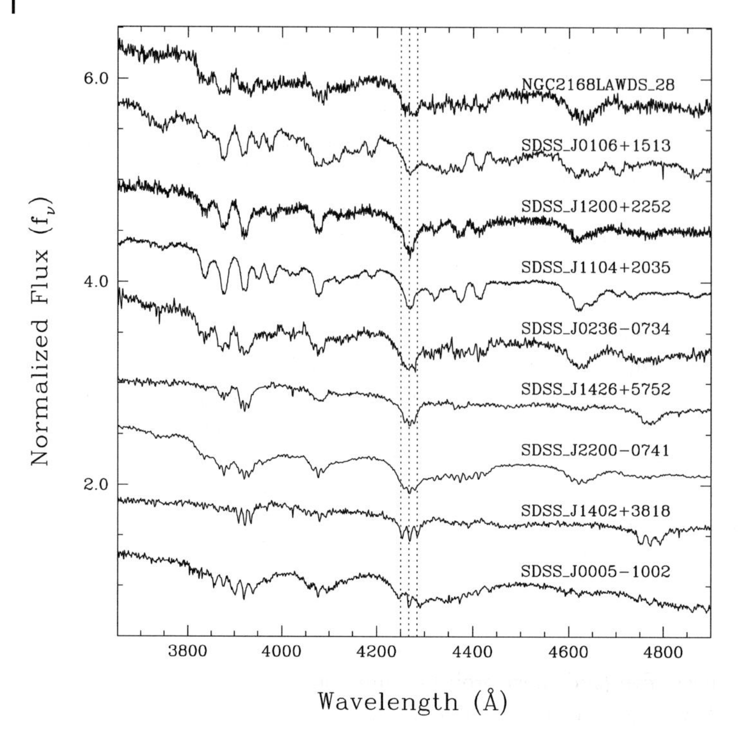

Figure 3.7 Optical spectroscopic observations of the magnetic (or suspected to be magnetic) Hot DQ white dwarfs. All spectra were obtained with the 6.5 m MMT telescope, except for LAWDS NGC 2168 28, SDSS J120027.73+225212.9, and SDSS J110406.08+203528.6, which were obtained with the Keck I telescope (K.A. Williams, private communication). The dotted lines indicate the expected location, for C II $\lambda 4267$, of the π (unshifted) and σ components for a magnetic field of 2 MG.

able for the 14 known Hot DQ white dwarfs are presented in Figures 3.6 and 3.7. What has been learned from these new observations, as well as from time series photometry of the Hot DQ stars, is described in the following sections.

3.4.2
Magnetism

The presence of strong stellar magnetic fields can usually be assessed quite easily from the presence of Zeeman line splitting in spectroscopic observations. Based on the rather noisy SDSS spectra of the first discovered Hot DQ white dwarfs (Dufour *et al.*, 2008a), only one object, SDSS J000555.90−100213.3, clearly showed signs of line splitting, with the separation of the components corresponding to a mean surface field strength of $\sim$ 1.5 MG. Furthermore, based on the unusually broad appearance of some carbon lines (presumably due to unresolved split line components), a weak magnetic field was suspected for two other stars, SDSS J010647.92+151327.8

and SDSS J220029.09−074121.5. However, the low S/N and resolution of the available spectra prevented confirmation of this.

As higher S/N spectroscopic observations became available, it was quickly realized that the presence of magnetic fields in these objects was not serendipitous but, rather, a common property among Hot DQ stars. Indeed, of the 14 known Hot DQ white dwarfs, five have now been shown to have clear signs of magnetic line splitting while four others appear to have unusually broadened lines, possibly the result of unresolved components of lines slightly split by a weak magnetic field. Figure 3.6 shows the optical spectra of the five objects that do not appear to show any trace of a magnetic field. Given the spectroscopic resolution of these observations, only upper limits of the order of 200–300 kG can be obtained for these stars. Only spectropolarimetric measurements which are, unfortunately, not yet available for these faint stars ($g \sim 17.7$–19.7) would reveal the presence of weaker fields. In contrast, Figure 3.7 shows the objects with resolved Zeeman components as well as those suspected to have a weaker field. While the presence of a magnetic field cannot be claimed with certainty for the latter objects, a visual comparison of their spectra in Figure 3.7 with the spectra of the nonmagnetic stars shown in Figure 3.6 seems to support this hypothesis (note in particular how the cores of the lines are sharper and deeper in the spectra of the nonmagnetic stars). Further observations are warranted for confirmation.

As noted above, five (and most probably nine) out of the 14 known Hot DQ white dwarfs have been shown to be magnetic. This fraction of magnetic objects among this spectral class (35–64%) is incredibly high considering that the fraction of magnetism found in samples of nearby white dwarfs is only about 10–15% (Liebert *et al.*, 2003a). Additionally, it is possible that even the stars that appear nonmagnetic will turn out to have weak fields below 200 kG when spectropolarimetric observations eventually become available. Hence, the atmospheric parameter determination based on nonmagnetic models presented in Dufour *et al.* (2008a) are, at least for the magnetic objects, unreliable. A new generation of magnetic model atmospheres are currently being developed that will include the effects of magnetically split line blanketing. These new models, which will be presented in due time, should provide much improved atmospheric parameters for these objects.

To conclude, such a high frequency of magnetic white dwarfs among the Hot DQ class raises many new interesting questions: are all Hot DQ white dwarfs magnetic at some level? If so, does magnetism have anything to do with the evolution of these stars (mass loss, convective mixing, diffusion, etc.)? Are the observed magnetic fields fossil fields (i. e., the result of magnetic flux conservation as a magnetic main sequence star shrinks in radius and becomes a white dwarf) or are they self-generated from a dynamo-type mechanism? Close to nothing is known about all these issues and future investigations should be aimed at resolving these questions in order to reach a better understanding of the evolution of Hot DQ white dwarfs.

3.4.3 Oxygen

Oxygen, along with carbon, is expected to be present in large proportion in the cores of white dwarf stars. Given it is believed that Hot DQ stars probably represent bare stellar cores (see Section 3.4.5), it would not be surprising to also find a significant quantity of oxygen in the atmospheres of these objects. In their analysis of low S/N SDSS observations, Dufour *et al.* (2008a) concluded, based on the absence of clear oxygen features in the spectra they analyzed, that the oxygen abundances had to be at least close to an order of magnitude lower than those of carbon. However, since that analysis, new atmospheric models, as well as higher S/N spectroscopic observations, have become available. Of particular importance to the present discussion is the discovery that there was a small error in the implementation of the oxygen continuum opacity used in the model calculations presented in Dufour *et al.* (2008a). Hence, the conclusion concerning the amounts of oxygen in Hot DQ stars was perhaps premature and, consequently, it must be reconsidered (see below).

Moreover, since the Dufour *et al.* (2008a) analysis, one of the newly discovered Hot DQ stars, SDSS J081839.23+010227.5, has been found to show strong oxygen features (see Figure 3.6). Thanks to new high S/N spectroscopic observations, there is also a clear detection of weak oxygen lines in three other objects (SDSS J010647.92+151327.8, SDSS J115305.54+005646.2, and SDSS J133710.19–002643.6). In addition, very weak oxygen features are possibly present in five more stars (see question marks in Table 3.3), but it would be preferable to wait for a detailed analysis using models that include magnetic blanketing before claiming any detection. A preliminary model atmosphere analysis of these white dwarfs with oxygen lines (Dufour *et al.* 2011, in preparation) indicates that the amount of oxygen can be quite significant, with abundances as high as 50% in some cases.

The presence of a significant amount of oxygen in some Hot DQ white dwarfs also considerably complicates the analysis of these objects. Indeed, while determining the atmospheric parameters of mono-elemental atmosphere white dwarfs such as DA or DB stars only requires navigating in a two-dimensional parameter space ($\log g$ and $T_{\rm eff}$), fitting models to carbon atmosphere white dwarfs now seems to necessitate the consideration of at least five parameters ($\log g$, $T_{\rm eff}$, and the hydrogen, helium and oxygen abundances). On top of that, magnetism also has to be considered for most objects, increasing the complexity in the analysis.

3.4.4 Pulsations

The extreme chemical composition of Hot DQ white dwarfs, as well as their magnetic properties, were more than enough to make these objects extremely intriguing. Nevertheless, some of them had still more remarkable characteristics waiting to be discovered. Indeed, a search for photometric variability among this new class of objects has revealed that four Hot DQ stars are also nonradial pulsators (Barlow *et al.*, 2008; Dunlap *et al.*, 2010; Montgomery *et al.*, 2008), adding a fourth class to the variable white dwarf family after the ZZ Ceti, V777 Herculis, and GW Virginis stars (see Fontaine and Brassard, 2008, for a recent review).

The existence of pulsating carbon atmosphere white dwarfs opens an exciting new astroseismological window through which further study of the physical properties of Hot DQ white dwarfs will be possible. However, the full potential of astroseismology analysis will only be possible if multiple modes are identified in the light curves of these variables. Unfortunately, the pulsating Hot DQ stars discovered so far are not very rich in periodicities. The first discovered pulsating Hot DQ, SDSS J142625.70+575218.4, was found to possess, in addition to the fundamental mode (417.71 s) and its first harmonic (208.85 s), only one additional faint periodicity at 319.72 s. Furthermore, the second mode uncovered in this faint white dwarf ($g = 19.16$) revealed itself only through some 106.4 h of useful photometry collected over a 40 day period (Green *et al.*, 2009). Clearly, long observational campaigns will be needed to uncover, if present, additional vibrational modes in this object. The brighter ($g = 17.7$) Hot DQ white dwarf SDSS J220029.29+074121.5 offers a little more potential since follow-up observations by Dufour *et al.* (2009a) revealed the possible presence of three unrelated oscillations (654.4, 577.6, and 254.7 s) as well as the first harmonic of the main periodicity at 327.2 s. However, the false alarm probability for two of these periodicities is relatively high and more sensitive observations are needed to confirm their existence. Dufour *et al.* (2009a) also report two periods (1044.2 and 416.9 s) and no harmonics above the noise level in the observations of SDSS J234843.30−094245.2 ($g = 19.0$). Finally, Dunlap *et al.* (2010) have recently found two harmonically related periodicities (169 and 339 s) in SDSS J133710.19−002643.7.

Although significant progress on the observational front will be needed before a full astroseismological analysis becomes possible, the pulsations observed so far already present some remarkable characteristics worthy of the attention of pulsation theorists. Most noteworthy is the amplitude of the first harmonic relative to the dominant oscillation for SDSS J142625.70+575218.4 and SDSS J220029.29+074121.5 (30 and 82%, respectively; see Figure 3.8). Such a large amplitude for harmonics is unusual for white dwarfs (Fontaine and Brassard, 2008). Moreover, the first harmonic for these two objects is found to be in antiphase with the main oscillation, which produces a highly nonlinear pulse shape with a flat maximum and a sharp minimum when the light curve is folded on the main period (Barlow *et al.*, 2008; Dufour *et al.*, 2009a; Green *et al.*, 2009; Montgomery *et al.*, 2008). High S/N observations also reveal (see Figure 3.7) that both of these objects are magnetic white dwarfs. These two stars are the only known white dwarfs to possess both large magnetic fields and nonradial pulsations. From a theoretical point of view, the rarity of pulsating white dwarfs with high magnetic fields is not very surprising since convective motions of charged particles, which are an important ingredient in the pulsation driving mechanism, are probably partially quenched by magnetic field lines. It is interesting to note that if one artificially turns off the convection in ZZ Ceti white dwarf models, pulsations are completely extinguished (G. Fontaine, private communication). This is not the case for Hot DQ white dwarfs since the

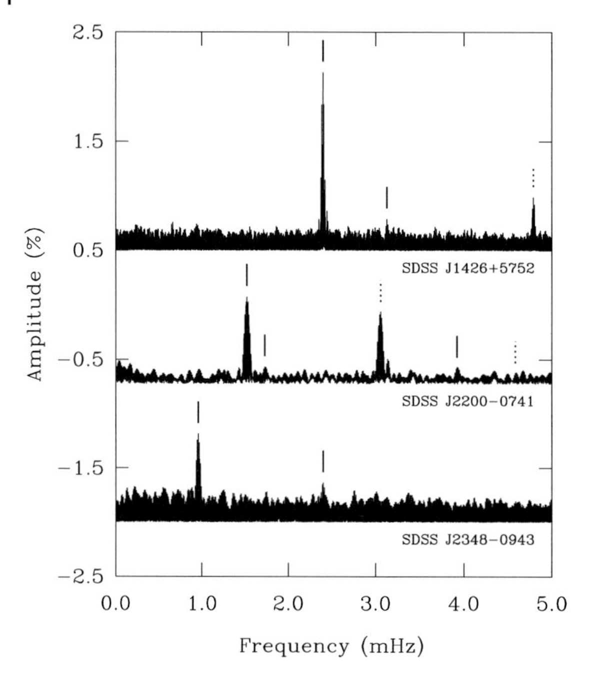

Figure 3.8 Fourier transforms for three of the four known pulsating Hot DQ white dwarfs. Solid tick marks indicate the location of the periodicities found in these objects, while dotted marks represent harmonics of the fundamental frequency.

huge opacity peak in the envelope (due to partial ionization of carbon and oxygen) is, by itself, sufficient to drive pulsational instabilities via the usual κ mechanism.

Could the strong magnetic fields observed in pulsating Hot DQ white dwarfs also be responsible for the strange amplitudes (and strange light curve shapes) found in the two objects discussed above? The absence of a detectable harmonic in the Fourier transform of the nonmagnetic white dwarf SDSS J234843.30−094245.2 (see Figure 3.8), as opposed to the strong harmonics found in the magnetic pulsators, certainly hints at that possibility (Dufour *et al.*, 2008b). However, the recently discovered pulsating Hot DQ star SDSS J133710.19−002643.7 (Dunlap *et al.*, 2010), which shows a strong first harmonic but no signs of magnetism, calls this interpretation into question.

The pulsational properties of Hot DQ white dwarfs are, thus, more mysterious than ever and much more work will be required before a detailed understanding of these objects emerges. To that effect, the vast body of knowledge gathered on the rapidly oscillating Ap stars, the main sequence analogs of the pulsating magnetic Hot DQ white dwarfs, might prove to be an extremely useful guide for future investigations. Meanwhile, on the observational side, it is important to find out if the subtle hints for the possible presence of amplitude modulations associated with rotation presented in the analyses of Green *et al.* (2009) and Dufour *et al.* (2009a) can be confirmed, as this would make these Hot DQ stars true white dwarf equivalents to rapidly oscillating Ap stars. Much more work remains to be done on the theoretical front as well since the nonadiabatic calculations presented in Fontaine *et al.* (2008) and Dufour *et al.* (2008a) relied on stellar parameters obtained from the first generation of model atmospheres and equilibrium models that had no oxygen and convection zones undamped by magnetic field lines.

3.4.5
Formation and Origin

How can white dwarfs with such extreme surface compositions form? What is the past (and future) evolutionary history of such objects? Answering these questions is key to a full understanding of post-AGB evolution, particularly that of stars massive enough to ignite carbon and heavier elements in their cores (see below). Empirical evidence now indicates that an explanation for the existence of such carbon/oxygen-rich objects must also necessarily involve a spectral evolution scheme as well. The reason for this is that all the Hot DQ white dwarfs that have been uncovered so far are found in a rather narrow range of effective temperature ($\sim$ 18 000–23 000 K). Despite an extensive search through the latest SDSS data release, no Hot DQ star outside this small window has ever been found (note that if such an object was present in the SDSS archives, it would have been found easily thanks to its very strong and recognizable spectral characteristics; e. g., see Dufour *et al.*, 2008a, 2009b). Hence, it appears that the Hot DQ phenomenon is only a short-lived phase in the cooling history of some peculiar white dwarfs. They must show a different surface composition, probably helium-rich (see below), outside the 18 000–23 000 K temperature range.

A spectral evolution scenario with the characteristics mentioned above was originally presented in Dufour *et al.* (2007a). The central idea of that scenario is inspired by the very special characteristics of H1504+65, the hottest member of the hydrogen-deficient PG 1159 class ($T_{\rm eff} \sim$ 200 000 K; Werner *et al.*, 2004). In addition to being the hottest, H1504+65 is also unique among the PG 1159 stars in that it is not only hydrogen deficient, but helium deficient as well. Its surface composition, by mass fraction, has been determined to be about 48% C, 48% O, 2% Ne, and 2% Mg (limits for helium are less than 1%; see Werner *et al.*, 2004, and references therein). This is very different from the surface composition of a normal PG 1159 star, where abundances are found in the range $\sim$ 30–85% He, 15–60% C, and 2–20% O (Werner and Herwig, 2006). The strange surface composition of H1504+65 is believed to be the result of a late helium-shell flash that occurred at the post-AGB or white dwarf stage. It is conjectured that this flash induced envelope mixing, which has almost completely eliminated all the remaining hydrogen and helium. Thus, H1504+65 could be a bare stellar core. It is also possible that H1504+65 had a different evolutionary history than normal PG 1159 stars and that it has gone through carbon burning in its past. In this scenario, it would presumably have an O/Ne/Mg core, on top of which a C/O-rich envelope is directly visible. If so, H1504+65 may be the result of the evolution of a "heavy-weight" intermediate-mass star near $8\,M_{\odot}$ (Werner *et al.*, 2004).

H1504+65 is the only other known post-AGB object, along with the Hot DQ stars, to be both H- and He-deficient. As such, it is only natural to consider it as a candidate progenitor for the Hot DQ white dwarfs. However, a simple cooling evolutionary path linking H1504+65 to the Hot DQ stars runs into the problem of explaining why no carbon/oxygen atmosphere white dwarfs are found at intermediate effective temperatures (between 23 000 and 200 000 K). As mentioned above, the absence of such objects necessarily implies that cooled down versions of H1504+65 do not have C/O surface compositions. The original idea proposed by Dufour *et al.* (2007a) is that a small and undetectable residual amount of helium present in the atmosphere of H1504+65 eventually diffuses upward to form a thin, but optically thick, layer above the C/O envelope (note that $\sim 10^{-14}$ to $10^{-15} M_\star$ would be sufficient to disguise the star as a white dwarf with a helium-rich atmosphere). As a consequence, such objects would simply appear, at least spectroscopically, as normal DO or DB white dwarfs along the cooling track. With further cooling, a convection zone develops in the subphotospheric mantle due to the recombination of carbon (and oxygen). Eventually, this convective zone meets the overlaying helium layer and completely dilutes it from below since the mass of the C/O convective zone is orders of magnitude larger than the mass of the helium layer. As a result of this mixing episode, such stars would presumably suffer a dramatic spectral change, transforming themselves into carbon (and oxygen) dominated atmospheres with abundances roughly in agreement with observational determinations. A more quantitative formulation of this hypothesis presented by Althaus *et al.* (2009) recently lent strong support to the validity of this scenario, although details linking the various possible initial conditions at the PG 1159 stage with the properties of individual Hot DQ white dwarfs (detailed composition, mass, etc.) still need to be worked out. Nevertheless, with an evolutionary connection between the hot PG 1159 star H1504+65 and the cooler Hot DQ white dwarfs firmly established, a new window on the core composition and evolution of post-AGB stars has been opened.

Similarly, since no white dwarfs with carbon/oxygen dominated atmospheres are found below $\sim$ 18 000 K, Hot DQ white dwarfs must undergo another drastic spectral change with further cooling. Although exact details are still unknown due to a lack of proper models, it is conjectured that the helium, which has been convectively diluted in the massive carbon/oxygen envelope, finds its way back to the surface for effective temperatures below $\sim$ 18 000 K (Dufour *et al.*, 2007a). In that context, it is interesting to note that the two coolest Hot DQ stars, SDSS J140222.25+381848.8 and SDSS J142625.70+575218.4, both exhibit a weak He I λ4471 line, possibly indicating that they are in the process of converting back into a helium-rich surface composition. Making predictions about the exact spectral appearance of such stars with further cooling would be, without evolutionary calculations, only speculation at this point. However, it is certainly interesting to note that some helium-rich DQ white dwarfs have recently been found to have larger than expected amounts of carbon and oxygen (Gänsicke *et al.*, 2010; Liebert *et al.*, 2003b, also see Section 3.2.4).

To summarize, the Hot DQ white dwarfs, together with H1504+65, appear to represent a new evolutionary channel for the formation of hydrogen-deficient stars. Together, these objects have much to reveal about stellar evolution, especially if it can be confirmed that such stars have, as suspected, at one point in time ignited carbon in their core. If so, then they could have, under the observed superficial C/O envelope, an O/Ne/Mg core (or perhaps even heavier elements if their initial mass on the main sequence was very near the limiting mass – $\sim$ 8 to $10 M_{\odot}$ – to produce supernova explosions). Preliminary analysis of observations using the new generation of model atmospheres indicates that the Hot DQ stars have higher surface gravities than previously believed (mainly due to the fact that it is now necessary to include a significant amount of oxygen to reproduce the observations; see Dufour *et al.*, 2008a), indicating that their progenitors might indeed have been very massive.

3.4.6
Concluding Remarks on the Hot DQ Stars

The realization that the Hot DQ white dwarfs were not merely hotter helium-rich DQ stars but, rather, a new type of object entirely, was made only a few years ago. As can be expected for such a young field of research, a deep understanding of all the aspects associated with these mysterious stars has not yet been achieved, although tremendous progress has been made in a short period of time (e. g., discovery of pulsations, magnetism, and oxygen).

It now appears that before further progress in this field can be made, more accurate atmospheric parameters will be needed. In particular, obtaining precise surface gravities (or masses) and chemical compositions (of utmost importance is oxygen) of Hot DQ white dwarfs is critical to test the details of post-AGB evolutionary theories. These parameters are also essential to reassess the predictive ability of pulsational models. The most challenging aspect will certainly be to successfully extract these important parameters despite the presence of a strong magnetic field in some objects. On the observational front, obtaining high resolution spectropolarimetric measurements, a difficult task for such faint objects is desirable in order to establish the true fraction of magnetism among Hot DQ white dwarfs.

Finally, it should be noted that all of the observations available so far have only been obtained in the optical part of the electromagnetic spectrum. With effective temperatures in the 18 000–23 000 K range, most of the energy emitted by the Hot DQ white dwarfs is in the ultraviolet. According to the most recent models discussed herein, the far-ultraviolet part of the Hot DQ white dwarf spectra should be severely affected by the presence of numerous carbon absorptions features. This absorption causes an important flux redistribution toward longer wavelength and, as a consequence, Hot DQ models appear much bluer in the optical than those of DB stars of the same effective temperature. Thus, it is of utmost importance to ensure that the ultraviolet regions of the spectrum are modeled correctly in order to be confident that the atmospheric parameters based on optical data alone are reliable. The author's program of ultraviolet spectroscopic observations of Hot

DQ white dwarfs with the Cosmic Origins Spectrograph on HST has recently been completed for five targets, and preliminary analysis indicates that the new models reproduce the observations quite well. The full analysis of these data is currently under way and will be presented in due time.

3.5 Conclusion

This brief tour of cool white dwarfs with traces of carbon and oxygen reveals that although tremendous progress has been made in the last four decades, there are still many unresolved problems. While many of these problems may only appear as minor details to non-specialists, it is, in fact, only through a close examination of those details that many of the hidden secrets of stellar evolution will be empirically revealed. Hopefully, this short review succeeded in demonstrating that the study of even relatively marginal spectral types of white dwarfs with traces of carbon (and sometimes oxygen), namely the DQ, DBQ, and Hot DQ stars, can be very rich in information about stellar evolution in general. Hopefully, a better quantitative understanding of the origin and evolution of these objects will emerge in the next few years. However, until all the problems noted above are properly addressed, no claim can be made that post-AGB evolution and white dwarf spectral evolution are fully understood.

This chapter concludes with a short list, not exhaustive by any means, of questions and outstanding problems which should be the focus of future research efforts in this area. Once resolved, they are the most likely to reveal the most about white dwarf evolution, in particular, and post-AGB evolution, in general.

1. It is now well established that the carbon pollution found in many DQ white dwarfs cannot be explained using only the canonical helium envelope thickness predicted by post-AGB evolution. How nature produces such an apparent diversity in helium layer thicknesses (assuming this is the correct interpretation) is not well understood at this point. Since variation in the mass distribution of these objects is not found to be a convincing explanation, what is it that makes a star evolve to a given carbon pollution sequence? Why is there a dichotomy (and perhaps even a trichotomy if the results from ultraviolet analyses is considered) in the distribution of carbon abundances, instead of a continuous distribution between the observed sequences? Would ultraviolet observations of more stars uncover a new, well-defined sequence or will there just be a new scattering of points to add to Figure 3.3? The answers to these questions might hold the key to a better understanding of the late phases of stellar evolution.
2. What is the nature of the DQpec white dwarfs? Are they the result of high atmospheric pressure effects (an explanation favored by the author), or simply a species of molecular absorption that has not yet been considered?
3. Why are so few DQ stars found with traces of hydrogen or heavy elements (DQA and DQZ types, respectively)? From over 200 cool DQ stars examined in

the latest SDSS data release (DR7), only five DQZ (H and K Ca II lines) and eight DQA (Hα) stars are found (Dufour *et al.* 2011, in preparation). On the other hand, DZ white dwarfs are numerous and about $\sim$ 25% of them also have hydrogen (Dufour *et al.*, 2007b). Given that it has now been established that the metals observed in many white dwarf stars most probably originate from the accretion of asteroids or planets that orbited the star and survived the post-AGB evolution (see Chapters 5 and 6 by J. Farihi and J.H. Debes), does the rarity of DQZ white dwarfs imply that planetary debris systems, if present, are more effectively destroyed in the late phases of evolution of these "born again" objects (or, similarly, does the rarity of DQA white dwarfs imply no accretion of watery/icy planetesimals)?

4. One major problem with the simple hypothesis outlined in the previous item is that heavy elements and hydrogen are found frequently in the hotter DB white dwarfs (i. e., DBZ and DBA types; see Zuckerman *et al.*, 2010), the most likely progenitors of the DQ stars. Alternatively, this might indicate that these elements are completely diluted in a massive, growing helium convection zone at lower effective temperature. However, as noted above, heavy elements and hydrogen are also frequently found in many cool helium-rich white dwarfs of spectral type DZ/DZA (Dufour *et al.*, 2007b). Are the convection zones of DZ and DC white dwarfs, which apparently did not dredge-up carbon in sufficient quantity to produce Swan bands, significantly different than those of DQ stars? If so, why? To summarize, what does the intriguing dearth of metals and hydrogen in cool DQ stars reveal about the helium convection zone sizes and the origin of various types of helium-rich white dwarfs (DQ, DZ/DZA, and DC)? Clearly, some of the physical mechanisms at work are not yet well understood.
5. What fraction of the cool helium-rich white dwarfs (DQ, DZ, DC) have DB stars as progenitors? What fraction are the result of convective mixing of DA white dwarfs with thin hydrogen layers?
6. What is the origin of the rare DQ white dwarfs with a large amount of oxygen found by Gänsicke *et al.* (2010) and Liebert *et al.* (2003b)? Are they O/Ne/Mg core white dwarfs?
7. How can the apparent plateau in carbon abundances observed in DB white dwarfs with temperatures between $\sim$ 31 000 and 21 000 K be explained? Is the weak wind interpretation the correct one?
8. Carbon abundances have been measured in only two DB stars in the 14 000–20 000 K temperature range, barely enough to test the theory on the "hot side" of the dredge-up maximum (i. e., above $\sim$ 12 000 K). How good is the dredge-up theory on the "hot side"? Clearly, more ultraviolet observations are needed before that question can be answered.
9. What is the true fraction of Hot DQ white dwarfs with magnetic fields, and what is the origin of those fields? Does magnetism play any significant role in the evolution of these objects?
10. What is the significance of the presence of large amounts of oxygen in a few Hot DQ white dwarfs? What initial conditions (chemical composition, mass, etc.) at the PG 1159 stage would produce Hot DQ white dwarfs with high oxygen abun-

dance? Are all the Hot DQ white dwarfs cooled down versions of H1504+65? Do they have an O/Ne/Mg (or perhaps even heavier elements) core?

11. Four Hot DQ stars are now known to be variable (pulsating) stars. Since the amplitudes of the modes are rather small, it is quite possible that other objects in this class are also pulsating but have gone undetected. Apart from obtaining better observations, which is not an easy task given the faintness of the objects and the relatively small amplitudes of the modes, the two main challenges to address are:
    – To explain why the amplitudes of the first harmonics of the two magnetic Hot DQ white dwarfs are so large. Does it have anything to do with the presence of the magnetic field? (In this regard, it should be noted that SDSS J133710.19−002643.7, which is not found to be magnetic, also shows this strange behavior.)
    – To accurately predict, based on measured atmospheric parameters, which stars should or should not be variable. The first attempt by Dufour *et al.* (2008a) clearly failed to do that, but only because the predictions were based on atmospheric parameters obtained from the analysis of noisy SDSS spectra with first generation model atmospheres that did not include the effects of oxygen and magnetism.
12. Is there an evolutionary connection between the Hot DQ stars, the oxygen-rich DQ white dwarfs (Gänsicke *et al.*, 2010), and some of the various "families" of cooler DQ stars found throughout the T_{eff} versus log C/He diagram?

As answers to some of these questions are gradually found, key pieces of the white dwarf evolutionary puzzle should fall into place. This will certainly provide a much more detailed understanding of the evolution of stars in general, of the various formation channels of white dwarfs, and also of the spectral evolution of white dwarfs as they cool. While answering these questions (as well as many others not mentioned here, or new ones that will emerge) will require a lot of progress on both the theoretical (evolutionary models, model atmospheres, etc.) and observational (ultraviolet observations, spectropolarimetry, spectroscopy, etc.) fronts, the time and effort invested in such an endeavor will be quite rewarding.

Acknowledgments

I wish to thank Alexandros Gianninas and Amélie Simon for carefully reading this manuscript, and Pierre Bergeron and Gilles Fontaine for many useful discussions. Patrick Dufour is a CRAQ postdoctoral fellow.

References

Althaus, L.G., García-Berro, E., Córsico, A.H., Miller Bertolami, M.M., and Romero, A.D. (2009) *Astrophys. J. Lett.*, **693**, L23.

Beauchamp, A. (1995) Ph.D. Thesis, Université de Montréal.

Barlow, B.N., Dunlap, B.H., Rosen, R., and Clemens, J.C. (2008) *Astrophys. J. Lett.*, **688**, L95.

Bergeron, P., Wesemael, F., and Fontaine, G. (1991) *Astrophys. J.*, **367**, 253.

Bergeron, P., Ruiz, M.-T., Leggett, S.K., Saumon, D., and Wesemael, F. (1994) *Astrophys. J.*, **423**, 456.

Bergeron, P., Ruiz, M.T., and Leggett, S.K. (1997) *Astrophys. J. (Suppl.)*, **108**, 339.

Bergeron, P., Leggett, S.K., and Ruiz, M.T. (2001) *Astrophys. J. (Suppl.)*, **133**, 413.

Bues, I. (1973) *Astron. Astrophys.*, **28**, 181.

Bues, I. (1979) in *White Dwarfs and Variable Degenerate Stars.* Proceedings of IAU Colloq. 53 (eds H.M. van Horn and V. Weidemann), University of Rochester, Rochester, p. 186.

Carollo, D., Koester, D., Spagna, A., Lattanzi, M.G., and Hodgkin, S.T. (2003) *Astron. Astrophys.*, **400**, L13.

de Martino, D., Koester, D., Treves, A., Sbarufatti, B., and Falomo, R. (2007) in *ASP Conf. Ser.* Proceedings of the 15th European Workshop on White Dwarfs, vol. 372 (eds R. Napiwotzki and M.R. Burleigh), Astronomical Society of the Pacific, San Francisco, p. 273.

Desharnais, S., Wesemael, F., Chayer, P., Kruk, J.W., and Saffer, R.A. (2008) *Astrophys. J.*, **672**, 540.

Dufour, P. (2007) Ph.D. Thesis, Université de Montréal.

Dufour, P., Bergeron, P., and Fontaine, G. (2005) *Astrophys. J.*, **627**, 404.

Dufour, P., Liebert, J., Fontaine, G., and Behara, N. (2007a) *Nature*, **450**, 522.

Dufour, P. *et al.* (2007b) *Astrophys. J.*, **663**, 1291.

Dufour, P., Fontaine, G., Liebert, J., Schmidt, G.D., and Behara, N. (2008a) *Astrophys. J.*, **683**, 978.

Dufour, P., Fontaine, G., Liebert, J., Williams, K., and Lai, D.K. (2008b) *Astrophys. J. Lett.*, **683**, L167.

Dufour, P., Green, E.M., Fontaine, G., Brassard, P., Francoeur, M., and Latour, M. (2009a) *Astrophys. J.*, **703**, 240.

Dufour, P., Liebert, J., Swift, B., Fontaine, G., and Sukhbold, T. (2009b) *J. Phys. Conf. Ser.*, **172**, 012012.

Dufour, P. *et al.* (2010) *Astrophys. J.*, **718**, 647.

Dunlap, B.H., Barlow, B.N., and Clemens, J.C. (2010) *Astrophys. J. Lett.*, **720**, L159.

Eisenstein, D.J. *et al.* (2006) *Astrophys. J. (Suppl.)*, **167**, 40.

Fontaine, G., Villeneuve, B., Wesemael, F., and Wegner, G. (1984) *Astrophys. J. Lett.*, **277**, L61.

Fontaine, G. and Brassard, P. (2005) in *ASP Conf. Ser.* Proceedings of the 14th European Workshop on White Dwarfs, vol. 334 (eds D. Koester and S. Moehler), Astronomical Society of the Pacific, San Francisco, p. 49.

Fontaine, G. and Brassard, P. (2008) *Publ. ASP*, **120**, 1043.

Fontaine, G., Brassard, P., and Dufour, P. (2008) *Astron. Astrophys.*, **483**, L1.

Gänsicke, B.T., Koester, D., Girven, J., Marsh, T.R., and Steeghs, D. (2010) *Science*, **327**, 188.

Green, E.M., Dufour, P., Fontaine, G., and Brassard, P. (2009) *Astrophys. J.*, **702**, 1593.

Greenstein, J.L. and Liebert, J.W. (1990) *Astrophys. J.*, **360**, 662.

Greenstein, J.L. and Matthews, M.S. (1957) *Astrophys. J.*, **126**, 14.

Grenfell, T.C. (1972) Ph.D. Thesis, University of Washington.

Grenfell, T.C. (1974) *Astron. Astrophys.*, **31**, 303.

Hall, P.B. and Maxwell, A.J. (2008) *Astrophys. J.*, **678**, 1292.

Hummer, D.G. and Mihalas, D. (1988) *Astrophys. J.*, **331**, 794.

Herzberg, G. (1950) *Molecular Spectra and Molecular Structure I: Spectra of Diatomic Molecules*, vol. 2, Van Nostrand Reinhold, New York.

Kawka, A. and Vennes, S. (2006) *Astrophys. J.*, **643**, 402.

Kilic, M., Winget, D.E., von Hippel, T., Lester, D.F., and Saumon, D. (2003) in *White Dwarfs, NATO Science Series II: Mathematics, Physics and Chemistry*, vol. 105 (eds D. de Martino, R. Silvotti, J.-E. Solheim, and

R. Kalytis), Kluwer Academic Publishers, Dordrecht, p 169.

Koester, D. and Knist, S. (2006) *Astron. Astrophys.*, **454**, 951.

Koester, D., Weidemann, V., and Zeidler, E.-M. (1982a) *Astron. Astrophys.*, **116**, 147.

Koester, D., Weidemann, V., Zeidler, E.M., and Vauclair, G. (1982b) *Astron. Astrophys.*, **113**, L13.

Kowalski, P.M. (2010) *Astron. Astrophys.*, **519**, L8.

Kowalski, P.M., Saumon, D., and Mazevet, S. (2005) in *ASP Conf. Ser.* Proceedings of the 14th European Workshop on White Dwarfs, vol. 334 (eds D. Koester and S. Moehler), Astronomical Society of the Pacific, San Francisco, p. 203.

Liebert, J., Bergeron, P., and Holberg, J.B. (2003a) *Astron. J.*, **125**, 348.

Liebert, J. and Dahn, C.C. (1983) *Astrophys. J.*, **269**, 258.

Liebert, J. *et al.* (2003b) *Astron. J.*, **126**, 2521.

McCook, G.P. and Sion, E.M. (1987) *Astrophys. J. (Suppl.)*, **65**, 603.

Montgomery, M.H., Williams, K.A., Winget, D.E., Dufour, P., De Gennaro, S., and Liebert, J. (2008) *Astrophys. J. Lett.*, **678**, L51.

Oswalt, T.D., Sion, E.M., Hammond, G., Vauclair, G., Liebert, J.W., Wegner, G., Koester, D., and Marcum, P.M. (1991) *Astron. J.*, **101**, 583.

Pelletier, C., Fontaine, G., Wesemael, F., Michaud, G., and Wegner, G. (1986) *Astrophys. J.*, **307**, 242.

Petitclerc, N., Wesemael, F., Kruk, J.W., Chayer, P., and Billères, M. (2005) *Astrophys. J.*, **624**, 317.

Provencal, J.L. and Shipman, H.L. (1999) in *ASP Conf. Ser.* Proceedings of the 11th European Workshop on White Dwarfs, vol. 169 (eds S.-E. Solheim and E.G. Meistas), ASP, San Francisco, p. 293.

Provencal, J.L., Shipman, H.L., Wesemael, F., Bergeron, P., Bond, H.E., Liebert, J., and Sion, E.M. (1997) *Astrophys. J.*, **480**, 777.

Provencal, J.L., Shipman, H.L., Thejll, P., and Vennes, S. (2000) *Astrophys. J.*, **542**, 1041.

Provencal, J.L., Shipman, H.L., Koester, D., Wesemael, F., and Bergeron, P. (2002) *Astrophys. J.*, **568**, 324.

Schmidt, G.D., Bergeron, P., and Fegley, B. (1995) *Astrophys. J.*, **443**, 274.

Shipman, H.L. (1972) *Astrophys. J.*, **177**, 723.

Strittmatter, P.A. and Wickramasinghe, D.T. (1971) *Mon. Not. R. Astron. Soc.*, **152**, 47.

Subasavage, J.P., Henry, T.J., Bergeron, P., Dufour, P., Hambly, N.C., and Beaulieu, T.D. (2007) *Astron. J.*, **134**, 252.

Subasavage, J.P., Jao, W.-C., Henry, T.J., Bergeron, P., Dufour, P., Ianna, P.A., Costa, E., and Méndez, R.A. (2009) *Astron. J.*, **137**, 4547.

Thejll, P., Shipman, H.L., MacDonald, J., and Macfarland, W.M. (1990) *Astrophys. J.*, **361**, 197.

Tremblay, P.-E. and Bergeron, P. (2008) *Astrophys. J.*, **672**, 1144.

Vauclair, G. and Fontaine, G. (1979) *Astrophys. J.*, **230**, 563.

Vornanen, T. Berdyugina, S.V., Berdyugin, A.V., and Piirola, V. (2010) *Astrophys. J. Lett.*, **720**, L52.

Wegner, G. (1983a) *Astron. J.*, **88**, 1034.

Wegner, G. (1983b) *Astron. Astrophys.*, **128**, 258.

Wegner, G. (1983c) *Astrophys. J.*, **268**, 282.

Wegner, G. and Koester, D. (1985) *Astrophys. J.*, **288**, 746.

Wegner, G. and Nelan, E.P. (1987) *Astrophys. J.*, **319**, 916.

Wegner, G. and Yackovich, F.H. (1983) *Astrophys. J.*, **275**, 240.

Wegner, G. and Yackovich, F.H. (1984) *Astrophys. J.*, **284**, 257.

Wegner, G., Yackovich, F.H., Green, R.F., Liebert, J., and Stocke, J.T. (1985) *Publ. ASP*, **97**, 575.

Weidemann, V. and Koester, D. (1995) *Astron. Astrophys.*, **297**, 216.

Werner, K. and Herwig, F. (2006) *Publ. ASP*, **118**, 183.

Werner, K., Rauch, T., Barstow, M.A., and Kruk, J.W. (2004) *Astron. Astrophys.*, **421**, 1169.

Williams, K.A., Liebert, J., Bolte, M., and Hanson, R.B. (2006) *Astrophys. J. Lett.*, **643**, L127.

Yackovich, F.H. (1982) Ph.D. Thesis, Pennsylvania State University.

Zeidler-K.T., E.M., and Koester, D. (1982) *Astron. Astrophys.*, **113**, 173.

Zuckerman, B., Melis, C., Klein, B., Koester, D., and Jura, M. (2010) *Astrophys. J.*, **722**, 725.

4
Planets Orbiting White Dwarfs

Rosanne Di Stefano

4.1
Introduction

The search for exoplanets, planets orbiting stars other than the Sun, has had a long and interesting history. Despite the challenges, the quest for planets was considered important because it was linked to the idea of searching for life outside the Solar System. That the extreme technical challenges and several false starts did not halt the search is undoubtedly due to the importance of the problem, combined with the near certainty on the part of many astronomers that exoplanets are common. Even before the first exoplanet was discovered, it seemed clear to many astronomers that if the Sun, an ordinary star, harbors a system of planets, then other stars must as well. Exoplanets would be discovered as soon as observational techniques become sensitive enough – and, indeed, they were.

To date, more than 500 planets orbiting main sequence stars have been discovered. Most have been found through studying periodic variations in the radial velocity of the host star associated with the planetary orbits. These discoveries were made by teams that had been engaged in long-term programs to monitor the radial velocities of large numbers of stars. The first announced planet to be discovered through radial velocity techniques was 51 Peg b (Mayor and Queloz, 1995). This system was identified from a group of 142 G and K dwarf stars that had been studied for 18 months. This group of stars was itself selected based on long-term radial velocity stability from a much larger set that had been monitored for more than a decade. Radial velocity studies are best suited to bright stars. For example, the host star in 51 Peg has an absolute V magnitude of 5.49. Indeed, most planets discovered through radial velocity measurements orbit stars brighter than tenth magnitude, although ongoing improvements in sensitivity now make it possible to measure radial velocity variations capable of detecting planets in dimmer stars. White dwarfs are intrinsically dim, and their luminosity function is not well-populated in the magnitude range best studied with radial-velocity techniques (Harris *et al.*, 2006).

The planet 51 Peg b is a gas giant in a 4.2 day orbit. Its discovery ran counter to most expectations at the time of its discovery (e. g., Boss, 1995), and provid-

White Dwarf Atmospheres and Circumstellar Environments. First Edition. Edited by D. W. Hoard.

ed a stark contrast with the distribution of planets in the Solar System. This suggests that it is important to search for planets of any mass in orbits of all types; that is, to search every accessible portion of the parameter space. This message is driven home even more sharply by the very first planets to be discovered: the largely unexpected planets around pulsars (Wolszczan and Frail, 1992). Furthermore, the pulsar planet discoveries are directly relevant to the search for white dwarf planets. This is not only because, like white dwarfs, the pulsars are stellar remnants, but also because of the fact that the timing technique used to discover these planets can be applied to white dwarfs, a significant number of which are stable pulsators. Interestingly enough, timing measurements have led to the discovery of two planets orbiting white dwarfs. However, in these cases, the measurements have not been of the white dwarf's pulsations. In addition, each of the two systems is an example of a rare stellar multiple. The first is a globular cluster system in which the planet orbits a millisecond pulsar + helium white dwarf binary which was once a low mass X-ray binary. The discovery was made by measuring the neutron star's pulse arrival times. Eclipse timing has yielded the first potential discovery of a planet orbiting a cataclysmic variable, a binary in which mass from a low mass stellar companion is falling onto a white dwarf, producing X-rays. (See Section 4.5 for additional discussion of these planet detections.)

As the Galaxy ages, white dwarfs will become the most common stars. Hence, the rare systems described above must represent a minuscule fraction of all current and future white dwarf planetary systems. A large fraction of stars have planets, and roughly 98% of all stars will become white dwarfs. If planetary systems survive the evolution of their low- and intermediate-mass parent stars, or if some planets can be formed in the aftermath of this evolution, then most planets will eventually be orbiting white dwarfs. Therefore, it should be expected that "garden-variety" white dwarf planetary systems in which planets orbit single white dwarfs or white dwarfs with distant stellar or substellar companions are common. Because white dwarfs are 10^2–10^5 times dimmer than the main sequence stars already found to have planets, search methods that do not rely on collecting a lot of light from the white dwarf are more likely to be successful. In fact, the low luminosities of white dwarfs mean that many search methods are potentially easier to apply to white dwarf systems. These methods will be the focus in this chapter.

Section 4.2 presents a discussion of expectations for the presence and characteristics of planets around white dwarfs. These are based partly on theory, partly on observations of related astrophysical systems, and partly on the possible effectiveness of the variety of search techniques that are being implemented. While it is important to be wary of concrete predictions, all signs point to the likelihood that white dwarfs typically host planetary systems and that it should be possible to discover large numbers of white dwarf planets in the near future.

Section 4.3 discusses the possibility of detecting radiation from planets orbiting white dwarfs, and summarizes the results of past searches. Although these searches have not yet discovered planets, they have produced upper limits on the masses of planets that could be orbiting the white dwarfs that have been studied. The lack of detections might indicate that typical white dwarf planets have masses comparable to or smaller than that of Jupiter.

Although there is not yet direct evidence of large planets orbiting white dwarfs, there is a great deal of evidence that white dwarfs host large systems of minor planets; that is, asteroids and comets. The support for this scenario is sketched in Section 4.4. The direct detection of minor planets around white dwarfs may be difficult, but photometric monitoring by the Kepler Mission may make it possible to detect asteroids with radii of 100 km or larger.

The work discussed in Section 4.2 shows that it is possible to detect large planets around white dwarfs, but that they appear to be rare. The work discussed in Section 4.3 shows that minor planets are common, but that it is difficult to detect them. These complementary results point to the importance of improving the detection sensitivities to make it possible to detect planets with masses smaller than that of Jupiter. Timing techniques can achieve this. These techniques are discussed in Section 4.5, along with the challenges researchers face in using them to discover planets orbiting white dwarfs.

Programs that monitor large portions of the sky are becoming more important in astronomical studies. Microlensing monitoring programs, for example, have been active for almost 20 years and have undergone several upgrades. They have demonstrated the ability to discover planets. Section 4.6 entertains the possibility that microlensing will be used to discover planets orbiting white dwarfs. Programs designed specifically for microlensing studies are being joined by several other large survey programs that will return to portions of the sky at regular intervals over a period of years. The Panoramic Survey Telescope and Rapid Response System (Pan-STARRS)[1] provides one such program, and it is already beginning to obtain science data. The Large Synoptic Survey Telescope (LSST)[2] will undertake another such survey program, and is scheduled to begin within the next decade. These programs have the potential to discover planetary lensing events. Planets discovered through their action as gravitational lenses can have masses significantly smaller than that of Jupiter and can also lie within $\approx 1\,\text{AU}$ of the star they orbit. The use of lensing to discover planets around nearby (within $\sim 1\,\text{kpc}$) white dwarfs will provide the opportunity to explore an important part of the parameter space.

Large-scale photometric monitoring programs have also demonstrated the ability to discover planetary transits across (or behind) the parent star. Although the microlensing programs monitor fields too bright to easily allow transits of white dwarfs to be detected, Pan-STARRS and LSST will be able to detect transits of white dwarfs by planets. Although the probability of detection per object is low, the number of objects to be monitored is large enough that some detections are expected. This method favors close-in planets and will be sensitive to even terrestrial-size planets. The possible use of transits to discover white dwarf planets is discussed in Section 4.7. Finally, the prospects for studying the architecture of white dwarf planetary systems are discussed in Section 4.8.

1) http://pan-starrs.ifa.hawaii.edu/public/ (9 May 2011)
2) http://www.lsst.org (9 May 2011)

4.2 Expectations

The first lesson learned about exoplanets was that the characteristics of real systems can be very different from the characteristics that might have been predicted through simple applications of basic physics and extrapolations from the properties of the Solar System. Therefore, effective strategies to search any part of the parameter space should be implemented. Predictions are useful, however, because they set the stage for the eventual discoveries of white dwarf planets. After these discoveries, comparing the properties of ensembles of real systems with those predicted will allow the determination of the relative roles of the physical processes that come into play. These processes include stellar evolution, tidal interactions, late-stage stellar winds, and motion through a planetary nebula, or even through the envelope of the parent star itself.

The most certain prediction is that white dwarfs likely harbor planets in wide orbits. It is known that main sequence stars, including the Sun, have planets in wide orbits. There is general agreement that in the Solar System, Mars and the planets beyond it will survive the Sun's post-main sequence evolution. Indeed, the discovery of a planet orbiting V391 Pegasi, a post-red giant star, provides evidence that planets in orbits with semimajor axes smaller than 2 AU can survive through the epoch of maximum stellar expansion (Silvotti *et al.*, 2007). The planet has $M \sin i$ equal to 3.2 times the mass of Jupiter (M_J), and is at a distance of 1.7 AU from the star. It is estimated that its original orbital separation was roughly 1 AU. This system is well on its way to becoming an ordinary white dwarf with a planet in an orbit similar to that of Mars.

It is important to note that 145 of the known exoplanets, more than one-fourth of the current total, are in orbits with semimajor axes larger than 1.5 AU, the approximate size of the orbit of Mars. Several of these planets have large enough angular separations from their parent star that it has been possible to image the planets, even though the stars are brighter than the Sun. The star HR 8799 is 40 pc away, and its mass is approximately $1.5 M_\odot$. It is orbited by three planets that have been imaged. The planets have masses between $5 M_J$ and $13 M_J$ and lie at orbital radii between 24 and 68 AU (Lafrenière *et al.*, 2009; Marois *et al.*, 2008). The star Fomalhaut has a mass of about $2 M_\odot$, and is 7.7 pc away. It is orbited by a planet of $\sim 3 M_J$ at an orbital separation of roughly 120 AU, orbiting just inside a belt of dust surrounding the central star (see Figure 4.1; Kalas *et al.*, 2008). A giant planet has been imaged in the disk of β Pictoris, which has a mass of $1.8 M_\odot$ and is at a distance of 19 pc. The planet has a mass of approximately $9 M_J$ and is located 8–15 AU from its parent star (Lagrange *et al.*, 2010).

Wide-orbit planets will move to even wider orbits when the central star evolves off the main sequence, loses mass, and becomes a white dwarf. The orbital separation will increase by a factor that could be as large as the ratio of the progenitor to white dwarf masses, M_* / M_{wd}. Tidal interactions with the evolved giant star, interactions with the giant's substantial late-stage wind, and with the subsequent planetary nebula, may limit the expansion of the semimajor axes of planets that

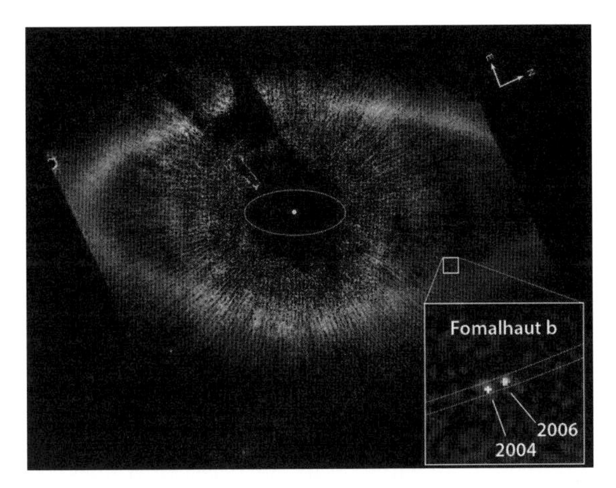

Figure 4.1 Coronagraphic image of Fomalhaut at 0.6 μm from HST. The position of the planet, Fomalhaut b, is shown in the small square located just within the inner boundary of the dust belt. The small filled circle at the center of the image marks the location of the star behind the occulting spot of the coronagraph, and the ellipse has a semimajor axis of 30 AU (3.9″), comparable to the orbit of Neptune. The inset is an expanded composite image showing the locations of Fomalhaut b in 2004 and 2006. The arcs in the inset are sections of two ellipses with semimajor axes of 114.2 and 115.9 AU that are concentric with the dust ring around Fomalhaut. From Kalas *et al.* (2008), reprinted with permission from AAAS.

start within several AU of a main sequence parent star (Rasio *et al.*, 1996; Villaver and Livio, 2007). It has been conjectured, for example, that planetary companions can play an important role in the release and shaping of planetary nebulae.[3] If this is so, then the planetary nebulae would also influence the evolution of the planetary orbits.

The closer the starting separation, the more difficult it is to predict the fate of the planet. While Mars and all of the outer planets are far enough from the Sun to avoid being engulfed even when the Sun achieves its largest post-main sequence radius, the fate of the Earth is controversial. Whether or not it is engulfed by the Sun depends on the details of the solar wind's evolution, the slowing of the Sun's spin, and the effects of the Sun's tidal bulge (e. g., Sackmann *et al.*, 1993, Schröder and Connon Smith, 2008). Whatever eventually happens to the Earth, it is clear that some planets will be engulfed by the giant envelope of their evolving stars. Some of these may spiral toward the giant's core, but nevertheless survive, achieving very close orbits. The discovery of a few binaries containing a brown dwarf in an $\sim 0.6 R_\odot$ orbit with a white dwarf establishes that substellar objects can survive a common

3) The term "planetary nebula" was coined in the eighteenth century because the appearance of these nebulae was considered similar to the appearance of the giant planets Jupiter and Saturn. When it was later understood that these nebulae are the escaping envelopes of dying stars, the name seemed somewhat inappropriate. It is, therefore, interesting that some researchers are now considering the conjecture that the genesis and features of at least some planetary nebulae may be determined by the star's interaction with its own planetary system.

envelope (e. g., WD 0137−349, Maxted *et al.*, 2006; SDSS J121209.31+013627.7, Schmidt *et al.*, 2005, Farihi *et al.*, 2008). This may well indicate that planets can survive the post-main sequence evolution of their parent star and spiral in to close orbits.

Since initially wide orbits are expected to expand while several mechanisms can shrink closer orbits, it might seem as if the distribution of orbital periods must be bimodal (e. g., Nordhaus *et al.*, 2010). However, this need not be the case. For example, interactions among planets during and after the evolution of the central star can alter the orbital separations. It has also been suggested that planets can form after the host star evolves. For example, when two white dwarfs initially in a close binary merge via loss of orbital angular momentum through gravitational radiation, planets might form from the resultant debris disk (Livio *et al.*, 2005; Villaver and Livio, 2007). The locations of these "second generation" planets could be spread out in a region surrounding the white dwarf. Searches for planets orbiting white dwarfs are, therefore, exploring the full ranges of possible planetary masses and orbital separations that are accessible to modern technology.

The goal is to discover large numbers of white dwarf planetary systems. As discussed in the following sections, this can be accomplished by using several different, but complementary, search techniques. Each technique has its own selection biases. As more systems are discovered, the biases can be understood in a quantitative way, allowing the true distributions of system properties to be computed from the ensemble of observed systems. The distributions can be compared with the distributions computed for the known main sequence planetary systems. This will allow information to be derived about interactions between a star and its planets during the late stages of stellar evolution.

4.3
Detecting Radiation from the Planets

The imaging of planets around HR 8799, Fomalhaut, and β Pictoris demonstrates the feasibility of direct detection of exoplanets. Because white dwarfs are dim compared with main sequence stars and are generally very hot ($T_{\text{eff}} \gtrsim 10\,000$ K), while, at the same time, self-luminous planets can be relatively large and are brightest in the infrared, planet searches around white dwarfs have an advantage. Burleigh *et al.* (2002) showed that ground-based imaging could detect planets around nearby white dwarfs and developed a "top ten" list of candidates to search. Ignace (2001) showed that infrared excesses could signal the presence of planets that cannot be spatially resolved from the white dwarfs they orbit. The advantages of searching for substellar objects around white dwarfs was highlighted by the discovery of a brown dwarf, GD 165B, orbiting a white dwarf (see Figure 4.2; Becklin and Zuckerman, 1988). Despite the great promise of infrared studies to discover massive planets around white dwarfs, progress has been slow. Currently, there is no evidence that photons have been collected from planets orbiting white dwarfs. Some of the observational limits imposed on this conclusion are described below.

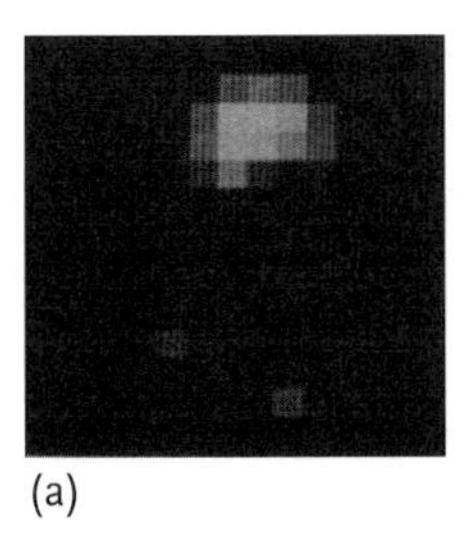
(a)

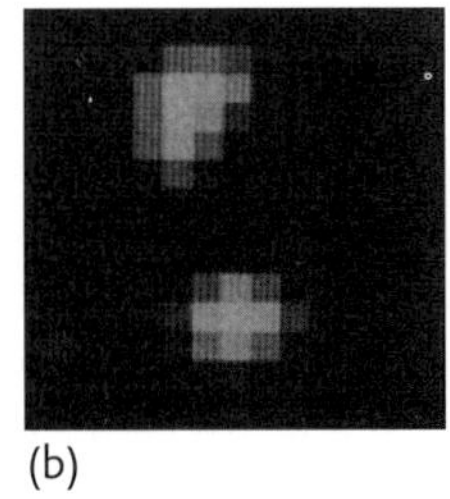
(b)

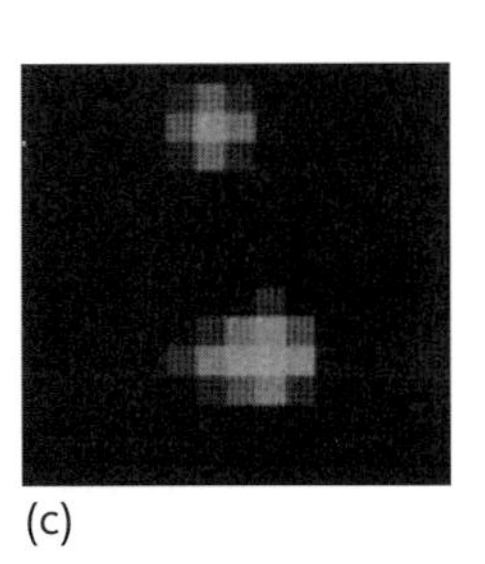
(c)

Figure 4.2 Infrared images of the white dwarf + brown dwarf binary GD 165/GD 165B at wavelengths of (a) 1.25 μm (*J* band), (b) 1.65 μm (*H* band), and (c) 2.2 μm (*K* band). Each frame is ∼ 7″ square. The white dwarf is the upper (northern) object in each frame. From Becklin and Zuckerman (1988), reprinted with the permission of Macmillan Publishers Ltd: Nature, ©1988.

Farihi *et al.* (2005) reported on more than 17 years of work by astronomers to find substellar companions to white dwarfs. Despite the successes for brown dwarfs, including GD 165B and a spectrally-confirmed brown dwarf, GD 1400B, no planets had been discovered. This null result was obtained even though planets with masses in the range 2–13 M_J could have been detected for 14 white dwarfs, and planets with masses in the range 14–20 M_J could have been detected for ten additional white dwarfs. Subsequently, Farihi *et al.* (2008) ruled out substellar companions to any white dwarf in the Hyades and Pleides clusters, with a handful of young field white dwarfs having limits on the companion mass of $\sim 5 M_J$.

Kilic *et al.* (2009) reported on a Spitzer Space Telescope search of 14 white dwarfs. The mass range was high enough to select remnants of intermediate mass stars (3–5 $M_\odot$). Because it is difficult for radial velocity or transit searches for planets to be successful for main sequence stars in this mass range, discovering planets orbiting the white dwarfs that descend from these stars may be the best way to study their planetary systems. Kilic *et al.* ruled out planets around the white dwarfs with masses larger than the following limits: $> 5 M_J$ for five white dwarfs, $> 7 M_J$ for four white dwarfs, $> 10 M_J$ for four white dwarfs, and $> 15 M_J$ for one white dwarf. They also pointed to additional null results, including those of Debes *et al.* (2005), Friedrich *et al.* (2007), and Mullally *et al.* (2007a). Kilic *et al.* (2010) used Spitzer to search for substellar companions to low mass white dwarfs ($M < 0.45 M_\odot$). They placed limits on the companion masses that extended down to $2 M_J$ for young assumed ages (10^8 yr), but were generally greater than $10 M_J$. In addition to these, there are several other sets of null results for detection of planets around white dwarfs; for example, see Burleigh *et al.* (2008), Clarke and Burleigh (2004), Mullally (2009), Mullally *et al.* (2007b, 2009). The models against which the data have been tested include blackbody models as well as more sophisticated models of substellar spectra, which, themselves, have been tested against observations of substellar objects (Baraffe *et al.*, 2003; Burrows *et al.*, 2003).

When cutting edge technology fails to find planets, it is reasonable to question if the planets were there, but the search techniques were simply ineffective. In the case of white dwarf planetary systems, there are good reasons to believe that planets could have been discovered by the surveys described above had they been there.

For example, as described in the next section, modest infrared excesses have, in fact, been discovered for some white dwarfs. These data have been convincingly modeled by debris disks. The ability to discover and model disks is a strong indication that the technology and modeling should also be able to work in concert to discover planets around white dwarfs.

In the meantime, one must ask what can be learned from the null results. Before attempting to answer this question, it is worth noting that the sample of white dwarfs that has been studied thus far is inhomogeneous. The white dwarfs display a wide range of properties: masses, ages, binary histories. They include those in which there are null results for orbiting planets as well as those with positive detections of dusty debris disks. Furthermore, the observational techniques and analysis methods differ from project to project, and thus care must be taken when comparing and combining results. While such an ambitious task is not attempted here, two directions for further research are clearly suggested by the results summarized above.

1. The planet-detection mass limits of the existing surveys tend to be high, larger than a few Jupiter masses in most cases. The null results, therefore, suggest that high mass self-luminous planets in wide orbits may not be common. In fact, evidence from another direction, gravitational microlensing, seems to indicate that the most common planets may be cool Neptune-mass objects (Sumi *et al.*, 2010). It is clearly important to implement methods, such as those discussed in the rest of this chapter, that can discover less massive, less luminous planets in orbit around white dwarfs.
2. It is important to compare the nondetections of high-mass planets orbiting white dwarfs with the many discoveries of such planets orbiting main sequence stars. Even as the number of known exoplanets has steadily increased following the first discoveries, their mass distribution by number has been consistently quantifiable (without regard for the homogeneity or completeness of the sample) as $dN/dM_p \propto M_p^{-\alpha}$, where $\alpha \approx 1.05$–1.20 (Butler *et al.*, 2006; Jorissen *et al.*, 2001; Marcy *et al.*, 2005, 2008). As of early January 2011, the NASA/IPAC/NExScI Star and Exoplanet Database (NStED)[4] contained 506 exoplanet mass estimates; of these, 59% have masses greater than $1\,M_J$, 38% have $M_p > 2M_J$, 9.5% have $M_p > 8M_J$, and 2.5% have $M_p > 15M_J$. When the parent stars of these planets evolve into white dwarfs, planets that start close to the star might be destroyed or lose much of their mass, while planets that start in wide orbits should survive intact. Furthermore, the descendants of the wide-orbit massive planets known today should remain bound after the star evolves. In fact, 16% of the known exoplanets (as of the January 2011 NStED sample) have both mass greater than $5\,M_J$ and orbital semimajor axis greater than 1.5 AU, and, therefore, have a good possibility of survival following the post-main sequence evolution of their parent star. Consequently, if white dwarf planet searches were able to focus on the descendants of the main sequence

4) http://nsted.ipac.caltech.edu/index.html (15 June 2011)

stars that have been the subject of planet searches, then significant numbers of massive planets in wide orbits should be found.

Of course, it would also be interesting if this predicted population of white dwarf planets was *not* found. For example, most of the stars searched for planets have masses similar to that of the Sun and, therefore, also have main sequence lifetimes on the order of 10^{10} yr. Given the time needed for these stars to evolve, those that produced today's white dwarfs may have started with lower metallicities than the Sun. The failure to find planets around current white dwarfs could translate into a failure of planet formation at low metallicity. At present, however, the different selection criteria for main sequence and white dwarf planet searches are likely to mean that the white dwarfs being targeted are not exact analogues to the descendants of the main sequence stars with known exoplanets. Even so, as more systems are studied, it will be possible to determine if there is a genuine and meaningful mismatch between the number of planet nondetections around white dwarfs, and the number of planets detected around main sequence stars.

4.4 Evidence for Minor Planets

Comets and asteroids are the smallest bulk components of the Solar System. Yet, if the Sun was being observed from a distance of tens or hundreds of parsecs, it would be difficult to find direct evidence of the existence of individual minor planets, or of events involving just one or a small number of them. In contrast, the pristine atmospheres of cool white dwarfs as well as their low luminosities and small sizes make it easier to infer the presence of minor planets orbiting white dwarfs. The photospheric metal enrichment, often referred to as "pollution", of dozens of white dwarfs is well-studied (e. g., see Figure 4.3). There are good examples in which the elemental composition of the contaminants suggests that they come from accretion of a disrupted minor planet (Farihi *et al.*, 2011; Jura *et al.*, 2009a). A significant fraction ($> 50\%$) of white dwarfs with metal-enriched atmospheres also display infrared excesses (Farihi *et al.*, 2009). These excesses can be fit by dust disk models (e. g., see Figure 4.4); the origin of the dust is presumed to be a tidally disrupted minor planet. These sets of observations and analyses form an ever-more comprehensive web that stands as the best current evidence that white dwarfs have planetary systems.

4.4.1 Survivability

Theoretical studies of planet formation find that small bodies form first, potentially accumulating into masses as large as the terrestrial planets (e. g., Greenberg *et al.*, 1978, Ormel *et al.*, 2010, Süli, 2010). If local conditions are favorable, these cores can accrete gas, producing planets with masses comparable to that of Jupiter.

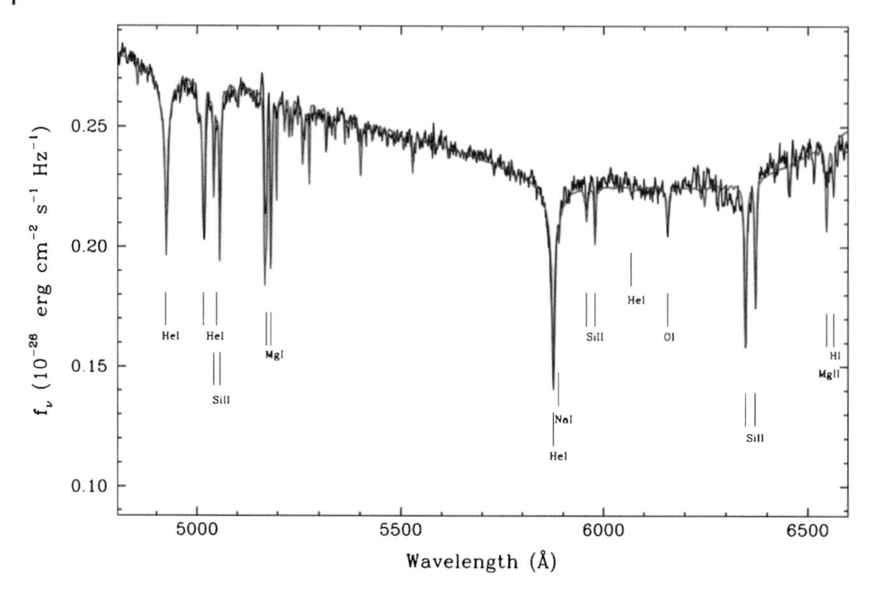

Figure 4.3 Optical spectrum (dark line) and best-fit model (light line) of SDSS J073842.56+183509.6, the most metal-rich white dwarf known. From Dufour *et al.* (2010), reproduced with the permission of the AAS.

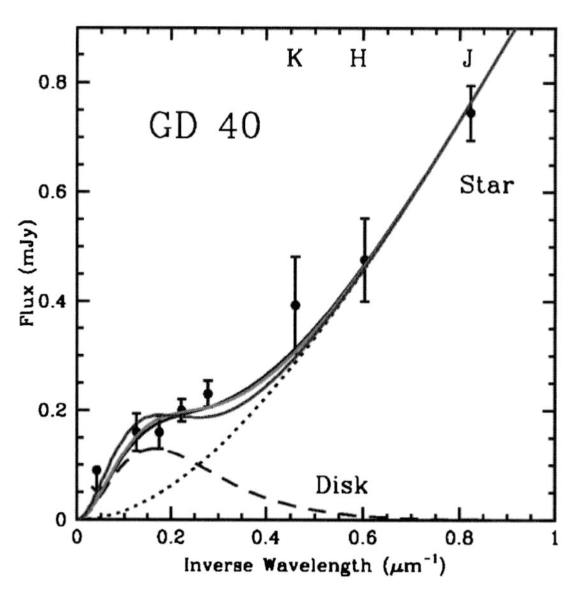

Figure 4.4 Photometric data with 2σ error bars for the dusty white dwarf GD 40, showing the contribution from the stellar photosphere (dotted line), a representative dust disk model (dashed line), and several combined models that utilize slightly different disk parameters (solid lines). From Jura *et al.* (2007a), reproduced with the permission of the AAS.

It is not yet known if the process of planet formation is successful for every star, although the diversity of exoplanets discovered to date seems to indicate that a large fraction of stars do have planetary systems. Even in cases in which the process of planet formation is truncated, however, one might expect that many small rocky or icy bodies would form. In the Solar System, the asteroid belt contains rocky minor planets that did not successfully aggregate into a larger mass. Likewise, the icy masses of the Kuiper belt and Oort Cloud are remnants of the early Solar System. When the Sun evolves, expands, and eventually loses a large fraction of its mass, the Solar System will evolve as well. For the sake of simplicity, all minor planets (i.e., bodies with radii smaller than that of Pluto) are, henceforth, referred to as "asteroids", regardless of their composition.

During the late stages of stellar evolution, planets and asteroids in wide orbits may be lost as the star loses mass and their orbits expand. Inner asteroids and planets may be engulfed by the star. Even when the star is on the main sequence, large planets may number only in the dozens, so it is conceivable that only a few, or even no, planets survive the post-main sequence evolution of some host stars. Asteroids, on the other hand, are much more numerous. With more than 10^{12} in the Solar System, dynamical studies show that asteroid or cometary systems can survive stellar evolution (Alcock *et al.*, 1986). It could even be the case that the perturbation of planetary orbits caused by the evolution of the host star could trigger collisions between planets, planets and their moons, or asteroids and planets, thereby producing even larger numbers of minor planets.

While asymmetries in the mass loss from the central star can influence survivability, some white dwarfs could experience asteroid impacts at a rate of $10^{-4}\,\mathrm{yr}^{-1}$ (Parriott and Alcock, 1998). Some asteroids may be in circular orbits, but many are likely to be in highly eccentric orbits. When an asteroid approaches the white dwarf, it can be tidally disrupted. A disk will form, and matter from the disk will gradually accrete onto the white dwarf. Heavy elements will sink through the atmosphere of a cool, hydrogen-dominated (DA) white dwarf on a timescale of days (for $T_{\mathrm{eff}} \lesssim 25\,000$ K) up to $\sim 10^4$ yr (for $T_{\mathrm{eff}} \lesssim 10\,000$ K; e. g., see Koester, 2009, and references therein). Even in the case of DB white dwarfs with helium-dominated atmospheres, the much longer ($\sim 10^6$ yr) metal diffusion timescales are still significantly shorter than white dwarf evolutionary timescales. Thus, when white dwarf atmospheres show long-term evidence of metal pollution, this signals ongoing contamination of the white dwarf, presumably by material falling onto its surface. In fact, the best direct evidence that white dwarfs have planetary systems has been obtained by studying the pollution of white dwarf atmospheres, and also by discovering evidence for debris disks around a significant fraction of white dwarfs. Beyond this, it is interesting to note that an interval of heavy bombardment of a white dwarf by asteroids may be a signal of more distant massive planets that can perturb the orbits of the system of minor planets (Debes and Sigurdsson, 2002; also see Chapter 6 by J.H. Debes).

4.4.2 Metal Enrichment and Disks

The atmospheres of cool white dwarfs provide a pristine astrophysical laboratory in which only small traces of elements heavier than hydrogen and helium can be detected. The metal-rich material that pollutes the atmospheres cannot be a remnant of the pre-white dwarf state since the evolutionary cooling timescales are many times longer than the time required for the heavy elements to sink below the outer atmosphere of the white dwarf where they can be detected.

It is valid to question if the pollution is being acquired as the white dwarfs move through the interstellar medium (ISM). This hypothesis has been tested in several ways (also, see the discussion of this topic in Chapter 5 by J. Farihi). First, the luminosity expected from accretion from the ISM has been compared to the observed luminosities of polluted white dwarfs. For example, one can compute the luminosity that should be generated from the capture of interstellar dust grains within the Bondi–Hoyle radius, and their accretion through Poynting–Robertson drag (Jura *et al.*, 2007a). Alternatively, one can consider the Poynting–Robertson drag on dust rings (Farihi *et al.*, 2010b). In either case, the 24 μm flux density should be at least ten times higher than is observed. In addition, by establishing upper limits on X-ray emission from polluted white dwarfs, one can establish upper limits on the mass infall rate. Observations from the X-ray Multi-Mirror Mission (XMM-Newton) have shown for both GD 362 and G29-38 that the rate of mass infall is much lower than it would be if the white dwarf was accreting hydrogen-rich matter from the ISM (Jura *et al.*, 2009b).

Second, if the accreted matter comes from the ISM and not from within the system itself, then it should be possible to find correlations between the positions and kinematics of white dwarfs and their states of metal enrichment. Yet, studies of modest sample sizes failed to establish meaningful correlations (Aannestad *et al.*, 1993; Kilic and Redfield, 2007). Continued work along these lines has been able to take advantage of a large sample of white dwarfs identified by the Sloan Digital Sky Survey (SDSS; Eisenstein *et al.*, 2006). Farihi *et al.* (2010a) considered a sample of 146 cool white dwarfs with atmospheres exhibiting significant metal enrichment, and found no evidence for a correlation between their positions in the sky and levels of enrichment. In fact, the positions and kinematics of roughly 2/3 of the enriched stars in their sample indicate that there has been little gas and dust in their environments for extended ($>$ Myr) intervals.

The presence of heavy elements that sink in the white dwarf's atmosphere on short time scales requires a source of replenishment. If this source is not the ISM, then it must be a circumstellar disk. Dust in a debris disk can produce an infrared excess. Such excesses have now been observed in ~ 20 white dwarfs with enriched atmospheres (e. g., see Becklin *et al.*, 2005; Brinkworth *et al.*, 2009; Jura *et al.*, 2007b; Kilic and Redfield, 2007; Kilic *et al.*, 2005, 2006; Zuckerman and Becklin, 1987; von Hippel *et al.*, 2007; Farihi *et al.*, 2008; Farihi *et al.*, 2009; Jura *et al.*, 2009a; Melis *et al.*, 2009; see Chapter 5 by J. Farihi for a complete census of known white dwarfs with dust disks). Farihi *et al.* (2009) found that there is a warm infrared excess in more than half of the single white dwarfs with implied metal accretion rates greater than $3 \times 10^8\,\mathrm{g\,s^{-1}}$. Equally important to establishing the presence of an infrared excess is that a range of detailed disk models are able to fit the data (e. g., see Figure 4.4). Yet, not all white dwarfs with polluted atmospheres exhibit infrared excesses. Farihi *et al.* (2010b) showed that even these white dwarfs could be accreting dust since the observable signature of the presence of dust depends on a number of parameters; for example, the thickness of the circumstellar disk, the size of its central hole (i. e., inner radius), and its maximum extent (i. e., outer radius). These factors appear to vary among the polluted white dwarfs that have been observed with Spitzer.

Interestingly, three of the white dwarfs initially thought to have no infrared excess were found to show spectroscopic evidence of the presence of *gaseous* disks. For example, Gänsicke *et al.* (2006) report observations of Ca II and Fe II double-peaked emission lines in the white dwarf SDSS J122859.93+104032.9, interpreted to be from a circumstellar, metal-rich gaseous disk. The white dwarf itself is hot enough ($T_{\mathrm{eff}} \sim 22\,000\,\mathrm{K}$) to burn off dust from the disk, which was presumed to account for the lack of an infrared excess. The disk itself is likely the remnant of a tidally disrupted rocky body of asteroid-sized mass. The case for a metal-rich disk is strengthened by observations of Mg II absorption lines in the stellar spectrum. Dynamical modeling of the system by Gänsicke *et al.* (2006) constrains the outer edge of the disk to be at about $1.2\,R_{\odot}$ and places the inner edge at approximately $0.64\,R_{\odot}$. Similar spectroscopic evidence was found in the other two gas disk systems, SDSS J084539.17+225728.0 (Gänsicke *et al.*, 2008) and SDSS J104341.53+085558.2 (Gänsicke *et al.*, 2007). Subsequent high sensitivity near- and mid-infrared measurements using Spitzer as well as ground-based infrared facilities have revealed the presence of weak infrared excess in two of these three systems (SDSS 1228, Brinkworth *et al.*, 2009; SDSS 0845, Melis *et al.*, 2010), indicating the additional presence of a dusty component in their circumstellar disks. The relatively large uncertainties on the Spitzer mid-infrared photometry of the faint third object (SDSS 1043) have so far confined any infrared excess, if present, to a $< 3\sigma$ confidence level; however, the presence of a dusty component of the gaseous disk around this white dwarf also has not been ruled out by the observations (Brinkworth *et al.*, 2011).

The bottom line is that there is a rich and rapidly evolving literature on metal-enriched white dwarfs and their associated dust disks. This body of work is developing a self-consistent and coherent picture in which the enrichment (or pollution) can be explained by the accretion of matter from tidally disrupted asteroids. In fact, this hypothesis is now largely accepted. This means that more attention can be devoted to using the data on enrichment and accretion to learn about the size and composition of the asteroids themselves as well as the frequency of their disruption. New investigations are increasingly focused on these questions.

4.4.3
Transits of Asteroids

Given the strong evidence summarized above, it is almost certain that asteroids orbit white dwarfs, very likely in large numbers. It will be shown in Section 4.7 that the small size of white dwarfs relative to main sequence stars means that the probability that a planet will transit a white dwarf is relatively small. Modest monitoring programs, like those used to detect planetary transits around main sequence stars, can succeed only if planets exist around most white dwarfs, and have orbits well within $\approx 1\,\mathrm{AU}$. If, however, white dwarfs have $\sim 10^9$–10^{12} asteroids, then monitoring for transits by asteroids can overcome the challenge posed by the low probability of planetary transits.

Unfortunately, however, it is not possible to detect asteroid transits from the ground. The depth of the transit scales as the ratio of the projected surface area of the asteroid to the projected surface area of the white dwarf, roughly 100 parts per million for a 100-km class asteroid. Serendipitously, Kepler has the requisite sensitivity. Kepler was designed to detect the transits of Sun-like stars by Earth-like planets. The ratio of the planetary to stellar surface areas are in exactly the range of asteroid to white dwarf surface areas. Thus, if most white dwarfs have asteroids, and if the orbital inclinations are not confined to a plane, but spread throughout a thick disk (~ 15–$30°$), then by monitoring roughly 10 white dwarfs for a year each, Kepler will have a good chance to discover white dwarf asteroids (Di Stefano *et al.*, 2010).

The time duration of an asteroid transit, τ_{cross}, is short,

$$\tau_{\mathrm{cross}} = \frac{2R_{\mathrm{wd}}}{v} \approx 9.4 \left(\frac{R_{\mathrm{wd}}}{7.5\times10^{8}\,\mathrm{cm}}\right)\left(\frac{a}{1\,\mathrm{AU}}\right)^{1/2}\left(\frac{0.8M_{\odot}}{M_{\mathrm{wd}}}\right)^{1/2}\ \mathrm{min}\,, \quad (4.1)$$

where R_{wd} is the radius of the white dwarf, v is approximately equal to the asteroid's orbital velocity, and a is the asteroid's orbital semimajor axis (Di Stefano *et al.*, 2010). Because Kepler has both a one-minute monitoring mode and the necessary sensitivity, it can detect or place meaningful limits on asteroid transits of white dwarfs. A small observing program monitoring two bright white dwarfs in the Kepler field is underway.

4.5
Timing

When a star emits periodic pulsations that are stable over long intervals, it is possible to search for the small variations in arrival times that would be expected if the star has a wobble associated with the orbit of a low-mass companion. The best targets for these studies are small stars: neutron stars, white dwarfs, and subdwarfs. Only three planetary systems have been discovered through pulsation variations. Each is important to the unfolding story of the discovery of white dwarf planets.

The timing of pulsations was the first method used to discover exoplanets (Wolszczan and Frail, 1992). The planets orbiting PSR 1257+12 are distinguished not only because of their place in history, but also because they have the smallest known masses ($7 \times 10^{-5} M_J$, $0.013 M_J$, and $0.012 M_J$) and orbital radii (0.19, 0.36, and 0.46 AU) of any exoplanets. Although the search for planets had been an ongoing enterprise, it was surprising to most astronomers that the first discoveries were of planets orbiting a stellar remnant. Numerous models have been developed to explain the presence of these planets (see Wolszczan, 2008, for a review.). In many of the models, they are proposed as "second generation" planets, formed after the remnant itself. There are suggestions that some white dwarfs could harbor second generation planetary systems. For example, when white dwarfs merge, the result is possibly a massive white dwarf with a high metallicity accretion disk (Livio *et al.*, 2005). Planets might be able to form in the disk (see Waters *et al.*, 1998, Wickramasinghe *et al.*, 2010, and Perets, 2010, for example, for an alternative scenario). However, a Spitzer survey of 14 white dwarfs that are possible merger remnants did not find the infrared excesses that might signal the presence of disks or planets (Hansen *et al.*, 2006).

The second planet to be discovered through timing variations orbits the millisecond pulsar PSR *B*1620−26 in the globular cluster M4. This is a $2.5 M_J$ planet in a 23 AU orbit around a binary (Sigurdsson *et al.*, 2003). The binary consists of the pulsar and a $0.34 M_\odot$ white dwarf. The white dwarf has been imaged using the Hubble Space Telescope (HST) (Richer *et al.*, 2003). PSR 1620−26b *is* a planet that orbits a white dwarf. Its formation history, however, is unusual (Ford *et al.*, 2000; Sigurdsson, 1992, 1993, 1995; Sigurdsson *et al.*, 2003). The white dwarf is thought to be the remnant of a star that replaced a less massive companion of the neutron star during a multiple-star interaction that occurred in the core of the globular cluster. In the aftermath of the interaction, the newly formed multiple system was catapulted from the core. When the captured star began to evolve, the inner binary became a low mass X-ray binary in which the neutron star could be spun up to millisecond frequencies. PSR 1620−26b can be viewed as the ultimate survivor among planets. It survived a complex stellar interaction in a globular cluster core, expulsion from the core, evolution of one of the stars it orbits, and a history of X-ray irradiation from a nearby accreting neutron star. It is a fascinating system, but it is very different from the garden-variety (i. e., future Solar System analog) white dwarf planetary systems that are the targets of most searches.

Pulsation timing has also been used to detect a planet orbiting a subdwarf star. V391 Peg b was discussed in Section 4.2. This planetary system was important not only because of its method of discovery, which makes it only the third planetary system to be discovered through the detection of periodic variations in pulse arrival times, but also because it signals the survival of a planet in a wide orbit through the post-main sequence evolutionary stage of maximum expansion of the host star (Silvotti *et al.*, 2007).

Like pulsars, pulsating white dwarfs can be stable astrophysical clocks. In particular, the class of hot, variable, hydrogen-atmosphere white dwarfs can exhibit stable pulsation modes that are suitable for long-term study and the reliable identification of periodic perturbations. Mullally *et al.* (2008) reported on a program monitoring 15 carefully-chosen white dwarfs. Of these, one exhibited variations in pulse arrival time that could be associated with an orbiting planet. The results for GD 66 are consistent with a $2 M_J$ planet in an orbit with a period of approximately 4.5 yr, corresponding to an orbital separation of 2.3 AU. Monitoring had not yet been conducted for a full orbit, although additional time series data indicate that the planet may be somewhat more massive and the orbit somewhat wider ($m_p \sin(i) = 2.36 M_J$ and $a = 2.75$ AU; Mullally *et al.*, 2009). These values highlight an important advantage of timing studies: they are capable of finding evidence of lower mass planets, closer to the parent star. This example demonstrates that timing allows the discovery of planets that cannot be directly imaged at present, either because their masses are too low or because they are too close to the host star. In the case of GD 66, Spitzer observations were made to search for an infrared excess that could confirm the presence of the planet (Mullally *et al.*, 2009). The observations did not reveal strong evidence for an infrared excess, but did provide an upper limit to the planet mass of 5–7 M_J, assuming an age of 1.2–1.7 Gyr.

The discovery of white dwarf planets through timing measurements can be difficult because the pulsation spectra can be complex and time-variable (e. g., see Kepler *et al.*, 2003). Winget *et al.* (1990) reported on an observation of the variable DA white dwarf G29-38 with the Whole Earth Telescope. A constant amplitude period of 615 s dominated the light curve, and displayed variations consistent with a planetary orbit. In fact, G29-38 is an interesting system within which to search for evidence of planets. It is enriched in Ca, Mg, and Fe (Koester *et al.*, 1997). In addition, it is an example of the type of system discussed in Section 4.4.2, which exhibits an infrared excess consistent with the presence of a dust disk (Jura, 2003; Reach *et al.*, 2005; Zuckerman and Becklin, 1987). The timing signature that indicated the possible presence of a planet around G29-38 has not, however, been stable since its initial discovery (Kleinman *et al.*, 1994; Winget *et al.*, 1990).

In addition to pulsation timing, any stable, periodic signal for which high-precision measurements are possible can be used to test for the presence of planets. For example, orbital period variations determined from eclipse timings of the central binary star have been used to infer the presence of planets and substellar (i. e., brown dwarf) companions in circumbinary orbits around a handful of cataclysmic variables and post-common envelope binaries (HU Aqr, Schwarz *et al.*, 2009; DP Leo, Beuermann *et al.*, 2011; NN Ser, Brinkworth *et al.*, 2006, Qian *et al.*, 2009, Beuermann *et al.*, 2010; QS Vir, Parsons *et al.*, 2010; HW Vir, Lee *et al.*, 2009). The planets are inferred to have minimum masses of a few M_J up to $\sim 10 M_J$, and orbital radii of $\lesssim 10$ AU. In all of these cases, the central binary consists of a white dwarf in a close orbit with a mass-losing, low mass stellar companion. As with PSR B1620–26, these would not be garden-variety planetary systems, but, instead, ones that have experienced an epoch of X-ray irradiation as mass was transferred to the white dwarf from the companion. However, it must be noted that the claims of the detection of planetary mass bodies in some of these systems (e. g., QS Vir, NN Ser) have recently been strongly disputed through very high precision, high speed eclipse timing measurements (Parsons *et al.*, 2010).

4.6 Mesolensing

This section assesses the likelihood that white dwarf planets will be discovered through their actions as gravitational lenses. The prospects might seem dim since currently no white dwarfs that have participated in lensing events have been identified. Nevertheless, it is estimated that 10–20% of all of the observed lensing events are caused by lenses located within 1–2 kpc. Of these events, roughly 15% are caused by white dwarfs (Di Stefano, 2008b). With approximately 6000 events known, it would be surprising if fewer than 100 were caused by white dwarf lenses. The information that can be gleaned from lensing events about planets orbiting white dwarfs, and the prospects of identifying lensing events in which white dwarfs are implicated, are discussed below.

4.6.1 Gravitational Lensing Basics

Gravitational lensing generally produces multiple distorted images of the lensed source. If the alignment between lens, source, and observer is perfect, the image is a ring with angular radius equal to the Einstein angle, $\theta_{\rm E}$. Let $M_{\rm L}$ be the lens (white dwarf) mass, $D_{\rm L}$ be the distance to the white dwarf, and $D_{\rm S}$ be the distance to the background star that is lensed. For most events, the lensed source is a member of the distant field of stars being monitored, and $D_{\rm L}/D_{\rm S} \approx 0$. The value of $\theta_{\rm E}$ is given by

$$\begin{aligned}\theta_{\rm E} &= \left[\frac{4GM_{\rm L}}{c^2 D_{\rm L}}\left(1-\frac{D_{\rm L}}{D_{\rm S}}\right)\right]^{1/2} \\ &\approx 7.0\left[\left(\frac{M_{\rm L}}{0.6M_\odot}\right)\left(\frac{100\,{\rm pc}}{D_{\rm L}}\right)\left(1-\frac{D_{\rm L}}{D_{\rm S}}\right)\right]^{1/2}\ {\rm mas}\,. \end{aligned} \qquad (4.2)$$

When there are multiple images, their angular separations are comparable to $\theta_{\rm E}$. Thus, the images associated with lensing of stars by stellar masses are not generally resolved from the ground. The signature of lensing is an enhancement in the amount of light received from the lensed source star (Einstein, 1936). The duration of the enhancement is comparable to the Einstein diameter crossing time, $\tau_{\rm E}$, given by

$$\begin{aligned}\tau_{\rm E} &= \frac{2\theta_{\rm E} D_{\rm L}}{v} \\ &\approx 81\left(\frac{30\,{\rm km\,s^{-1}}}{v}\right)\left[\left(\frac{M_{\rm L}}{0.6M_\odot}\right)\left(\frac{D_{\rm L}}{100\,{\rm pc}}\right)\left(1-\frac{D_{\rm L}}{D_{\rm S}}\right)\right]^{1/2}\ {\rm days}\,, \end{aligned} \qquad (4.3)$$

where v is the relative transverse velocity of the lens (see Figure 4.5). The shape of the lensing light curve is a well defined mathematical function, symmetric about the peak (Einstein, 1936; see the example light curve in Figure 4.6). Model fits to

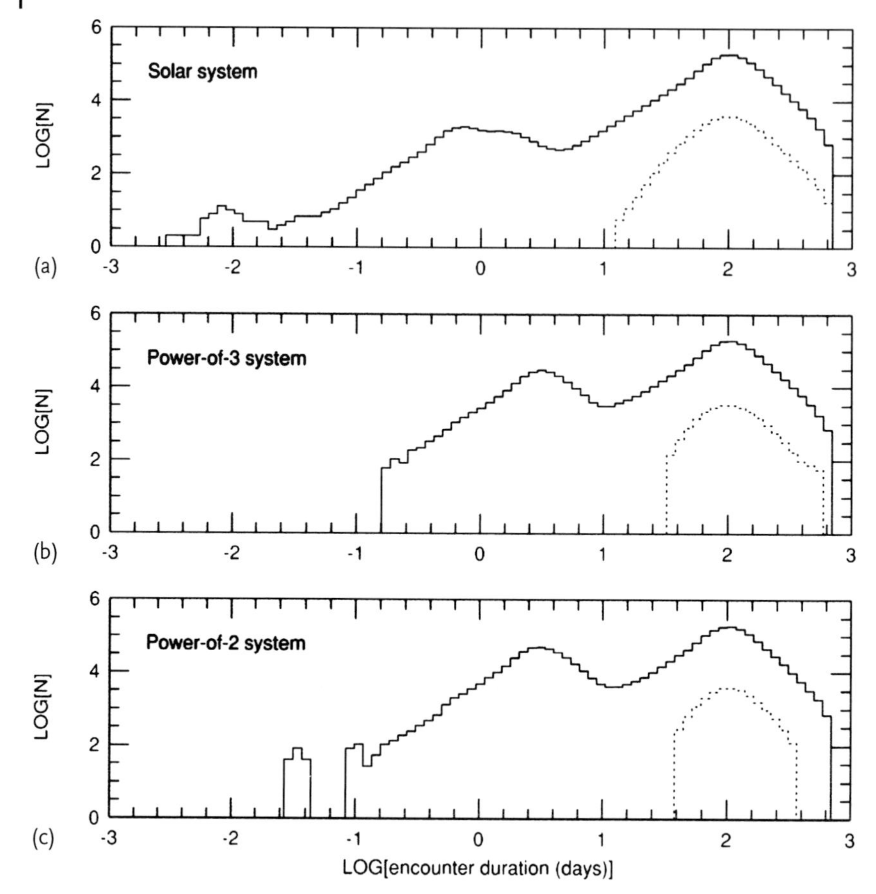

Figure 4.5 Distributions of durations for isolated (nonrepeating) lensing events from a set of Monte Carlo simulations in which three types of planetary systems were placed in the Galactic Bulge and served as lenses for more distant Bulge stars. The three planetary systems are the Solar System (a), a "power-of-3" system (b), and a "power-of-2" system (c). In the "power-of-*n*" models, the orbital separation between planets and the central star increases by a factor of *n* proceeding outward from the central star. The rightmost peak in each panel is due to the central star of the system, with overlap events involving the innermost wide planet shown as a dotted line; the peaks on the left are due to planetary encounters. The central feature in (a) is a superposition of peaks corresponding to Uranus, Neptune, and Saturn; the encounters due to Saturn are visible as a shoulder at about 1.6 days. In the Solar System, power-of-3, and power-of-2 models, there were 3.1, 30.4, and 55.9 isolated events of short-duration per year and 4.2, 3.6, and 4.0 overlap events, respectively. From Di Stefano and Scalzo (1999), reproduced with the permission of the AAS.

the light curve provide the value of τ_E. The duration of detectable deviations from the baseline can be a few times longer than τ_E.

The value of τ_E is the only parameter influencing the shape of the lensing light curve that is related to the lens mass. As (4.3) shows, however, the value of τ_E depends on the values of M_L, D_L, and v. In most lensing events, neither D_L nor v are known, making it impossible to uniquely determine the lens mass.

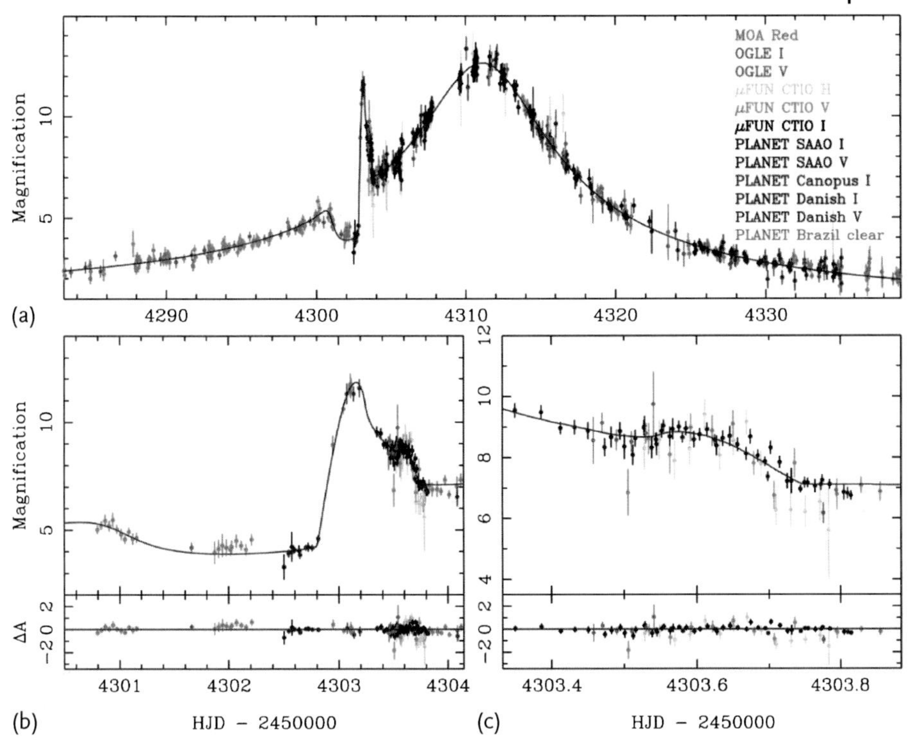

Figure 4.6 Light curve spanning the entire gravitational microlensing event OGLE-2007-BLG-368, which involved a planet-hosting lens star (a), and close-ups around the planetary deviation (b) and second caustic crossing (c), with the residuals from the best fit model. The solid line is the best fit model. The mass and distance of the planet-hosting K-dwarf lens star are $M_L = 0.64^{+0.21}_{-0.26} M_\odot$ and $D_L = 5.9^{+0.9}_{-1.4}$ kpc, respectively. The mass and orbital separation of its planet are $M_p = 20^{+7}_{-8} M_\oplus$ (i. e., Neptune-mass) and $a = 3.3^{+1.4}_{-0.8}$ AU, respectively. The lensed object (i. e., the source) is a mid-G star located in the Galactic Bulge with a mass of $M_S = 0.9 \pm 0.1 M_\odot$. From Sumi *et al.* (2010), reproduced with the permission of the AAS.

4.6.2 Nearby Lenses

When the lens is nearby, it is often possible to obtain more information about the lens and/or the lensing event, thereby breaking the degeneracy among the parameters that influence the observational signature of gravitational lensing. The gravitational mass of the lens is more likely to be measurable, and the presence of dark companions, such as, planets, can be more readily established. Because additional modes of study are possible for nearby lenses, with their greater accessibility, greater proper motions, and larger Einstein angles, they have been referred to as "mesolenses".

It might be possible, for example, to detect nearby lenses directly. This has been serendipitously achieved for several lenses (e. g., LMC-5, Alcock *et al.*, 2001a, 2001b,

Drake *et al.*, 2004, Gould, 2004, Gould *et al.*, 2004, Nguyen *et al.*, 2004; LMC-20, Kallivayalil *et al.*, 2006), and programs to routinely check for radiation from the lenses are being implemented. Direct detection of the lens confers several advantages. First, by measuring the flux from the lens in several filters and/or by obtaining a spectrum, it may be possible to determine its spectral type and then estimate its distance, D_L, via the method of photometric or spectroscopic parallax. In some cases, it may even be possible to determine the value of D_L by measuring the trigonometric parallax of the lens. Second, nearby lenses tend to have high proper motions, and thus a sequence of images taken over an interval of a few years might be usable to measure the angular speed of the lens across the sky. This provides another potential route toward determining D_L by comparing the space motion of the lens with Galactic mean motions along the relevant line-of-sight.

If direct detection of the lens yields estimates for D_L and v, then the value obtained for τ_E from the light curve fit provides an estimate of the gravitational mass of the lens. Another advantage of nearby lenses is that θ_E can be large enough that the centroid shift associated with lensing can be measured using high resolution imaging from space (e. g., using HST), providing a direct measurement of the Einstein angle. This allows the gravitational mass to be measured, even if only D_L is independently known. The discussion that follows focuses on the case in which the lens is either a white dwarf or a planetary companion to a white dwarf, and D_L is smaller than a few hundred pc, so that the lens can be detected and identified as a white dwarf.

4.6.3 Planet Detection via Lensing

Planets orbiting white dwarfs can be found through their actions as lenses, which translate into detectable light curve signatures. The characteristics of these signatures depend on the projected separation between the planet and white dwarf, expressed in units of the Einstein radius, R_E, which is given by

$$R_E = \theta_E D_L \approx 0.70 \left[\left(\frac{M_L}{0.6 M_\odot} \right) \left(\frac{D_L}{100\,\text{pc}} \right) \left(1 - \frac{D_L}{D_S} \right) \right]^{1/2} \text{AU}. \tag{4.4}$$

First, consider the case of wide orbits, as expected in the eventual descendant of the HR 8799 system (see Section 4.2). With $D_L = 40$ pc, and assuming the mass of the white dwarf remnant to be $0.6 M_\odot$, then $\theta_E = 11.1$ mas and $R_E = 0.44$ AU (assuming $D_L/D_S \approx 0$). The expanded post-main sequence orbits of the planets will have semimajor axes of roughly 100 AU, so if the system is oriented face-on, then the projected separation will be 2.5″. The proper motion of HR 8799 is $\mu = 0.12'' \text{yr}^{-1}$ (Perryman *et al.*, 1997), corresponding to $v = 23 \text{ km s}^{-1}$ at a distance of 40 pc. Therefore, if a planet with mass of $10 M_J$ in this system served as a lens for a background star, then it would produce an event with $\tau_E = 8.4$ days. This lensing event would be recognizable as being due to a planet because of its relatively short duration as well as the presence of a nearby white dwarf with a known proper motion. In the example outlined here, the planet interpretation could be easily verified because the planets would be detectable via their photometric signature and high resolution imaging. The lensing event would serve as a signal that planets should be looked for along this particular line-of-sight. Immediately after the event, the lensing planet might be detectable through an infrared signature, in excess of that produced by the lensed source. After several years, the planet will have moved far enough off the lensed star to be resolvable in images. The lensing event also allows the gravitational mass of the planet to be measured. This is important in itself and, in addition, provides a constraint that can be used to test models for planetary emission.

For special orientations of the track of the lens relative to the binary axis, the white dwarf can produce a separate lensing event. That is, the white dwarf and planet produce a sequence of two events, often referred to as a "repeating event". If the planet is closer to the white dwarf, or the orbital orientation is closer to edge-on, then the projected orbital separation will be smaller. With decreasing separation, the fraction of tracks producing repeating events increases. When both the star and planet contribute to the lensing signature, the physical connection between the planet and white dwarf becomes well-established, whether or not the planet can be directly imaged. When the projected separation between white dwarf and planet is $< 3.5 R_E$, any event in which the planet serves as a lens will also show evidence of the white dwarf. For separations $\lesssim 1.5 R_E$, the planet manifests as a short-lived deviation in a longer event in which the white dwarf serves as the primary lens. In addition, in such cases, information about D_L can translate into measurements of both masses.

The potential importance of gravitational lensing as a tool to discover white dwarf planetary systems lies in the fact that it is sensitive to low mass planets, including Earth mass planets. Observations of lensing events can also be used to discover planets across a wide range of semimajor axes, typically ranging from $\sim$ 0.1 AU out to hundreds of AU.

4.6.4
Identifying White Dwarfs in the Lens System

In order for lensing to become an effective tool for the discovery and study of white dwarf planets, it must be possible to determine which events are caused by either white dwarfs or low mass companions to white dwarfs. This has not yet been accomplished, but not because of fundamental limitations. Rather, since the first-generation monitoring programs assumed that the lenses would be dark or dim masses located more than a few kpc away, they did not attempt to match the positions of events to the positions of catalogued objects. This is beginning to change, however, as theoretical work has emphasized the importance of nearby lenses (Di Stefano, 2008a, 2008b), and as nearby lenses are beginning to be discovered. In fact, several brown dwarf (e. g., Gould *et al.*, 2009) and M dwarf events (e. g., Alcock *et al.*, 2001b) have been identified.

Today, the discovery of an interesting gravitational lensing event often triggers frequent monitoring by observers world-wide to achieve highly sensitive, almost continuous coverage of the light curves. The events deemed of sufficient interest to merit this sort of attention have generally been events whose evolving light curve fits show that they will reach high magnification, and thus the flux from the lensed source is predicted to reach hundreds or thousands of times more than its baseline level. The motivation is that monitoring during the high-magnification portion of an event will either discover evidence of orbiting planets or else will place limits on the existence of a planet in the region around the Einstein ring. If it is known that a possible white dwarf is located near the position of a just-discovered lensing event, then intensive monitoring should also be started, even if the magnification will not be very large. Note that planets can be discovered in events of only moderate magnification and in short-duration events of any magnification.

Halfway through 2010, the Optical Gravitational Lensing Experiment (OGLE; Udalski *et al.*, 1992)[5] is in the late stages of an upgrade that has the potential to increase the discovery rate of lensing events by a factor of a few, from the present rate of 600–700 events per year, mostly in Baade's window; the Microlensing Observations in Astrophysics (MOA)[6] monitoring program also discovers lensing events (e. g., see Figure 4.6). In addition, Pan-STARRS and LSST will monitor the entire sky at intervals of several days, with portions monitored more frequently, and have the potential to discover lensing events. Conservative expectations for the near future, therefore, are for the discovery of roughly two dozen white dwarf lens events per year. In order to have a good chance to identify and learn from these events, two additions to the current monitoring protocol are needed.

- When each event occurs, a check must be performed to determine if there is a catalogued white dwarf near the event location. The white dwarf catalogues compiled from the SDSS (e. g., Eisenstein *et al.*, 2006 and future versions) will be helpful when the monitoring program discovering the event is Pan-STARRS because there is overlap between the parts of the sky covered by SDSS and Pan-STARRS. It is, nevertheless, important, and for the Galactic Bulge, it will be essential to search in the vicinity of each event for catalogued counterparts of any type, and then to determine whether the multiband spectra are consistent with the interpretation that the lens is a white dwarf.
- If it is possible that the lens is a white dwarf, the event must be subjected to the type of intensive monitoring campaigns mentioned above. This means worldwide coverage several times per night, and even more frequently during any intervals when the light curve is predicted to exhibit high magnification. For every event monitored in this fashion, it will be possible to either discover planets or place limits on the existence of planets around the candidate white dwarf. The modeling of the event could be simple or complex. One particularly interesting case occurs when the lens is a planet located more than a few R_E from its white dwarf parent star. In this case, there will be a short event, and HST or other space-based observations obtained within months of the event should be able to resolve the angle separating the lensed source and the white dwarf, establishing the projected distance between the white dwarf and its planet.

5) http://ogle.astrouw.edu.pl/ (15 June 2011)
6) http://www.phys.canterbury.ac.nz/moa/ (15 June 2011)

It should be feasible to take the steps outlined above for 5–20 white dwarf lensing events per decade. This is a new approach with the potential to significantly contribute to the knowledge of white dwarf planetary systems.

4.7 Transits

4.7.1 Basics

Planetary transits can be rare. There is only a small probability, $\mathcal{P}$, that the orientation of the orbit with respect to the observer is favorable, given by

$$\mathcal{P} = \frac{R_{\rm wd} + R_{\rm p}}{a} \approx 6.7 \times 10^{-5} \left[\frac{(R_{\rm wd} + R_{\rm p})}{10^9\,{\rm cm}}\right] \left[\frac{1\,{\rm AU}}{a}\right] , \qquad (4.5)$$

where $R_{\rm wd}$ is the radius of a white dwarf, and $R_{\rm p}$ is the radius of a planet orbiting at a distance a from the white dwarf (Di Stefano *et al.*, 2010). Furthermore, detection requires observing the white dwarf during the short interval, typically minutes, when the planet passes in front of it. For circular orbits, the duty cycle is comparable to the value of $\mathcal{P}$.

These considerations show that the small size of white dwarfs relative to main sequence stars makes the probability of detecting a planetary transit so small that monitoring a fixed set of hundreds or thousands of white dwarfs for transits is not likely to yield positive results unless a large fraction of the white dwarfs have planets in orbits well within 1 AU. On the other hand, planetary transits of white dwarfs are an attractive avenue for research. Given the relatively small size of white dwarfs, Earth-size planets can completely eclipse high mass white dwarfs, and transits of even smaller planets can be detected with ground-based observations capable of 1% photometry.

4.7.2 SuperWASP

SuperWASP (Street and the SuperWASP Consortium, 2004)[7], the follow-up to the initial Wide Angle Search for Planets (WASP) project (Kane *et al.*, 2004), is a wide-field monitoring program designed to discover and study extrasolar planets orbiting main sequence stars. It has been very successful in its primary mission. Among the stars it monitors are 194 white dwarfs found by cross-correlating the SuperWASP catalogue with the McCook and Sion (1999) catalogue of white dwarfs. Faedi *et al.* (2009, 2010) were able to establish a low frequency ($< 10\%$) for the existence of brown dwarfs and gas giants in close orbits ($P_{\rm orb} < 0.1$–0.2 d) with the white dwarfs. It is, naturally, desirable to achieve more stringent limits. Of particu-

7) http://www.superwasp.org (9 May 2011)

lar interest are discoveries of, or limits on, terrestrial-size planets. Faedi *et al.* noted that larger sample size and more frequent observations are needed.

4.7.3
Wide-Field Monitoring

Fortunately, it should be possible to take advantage of larger wide-field monitoring programs; for example, Pan-STARRS and LSST. It is difficult to estimate the numbers of white dwarf transiting planets that could be discovered by Pan-STARRS and LSST. It is expected, however, that these programs will monitor a large fraction of the $\sim 10^6$ white dwarfs within a few hundred pc. Although the cadence will be far from ideal, these programs will nevertheless be sensitive to orbits with periods of tens of days. For a planet with a period of 10 days, for example, the duty cycle for transits is roughly 1%. Over a ten year period, these systems will be observed hundreds of times and there is a near certainty that one or more visits to the field containing the white dwarf will find it in eclipse during an ongoing planet transit. Such an observation should be used to trigger intensive monitoring of the white dwarf to search for signs of additional transits, and to provide continuous monitoring during transits. Because transits by planets in such close orbits will repeat, and can be observed in detail, a high level of confidence in the detection can be achieved. Wide-field monitoring programs will, therefore, be able to detect or to place limits on the fraction of white dwarfs with small planets orbiting within ~ 0.1 AU.

4.8
Prospects for the Future

White dwarfs represent an important frontier for planet discovery. There are many reasons to search for planets around white dwarfs. The most direct application is the study of planets around intermediate mass stars, for which all of the planet detection methods discussed here (radial velocity, timing, transits, and lensing) are difficult to apply. Finding planets orbiting white dwarfs descended from intermediate mass stars might be the most direct way to study these planetary systems. There are other important science returns as well. What happens to planets and planetary systems when the host star evolves? Can planets form after the central star has become a white dwarf? By finding an ensemble of white dwarf planetary systems, a number of questions of great interest to astronomers can be answered.

Finding the white dwarf planets will also answer a number of other questions of interest to biologists, philosophers, and the general public as well as to astronomers. What will happen to the Earth and the other terrestrial planets after the Sun evolves off the main sequence? What will happen to life in the universe after galaxies become so old that star formation has long since ceased? If life will exist in the far future universe, trillions of years from now, it will be because the conditions necessary to support life are common. It is known that life can thrive on planets. However, it is not yet known how different conditions can be from those on the Earth, and still support life. Many of the processes capable of producing enough energy, even in the absence of significant stellar luminosity, to support life, such as tidal interactions, are expected to continue far into the future. It may be that the most common environment in the universe to support life will eventually be provided by planets orbiting white dwarfs. It is interesting to note that some of the white dwarfs already in existence have been white dwarfs for longer than the time it has taken for intelligent life to evolve on the Earth (roughly 4 billion years). They will continue to be white dwarfs until some time in the indefinite future when the decay of fundamental particles might occur (perhaps 10^{45} years from now). If the present-day white dwarfs host planetary systems where life can develop, then there will be ample time for civilizations of all types to evolve, vanish, and evolve again.

On a much shorter time scale (i. e., within the next few years), planets around "garden-variety" white dwarfs should begin to be discovered. Exactly what types of systems will be discovered cannot be known now. Research on exoplanets has produced surprises; perhaps the discovery of white dwarf planets will do so as well. Therefore, rather than focusing on possible results, this chapter has concentrated on possible methods of discovery. It has been shown that a set of complementary methods exist which, taken together, can test for white dwarf planets with a wide range of masses and orbital periods. Some of these methods are already being aggressively pursued. Searches are being conducted for infrared signatures of massive planets, and timing studies are well underway. Great success has been achieved by groups studying metal pollution of white dwarfs and the dust disk signatures around those polluted white dwarfs. These studies have successfully established that many white dwarfs are very likely to host large asteroid systems. They have, in fact, graduated to using the enrichment profiles observed in the white dwarf atmospheres, and the composition of the disks, to learn about the asteroids themselves. Lensing and transit searches for planets around white dwarfs are in their infancy. Improvements in ongoing programs, and the development of new wide-field monitoring programs makes both lensing and transits promising avenues for future research.

The goal of these searches must be to discover large numbers of white dwarf planetary systems. It is only by doing this that the level of influence of stellar evolution on planetary systems can be established, and a determination can be made as to whether or not new planets can form around white dwarfs. The study of a large population of white dwarf planetary systems is needed if an understanding is to be developed of the characteristics that will be exhibited by the most common planets of the distant future.

References

Aannestad, P.A., Kenyon, S.J., Hammond, G.L., and Sion, E.M. (1993) *Astron. J.*, **105**, 1033.

Alcock, C., Fristrom, C.C., and Siegelman, R. (1986) *Astrophys. J.*1, **302**, 462.

Alcock, C. *et al.* (2001a) *Astrophys. J.*, **552**, 582.

Alcock, C. *et al.* (2001b) *Nature*, **414**, 617.

Baraffe, I., Chabrier, G., Barman, T.S., Allard, F., and Hauschildt, P.H. (2003) *Astron. Astrophys.*, **402**, 701.

Becklin, E.E. and Zuckerman, B. (1988) *Nature*, **336**, 656.

Becklin, E.E., Farihi, J., Jura, M., Song, I., Weinberger, A.J., and Zuckerman, B. (2005) *Astrophys. J. Lett.*, **632**, L119.

Beuermann, K. *et al.* (2010) *Astron. Astrophys.*, **521**, L60.

Beuermann, K. *et al.* (2011) *Astron. Astrophys.*, **526**, A53.

Boss, A.P. (1995) *Science*, **267**, 360.

Brinkworth, C.S., Marsh, T.R., Dhillon, V.S., and Knigge, C. (2006) *Mon. Not. R. Astron. Soc.*, **365**, 287.

Brinkworth, C.S., Gänsicke, B.T., Marsh, T.R., Hoard, D.W., and Tappert, C. (2009) *Astrophys. J.*, **696**, 1402.

Brinkworth, C.S., Gänsicke, B.T., Girven, J.M., Hoard, D.W., Marsh, T.R., and Parsons, S. (2011) *Astrophys. J.*, submitted.

Burleigh, M.R., Clarke, F.J., and Hodgkin, S.T. (2002) *Mon. Not. R. Astron. Soc.*, **331**, L41.

Burleigh, M.R. *et al.* (2008) *Mon. Not. R. Astron. Soc.*, **386**, L5.

Burrows, A., Sudarsky, D., and Lunine, J.I. (2003) *Astrophys. J.*, **596**, 587.

Butler, R.P. *et al.* (2006) *Astrophys. J.*, **646**, 505.

Clarke, F.J., and Burleigh, M.R. (2004) *Extrasolar Planets: Today and Tomorrow, ASP Conf. Ser.*, vol. 321 (eds J.-P. Beaulieu, A. Lecavelier des Etangs, and C. Terquem), ASP, San Francisco, p 76.

Debes, J.H. and Sigurdsson, S. (2002) *Astrophys. J.*, **572**, 556.

Debes, J.H., Sigurdsson, S., and Woodgate, B.E. (2005) *Astrophys. J.*, **633**, 1168.

Di Stefano, R. (2008a), *Astrophys. J.*, **684**, 46.

Di Stefano, R. (2008b), *Astrophys. J.*, **684**, 59.

Di Stefano, R., Howell, S.B., and Kawaler, S.D. (2010) *Astrophys. J.*, **712**, 142.

Di Stefano, R. and Scalzo, R.A. (1999) *Astrophys. J.*, **512**, 564.

Drake, A.J., Cook, K.H., and Keller, S.C. (2004) *Astrophys. J. Lett.*, **607**, L29.

Dufour, P., Kilic, M., Fontaine, G., Bergeron, P., Lachapelle, F.-R., Kleinman, S.J., and Leggett, S.K. (2010) *Astrophys. J.*, **719**, 803.

Einstein, A. (1936) *Science*, **84**, 506.

Eisenstein, D.J. *et al.* (2006) *Astrophys. J. (Suppl.)*, **167**, 40.

Faedi, F., West, R., Burleigh, M.R., Goad, M.R., and Hebb, L. (2009) *J. Phys. Conf. Ser.*, **172**, 012057.

Faedi, F., West, R.G., Burleigh, M.R., Goad, M.R., and Hebb, L. (2010) *Mon. Not. R. Astron. Soc.*, in press.

Farihi, J., Zuckerman, B., and Becklin, E.E. (2005) *Astron. Nachr.*, **326**, 964.

Farihi, J., Burleigh, M.R., and Hoard, D.W. (2008a) *Astrophys. J.*, **674**, 421.

Farihi, J., Zuckerman, B., and Becklin, E.E. (2008b) *Astrophys. J.*, **674**, 431.

Farihi, J., Becklin, E.E., and Zuckerman, B. (2009a) *Astrophys. J.*, **681**, 1470.

Farihi, J., Jura, M., and Zuckerman, B. (2009b) *Astrophys. J.*, **694**, 805.

Farihi, J., Barstow, M.A., Redfield, S., Dufour, P., and Hambly, N.C. (2010a), *Mon. Not. R. Astron. Soc.*, **404**, 2123.

Farihi, J., Jura, M., Lee, J.-E., and Zuckerman, B. (2010b) *Astrophys. J.*, **714**, 1386.

Farihi, J., Brinkworth, C., Gänsicke, B.T., Marsh, T.R., Girven, J., Hoard, D.W., Klein, B., and Koester, D. (2011) *Astrophys. J. Lett.*, **728**, L8.

Ford, E.B., Joshi, K.J., Rasio, F.A., and Zbarsky, B. (2000) *Astrophys. J.*, **528**, 336.

Friedrich, S., Zinnecker, H., Correia, S., Brandner, W., Burleigh, M., and McCaughrean, M. (2007) in *ASP Conf. Ser.* Proceedings of the 15th European Workshop on White Dwarfs, vol. 372 (eds R. Napiwotzki and M.R. Burleigh), ASP, San Francisco, p 343.

Gänsicke, B.T., Marsh, T.R., Southworth, J., and Rebassa-Mansergas, A. (2006) Science, **314**, 1908.

Gänsicke, B.T., Marsh, T.R., and Southworth, J. (2007) *Mon. Not. R. Astron. Soc.*, **380**, L35.

Gänsicke, B.T., Koester, D., Marsh, T.R., Rebassa-Mansergas, A., and Southworth, J. (2008) *Mon. Not. R. Astron. Soc.*, **391**, L103.

Gould, A. (2004) *Astrophys. J*, **606**, 319.

Gould, A., Bennett, D.P., and Alves, D.R. (2004) *Astrophys. J.*, **614**, 404.

Gould, A. *et al.* (2009) *Astrophys. J. Lett.*, **698**, L147.

Greenberg, R., Hartmann, W.K., Chapman, C.R., and Wacker, J.F. (1978) *Icarus*, **35**, 1.

Hansen, B.M.S., Kulkarni, S., and Wiktorowicz, S. (2006) *Astron. J.*, **131**, 1106.

Harris, H.C. *et al.* (2006) *Astron. J.*, **131**, 571.

Ignace, R. (2001) *Publicat. ASP*, **113**, 1227.

Jorissen, A., Mayor, M., and Udry, S. (2001) *Astron. Astrophys.*, **379**, 992.
Jura, M. (2003) *Astrophys. J. Lett.*, **584**, L91.
Jura, M., Farihi, J., and Zuckerman, B. (2007a) *Astrophys. J.*, **663**, 1285.
Jura, M., Farihi, J., Zuckerman, B., and Becklin, E.E. (2007b) *Astron. J.*, **133**, 1927.
Jura, M., Farihi, J., and Zuckerman, B. (2009a) *Astron. J.*, **137**, 3191.
Jura, M., Muno, M.P., Farihi, J., and Zuckerman, B. (2009b) *Astrophys. J.*, **699**, 1473.
Kalas, P. *et al.* (2008) *Science*, **322**, 1345.
Kallivayalil, N., Patten, B.M., Marengo, M., Alcock, C., Werner, M.W., and Fazio, G.G. (2006) *Astrophys. J. Lett.*, **652**, L97.
Kane, S.R., Horne, K., Lister, T., Cameron, A.C., Street, R.A., Pollacco, D.L., James, D., and Tsapras, Y. (2004) *Extrasolar Planets: Today and Tomorrow. ASP Conf. Proc.*, vol. 321 (eds J.-P. Beaulieu, A. Lecavelier des Etangs, and C. Terquem), ASP, San Francisco, p. 115.
Kepler, S.O. *et al.* (2003) *Astron. Astrophys.*, **401**, 639.
Kilic, M. and Redfield, S. (2007) *Astrophys. J.*, **660**, 641.
Kilic, M., von Hippel, T., Leggett, S.K., and Winget, D.E. (2005) *Astrophys. J. Lett.*, **632**, L115.
Kilic, M., von Hippel, T., Leggett, S.K., and Winget, D.E. (2006) *Astrophys. J.*, **646**, 474.
Kilic, M., Gould, A., and Koester, D. (2009) *Astrophys. J.*, **705**, 1219.
Kilic, M., Brown, W.R., and McLeod, B. (2010) *Astrophys. J.*, **708**, 411.
Kleinman, S.J. *et al.* (1994) *Astrophys. J.*, **436**, 875.
Koester, D. (2009) *Astron. Astrophys.*, **498**, 517.
Koester, D., Provencal, J., and Shipman, H.L. (1997) *Astron. Astrophys.*, **320**, L57.
Lafrenière, D., Marois, C., Doyon, R., and Barman, T. (2009) *Astrophys. J. Lett.*, **694**, L148.
Lagrange, A.-M. *et al.* (2010) *Science*, **329**, 57.
Lee, J.W., Kim, S.-L., Kim, C.-H., Koch, R.H., Lee, C.-U., Kim, H.-I., and Park, J.-H. (2009) *Astron. J.*, **137**, 3181.
Livio, M., Pringle, J.E., and Wood, K. (2005) *Astrophys. J. Lett.*, **632**, L37.
Marcy, G., Butler, R.P., Fischer, D., Vogt, S., Wright, J.T., Tinney, C.G., and Jones, H.R.A. (2005) *Prog. Theor. Phys. (Suppl.)*, **158**, 24.
Marcy, G.W. *et al.* (2008) *Phys. Scr.*, **T130**, 014001.
Marois, C., Macintosh, B., Barman, T., Zuckerman, B., Song, I., Patience, J., Lafrenière, D., and Doyon, R. (2008) *Science*, **322**, 1348.
Maxted, P.F.L., Napiwotzki, R., Dobbie, P.D., and Burleigh, M.R. (2006) *Nature*, **442**, 543.
Mayor, M. and Queloz, D. (1995) *Nature*, **378**, 355.
McCook, G.P. and Sion, E.M. (1999) *Astrophys. J. (Suppl.)*, **121**, 1.
Melis, C., Zuckerman, B., Song, I., Rhee, J.H., and Metchev, S. (2009) *Astrophys. J.*, **696**, 1964.
Melis, C., Jura, M., Albert, L., Klein, B., and Zuckerman, B. (2010) *Astrophys. J.*, **722**, 1078.
Mullally, F. (2009) *J. Phys. Conf. Ser.*, **172**, 012056.
Mullally, F., Kilic, M., Reach, W.T., Kuchner, M.J., von Hippel, T., Burrows, A., and Winget, D.E. (2007a) *Astrophys. J. (Suppl.)*, **171**, 206.
Mullally, F., von Hippel, T., and Winget, D.E. (2007b) in *ASP Conf. Ser.* Proceedings of the 15th European Workshop on White Dwarfs, vol. 372 (eds R. Napiwotzki and M.R. Burleigh), ASP, San Francisco, p. 355.
Mullally, F., Winget, D.E., De Gennaro, S., Jeffery, E., Thompson, S.E., Chandler, D., and Kepler, S.O. (2008) *Astrophys. J.*, **676**, 573.
Mullally, F., Reach, W.T., De Gennaro, S., and Burrows, A. (2009) *Astrophys. J.*, **694**, 327.
Nguyen, H.T., Kallivayalil, N., Werner, M.W., Alcock, C., Patten, B.M., and Stern, D. (2004) *Astrophys. J. (Suppl.)*, **154**, 266.
Nordhaus, J., Spiegel, D.S., Ibgui, L., Goodman, J., and Burrows, A. (2010) *Mon. Not. R. Astron. Soc.*, **408**, 631.
Ormel, C.W., Dullemond, C.P., and Spaans, M. (2010) *Icarus*, **210**, 507.
Parriott, J. and Alcock, C. (1998) *Astrophys. J.*, **501**, 357.
Parsons, S.G. *et al.* (2010) *Mon. Not. R. Astron. Soc.*, **407**, 2362.
Perets, H. (2010) *Bull. Am. Astron. Soc.*, **42**, 938.
Perryman, M.A.C. *et al.* (1997) *Astron. Astrophys.*, **323**, L49.
Qian, S.-B., Dai, Z.-B., Liao, W.-P., Zhu, L.-Y., Liu, L., and Zhao, E.G. (2009) *Astrophys. J. Lett.*, **706**, L96.

Qian, S.-B., Liao, W.-P., Zhu, L.-Y., Dai, Z.-B., Liu, L., He, J.-J., Zhao, E.-G., and Li, L.-J. (2010) *Mon. Not. R. Astron. Soc.*, **401**, L34.

Rasio, F.A., Tout, C.A., Lubow, S.H., and Livio, M. (1996) *Astrophys. J.*, **470**, 1187.

Reach, W.T., Kuchner, M.J., von Hippel, T., Burrows, A., Mullally, F., Kilic, M., and Winget, D.E. (2005) *Astrophys. J. Lett.*, **635**, L161.

Richer, H.B., Ibata, R., Fahlman, G.G., and Huber, M. (2003) *Astrophys. J. Lett.*, **597**, L45.

Sackmann, I.-J., Boothroyd, A.I., and Kraemer, K.E. (1993) *Astrophys. J.*, **418**, 457.

Schmidt, G.D., Szkody, P., Silvestri, N.M., Cushing, M.C., Liebert, J., and Smith, P.S. (2005) *Astrophys. J. Lett.*, **630**, L173.

Schröder, K.-P. and Connon Smith, R. (2008) *Mon. Not. R. Astron. Soc.*, **386**, 155.

Schwarz, R., Schwope, A.D., Vogel, J., Dhillon, V.S., Marsh, T.R., Copperwheat, C., Littlefair, S.P., and Kanbach, G. (2009) *Astron. Astrophys.*, **496**, 833.

Sigurdsson, S. (1992) *Astrophys. J. Lett.*, **399**, L95.

Sigurdsson, S. (1993) *Astrophys. J. Lett.*, **415**, L43.

Sigurdsson, S. (1995) *Astrophys. J.*, **452**, 323.

Sigurdsson, S., Richer, H.B., Hansen, B.M., Stairs, I.H., and Thorsett, S.E. (2003) Science, **301**, 193.

Silvotti, R. *et al.* (2007) *Nature*, **449**, 189.

Street, R.A. and the SuperWASP Consortium (2004) *Balt. Astron.*, **13**, 707.

Süli, Á. (2010) *J. Phys. Conf. Ser.*, **218**, 012004.

Sumi, T. *et al.* (2010) *Astrophys. J.*, **710**, 1641.

Udalski, A., Szymanski, M., Kaluzny, J., Kubiak, M., and Mateo, M. (1992) *Acta Astron.*, **42**, 253.

Villaver, E. and Livio, M. (2007) *Astrophys. J.*, **661**, 1192.

von Hippel, T., Kuchner, M.J., Kilic, M., Mullally, F., and Reach, W.T. (2007) *Astrophys. J.*, **662**, 544.

Waters, L.B.F.M. *et al.* (1998) *Nature*, **391**, 868.

Wickramasinghe, D.T., Farihi, J., Tout, C.A., Ferrario, L., and Stancliffe, R.J. (2010) *Mon. Not. R. Astron. Soc.*, **404**, 1984.

Winget, D.E. *et al.* (1990) *Astrophys. J.*, **357**, 630.

Wolszczan, A. (2008) *Phys. Scr.*, **T130**, 014005.

Wolszczan, A. and Frail, D.A. (1992) *Nature*, **355**, 145.

Zuckerman, B. and Becklin, E.E. (1987) *Nature*, **330**, 138.

5
White Dwarf Circumstellar Disks: Observations

Jay Farihi

5.1
Introduction

A circumstellar disk or ring is particulate matter that surrounds a star and is primarily confined to the plane of stellar rotation. Thus, disks distinguish themselves from spherical clouds or envelopes of gas (and dust) that typically surround protostellar objects and evolved giant stars. Circumstellar disks appear at every stage of stellar evolution, though the origin of the orbiting material is not always clearly understood. Pre-main sequence stars accrete material from a disk that is the flattened remnant of the cloud out of which they formed. Disks found around young stars in subsequent evolutionary stages are the likely site of planet formation, migration, and sometimes destruction. Mature main sequence stars exhibit dusty disks owing to recent or ongoing energetic collisions among orbiting asteroid or comet analogs, commonly referred to as the "Vega Phenomenon" (named after the first star observed to have orbiting, nonstellar material). Several first-ascent and asymptotic giant stars are also known to have circumstellar disks, though their origins are still debated, with hypotheses ranging from debris in a cold cometary cloud to consumed stellar companions. For a thorough (pre-Spitzer Space Telescope) review, see Zuckerman (2001).

White dwarfs are a relatively recent addition to the list of stellar objects with circumstellar disks, as their intrinsic faintness and the infrared-bright sky have conspired to keep them hidden. The last seven years have seen a profusion of white dwarf disk discoveries due primarily to the 2003 launch and unprecedented infrared performance of Spitzer, while ground-based projects with large sky coverage such as the Sloan Digital Sky Survey (SDSS; York *et al.*, 2000) and the Two Micron All Sky Survey (2MASS; Skrutskie *et al.*, 2006) have also played important roles. Given the small radii of white dwarfs, and because circumstellar material derives its luminosity from the central star, white dwarf disk observations are challenging.

White Dwarf Atmospheres and Circumstellar Environments. First Edition. Edited by D. W. Hoard.

5.2
History and Background

Developed in the fifties and sixties, the first photoelectric detectors for infrared[1] astronomy were revolutionary (McLean, 1997), but restricted to luminous sources, especially at wavelengths beyond 3 μm where the sky itself is bright and variable. Even so, the circumstellar disk around Vega, one of the brightest stars in the sky, was not discovered until 1983 with the launch of the Infrared Astronomical Satellite (IRAS). Despite the unprecedented sensitivity and all-sky coverage of IRAS, the detection of dust around white dwarfs had to await the development of more sensitive infrared detectors.

5.2.1
Early Searches

Excess infrared emission associated with a star can arise from heated circumstellar dust or from a self-luminous companion. Owing to their compact nature, white dwarfs can be easily outshone by low mass stellar and brown dwarf companions. This fact led Probst (1981) to search a large sample of nearby white dwarfs for near-infrared JHK photometric excess to study the luminosity function of the lowest mass stars and brown dwarfs (Probst, 1983; Probst and O'Connell, 1982). This was the first search for infrared excess associated with white dwarfs, and Probst deserves the credit for an insight that is now taken for granted and which fostered an abundance of subsequent infrared work on white dwarfs.

The first mid-infrared search for excess emission from white dwarfs was similarly motivated by the potential identification of brown dwarf companions. Shipman (1986) carried out a cross correlation of IRAS catalogue point sources with known, nearby white dwarfs, but the search did not produce any detections. This is perhaps unsurprising given that IRAS was only sensitive to point sources brighter than 500 mJy at its shortest wavelength bandpass of 12 μm, while the brightest white dwarf in the sky (Sirius B) should only be around 5 mJy at that wavelength.

5.2.2
The Discovery of Infrared Excess from G29-38

Inspired by the work of Probst and armed with a rapidly evolving set of infrared detectors atop Mauna Kea at the NASA Infrared Telescope Facility (IRTF), Zuckerman and Becklin (1987) detected the first white dwarf with infrared excess that was not associated with a stellar companion: G29-38. Photometric observations at three bandpasses longward of 2 μm revealed flux in excess of that expected for the stellar photosphere of this relatively cool, $T_{\text{eff}} = 11\,500$ K, white dwarf (see Figure 5.1). Because the prime motivation behind the observations was to search for substellar companions, the infrared excess was attributed to a spatially unresolved

1) The terms near-, mid-, and far-infrared are used herein to refer to the wavelength ranges 1–5, 5–30, and 30–200 μm; alternative divisions and definitions exist (McLean, 1997).

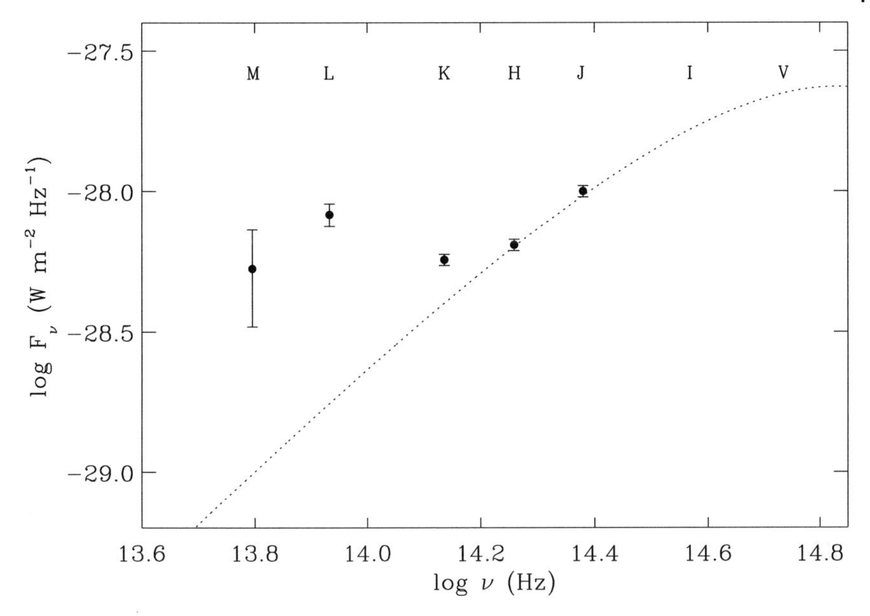

Figure 5.1 Discovery of infrared excess emission from the white dwarf G29-38 made at the IRTF, showing measurements with uncertainties in five infrared bands labeled in the figure: *J* (1.25 μm), *H* (1.65 μm), *K* (2.2 μm), *L* (3.5 μm), and *M* (4.8 μm). The dotted line is a blackbody plotted through the two shortest wavelength points, consistent with the white dwarf photosphere. This figure utilizes the photometric data for G29-38 provided in Zuckerman and Becklin (1987).

brown dwarf. Interestingly, circumstellar dust as warm as 1000 K was not favored due to the likelihood of rapid dissipation due to ongoing accretion and radiation drag. Zuckerman and Becklin (1987) prophetically noted that if a disk of material were orbiting close enough to reach such high temperatures, then spectral signatures of accretion should be seen (at the time, G29-38's atmospheric metals had not yet been detected).

Over the next few years, the infrared emission of G29-38 was studied intensely by many groups, and its unique properties sparked interest across many subfields of astrophysics research: brown dwarf and planet hunters, astroseismologists, infrared astronomers, and white dwarf pundits. Observational evidence gradually began to disfavor a brown dwarf as the source of infrared emission. First, some of the premier near-infrared imaging arrays revealed G29-38 to be a point source in several bandpasses. Second, near-infrared spectroscopy measured a continuum flux source (Tokunaga *et al.*, 1988), whereas a very cool (brown dwarf) atmosphere was expected to exhibit absorption features. Third, the detection of optical stellar pulsations echoed in the near-infrared was difficult to reconcile with a brown dwarf secondary (Graham *et al.*, 1990; Patterson *et al.*, 1991). Fourth, significant 10 μm emission was detected from G29-38, a few times greater than expected for a cool object with the radius of Jupiter, essentially ruling out the brown dwarf companion hypothesis (Telesco *et al.*, 1990; Tokunaga *et al.*, 1990).

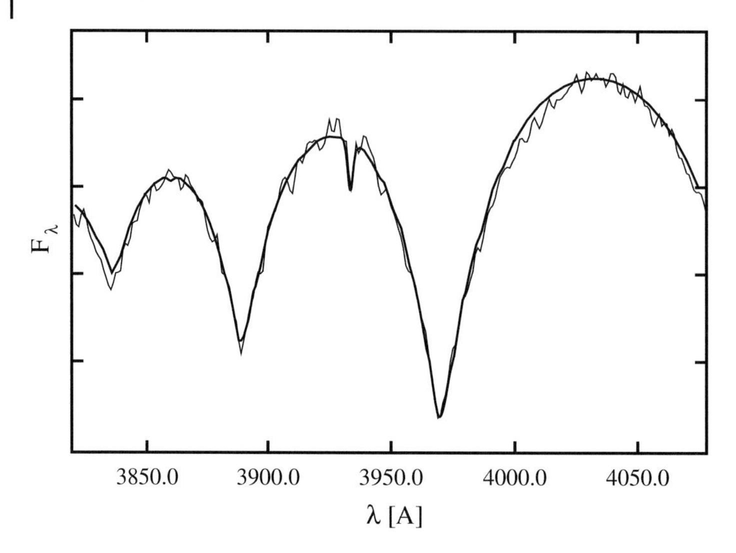

Figure 5.2 Discovery of photospheric calcium in the optical spectrum of G29-38, via the Fraunhofer K line. The thin line is the actual data while the thick line is the best model fit. It is noteworthy that the absorption line is quite weak despite the large calcium abundance (log [Ca/H] = −6.8; Zuckerman *et al.*, 2003), and is due to the relatively high opacity of hydrogen-rich white dwarf atmospheres. All else being equal, a similar calcium abundance produces a line equivalent width roughly 100 times greater in a helium-dominated atmosphere (see Figure 5.3). From Koester *et al.* (1997), reproduced with permission. © ESO.

Some lingering doubt remained that G29-38 was indeed surrounded by very warm dust, but variations seen in radial velocity (Barnbaum and Zuckerman, 1992) and pulse arrival times (Kleinman *et al.*, 1994) were never successfully attributed to an orbiting companion. A decade after the discovery of its infrared excess, the optical and ultraviolet spectroscopic detection of multiple metal species in the atmosphere of G29-38 (Koester *et al.*, 1997) made it clear that the star is currently accreting from its circumstellar environs (see Figure 5.2).

5.2.3
The Polluted Nature of Metal-Rich White Dwarfs

It would not be possible to tell the story of G29-38 and subsequent white dwarf disk detections without introducing the phenomenon of atmospheric metal contamination. The origin and abundances of photospheric metals in isolated white dwarfs has been an astrophysical curiosity dating back to the era when the first few white dwarfs were finally understood to be subluminous via the combination of spectra and parallax (van Maanen, 1919). In a half page journal entry, van Maanen (1917) noted that his accidentally discovered faint star with large proper motion had a spectral type of "about F0", almost certainly based on its strong calcium H and K absorption features (see Figure 5.3). Only four decades later did it become clear that vMa 2 was metal-poor with respect to the Sun (Weidemann, 1960). Over the next

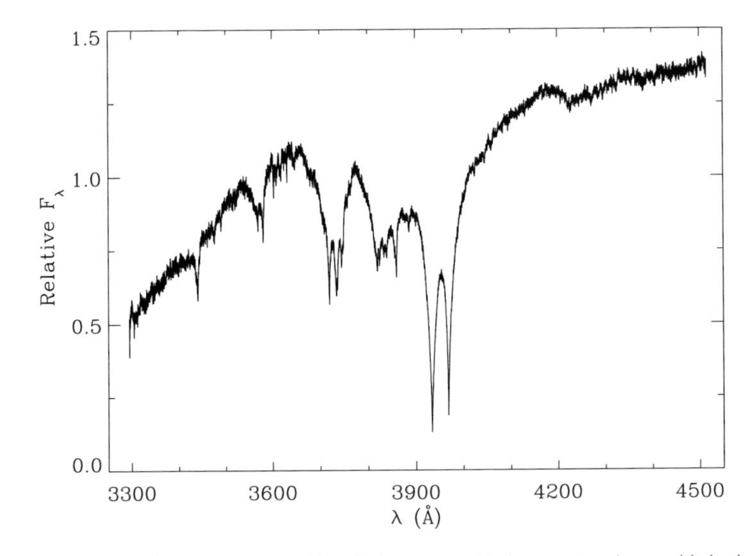

Figure 5.3 The prototype metal-lined white dwarf vMa 2 (van Maanen's star; van Maanen, 1917). In contrast to hydrogen-rich atmospheres, metal absorption features in helium-rich stars can be quite prominent, and often dominate their optical spectra as in this case. This is a previously unpublished spectrum obtained with the UVES spectrograph on the Very Large Telescope as part of the SPY survey (Napiwotzki *et al.*, 2003). All salient features are absorption due to iron, magnesium, or calcium.

decade and a half, it became gradually understood that white dwarfs (as a class) had metal abundances a few to several orders of magnitude below solar (Wegner, 1972; Wehrse, 1975).

Any primordial heavy elements in white dwarfs can only be sustained in their photospheres for the brief period during which the star is still rather hot and contracting, and then only to a certain degree (Chayer *et al.*, 1995). For $T_{\rm eff} < 25\,000$ K, gravitational settling is enhanced by the onset of convection and heavy elements sink rapidly in the high surface gravity atmospheres of white dwarfs (Alcock and Illarionov, 1980; Fontaine and Michaud, 1979; Vauclair *et al.*, 1979), leaving behind only hydrogen or helium. Downward diffusion timescales for heavy elements in cool white dwarfs are always orders of magnitude shorter than their evolutionary (cooling) timescales (Paquette *et al.*, 1986). Thus, external sources must be responsible for the presence of any metals within their photospheres.

The term "metal-rich" is used (somewhat ironically) to refer to cool white dwarfs that have trace abundances of atmospheric heavy elements. These are either hydrogen- or helium-rich atmosphere white dwarfs whose optical spectra exhibit the calcium K absorption line; the same atomic transition that is strongest in the Sun. While iron and magnesium absorption features are detected in the optical spectra for a substantial fraction of these stars, and additional elements are seen in a few cases, all currently known metal-rich white dwarfs display the calcium K line. (White dwarfs in binary systems can accrete heavy elements from a companion star via Roche lobe overflow or wind capture, e. g., cataclysmic variables, but the discussion here is restricted to white dwarfs that lack close stellar companions and excludes the topic of gaseous accretion disks resulting from binary mass transfer.)

5.2.4
Interstellar or Circumstellar Matter

There are two possible sources for the atmospheric metals seen in cool, single white dwarfs: accretion from the interstellar medium or from its immediate circumstellar environment. The latter case refers to material physically associated with the white dwarf and its formation (i. e., in simplest terms, a remnant planetary system) as opposed to a local accumulation of matter with distinct origins. In both cases, the accretion of heavy elements necessary to enrich the white dwarf atmosphere may be accompanied by the formation of a circumstellar disk. Thus, for cool white dwarfs, the phenomena of photospheric metals and circumstellar disks are likely to have a profound physical connection.

Historically, accretion from the interstellar medium was the most widely accepted hypothesis for the metals detected in cool white dwarfs. This was perhaps all, or for the most part, due to the fact that until 1983, all metal absorption features detected in cool white dwarfs were the result of relatively transparent, helium-dominated atmospheres (Sion *et al.*, 1990b). Such stars (referred to here as type DBZ) have relatively deep convection zones and commensurately long timescales for the downward diffusion of heavy elements, up to 10^6 yr (Paquette *et al.*, 1986). This allows for the possibility that their extant photospheric metals could be remnants of an interstellar cloud encounter several diffusion timescales prior (Dupuis *et al.*, 1992, 1993a, 1993b). However, the general lack of significant hydrogen in DBZ stars has been a continually recognized and glaring drawback for the interstellar accretion hypothesis (Aannestad *et al.*, 1993; Koester, 1976; Wesemael, 1979).

The confirmation of the first cool, hydrogen atmosphere white dwarf with metal absorption (G74-7, type DAZ; Lacombe *et al.*, 1983) presented a new challenge to the idea of interstellar cloud accretion. It took some time for robust stellar models to emerge, but it was basically understood that DA white dwarfs have relatively thin convection zones and correspondingly short metal diffusion timescales. These span a wide range from a matter of days in warmer stars like G29-38, all the way up to a few 10^3 yr for relatively cool stars such as G74-7 (Paquette *et al.*, 1986). Compared to all previously known metal-enriched white dwarfs, it was clear that the first DAZ star had experienced a recent accretion event. This led to the idea that comet impacts could be responsible for the photospheric metals in polluted white dwarfs (Alcock *et al.*, 1986).

The strength of the cometary impact model was that it capitalized on the hydrogen-poor nature of the accreted material in the numerous DBZ stars (Sion *et al.*, 1990a), yet it was difficult to reconcile with the lack of detected DAZ stars, as only G74-7 was known at the time (Alcock *et al.*, 1986). This apparent dearth of DAZ stars was eventually understood as an observational bias due to the relatively high opacity of hydrogen atmospheres compared to those composed primarily of helium (see Figures 5.2 and 5.3, and Dupuis *et al.*, 1993b). The eventual detection of atmospheric metals in numerous cool, hydrogen-rich white dwarfs required the combination of large telescopes and high resolution spectroscopy (Zuckerman and Reid, 1998). While these detections presented a challenge to the interstellar accretion hypothesis, they failed to breathe new life into the cometary impact model. For example, the second DAZ white dwarf to be found, G238-44 (Holberg *et al.*, 1997), has a metal diffusion timescale of only a few days, and an unlikely, continuous rain of comets would be needed to account for its metal abundance (Holberg *et al.*, 1997).

5.2.5
G29-38 and the Asteroid Accretion Model

Thus, did the study of metal-enriched white dwarfs come to somewhat of a historic crossroads circa 2003, a few months prior to the launch of Spitzer. With the sole exception of G29-38, there was a distinct lack of reliable (infrared) data on the circumstellar environments of white dwarfs, yet there was a growing profusion of stars contaminated by metals and problems with existing hypotheses (Zuckerman *et al.*, 2003). Because of their long metal dwell times, the contamination measured in DBZ stars with trace hydrogen could be made consistent with interstellar accretion models. However, the marked lack of dense interstellar clouds within 100 pc of the Sun (i. e., the Local Bubble; Welsh *et al.*, 1994, 1999) made this scenario difficult to reconcile with the existence of DAZ stars. In particular, the necessity for ongoing, high-rate accretion of heavy elements onto some DAZ stars (e. g., G238-44) rendered both the cometary impact and the interstellar cloud models unattractive and unlikely. Lacking a detailed model, Sion *et al.* (1990a) had speculated that asteroidal or planetary debris could be the ultimate source of the photospheric metals in hydrogen-poor DBZ white dwarfs.

In a short but seminal paper, Jura (2003) modeled the observed properties of G29-38 by invoking a tidally destroyed minor planet (i. e., asteroid, as suggested by Graham *et al.*, 1990) that generates an opaque, flat ring of dust analogous to the rings of Saturn. Rather than impacting the star, an asteroid perturbed into a highly eccentric orbit makes a close approach to the white dwarf, passes within its Roche limit, and is torn apart by gravitational tides. Ensuing collisions reduce the fragments to rubble and dust, and the resulting disk of material rapidly relaxes into a flat configuration owing to a range of very short ($P \sim 1$ h) orbital periods. The closely orbiting dust is heated by the white dwarf, producing an infrared excess, and slowly rains down onto the stellar surface, polluting its otherwise-pristine atmosphere with heavy elements. The bulk of a flat disk is shielded from the full light of the central star, allowing dust grains to persist for timescales longer than permitted by radiation drag forces. This model compared well to all the available infrared data on G29-38, including Infrared Space Observatory (ISO) 7 and 15 μm photometry (Chary *et al.*, 1999).

5.3
Pre-Spitzer and Ground-Based Observations

5.3.1
Photometric Searches for Near-Infrared Excess

The infrared excess and photospheric metals in G29-38 provided an empirical model to test on other white dwarfs prior to the development of the asteroid accretion model. Because the dust emission from G29-38 is prominent beginning at 2 μm (i. e., the *K* band), an obvious starting point would be to search for white dwarfs with similar near-infrared excess detectable with ground-based photometry. As previously mentioned, the G29-38 excess was found in just such a survey, but one aimed at identifying unevolved, low mass companions such as brown dwarfs. Although the authors did not state the result, Zuckerman and Becklin (1992) found no candidate analogs to G29-38 among *J H K* photometry of roughly 200 white dwarfs searched for near-infrared excess. Zuckerman and Becklin continued to obtain similar data for white dwarfs over the next several years, but very little was published on the topic until the Spitzer era.

Three photometric studies, again aimed at identifying low mass stellar and brown dwarf companions to white dwarfs, were published beginning in 2000. Green *et al.* (2000) surveyed around 60 extreme ultraviolet-selected white dwarfs in *J* and *K*, and identified a few stars with near-infrared excess consistent with stellar companions. These stars are too hot for any photospheric metals to be considered pollutants, and they did not search for mild *K*-band excesses that might be expected from circumstellar dust. Farihi *et al.* (2005) published the cumulative results of a decade and a half of near-infrared observations of white dwarfs begun by Zuckerman and Becklin. Among 371 white dwarfs, the study included over two dozen white dwarfs with a *K*-band excess from cool secondary stars, but no candidates for dusty white dwarfs. However, only roughly one third of the sample stars had independent *J K* photometry, while data for the remainder of the sample was taken from 2MASS.

Wachter *et al.* (2003) and later Hoard *et al.* (2007) published the results of a cross-correlation between the 2249 entries in the white dwarf catalogue of McCook and Sion (1999) and the 2MASS Point Source Catalogue Second and Final Data Releases. This enormous undertaking found a few dozen previously unidentified white dwarfs with infrared excess owing to cool stellar companions. Yet, although G29-38 satisfied their criteria for infrared excess, the search found no other candidates with similar colors; a startling result when taken at face value. However, the sensitivity limits of 2MASS, particularly in the K_S-band, often prevent reliable flux estimates for white dwarfs. A typical catalogue entry in McCook and Sion (1999) has $V \sim 15$ mag and near zero optical and infrared colors, and thus its predicted near-infrared magnitudes should be similar. Unfortunately, 2MASS K_S-band data becomes increasingly unreliable for sources fainter than $K_S = 14$ mag, severely limiting a robust search for dust emission from a large number of nearby white dwarfs (Farihi, 2009).

5.3.2
Metal-Polluted White Dwarf Discoveries

Contemporaneous with the launch of Spitzer, there were two major surveys of white dwarfs using high resolution optical spectrographs on the world's largest telescopes. Zuckerman *et al.* (2003) published a survey of nearly 120 cool DA white dwarfs with the High Resolution Echelle Spectrometer (HIRES) on the Keck I telescope. The study specifically aimed to detect photospheric metals in $T_{\rm eff} \lesssim$ 10 000 K, hydrogen-rich white dwarfs and was highly successful. Overall, 24 new DAZ stars were identified via the calcium K-line, including five that were known to have detached, low mass main-sequence companions. The study of Zuckerman *et al.* (2003) underscored the fact that detecting the weak optical absorption features from photospheric metals in DA white dwarfs requires high-powered instruments; nearly all of their detections had equivalent widths smaller than 0.5 Å.

Less than two years later, Koester *et al.* (2005b) published a subsample of stars from the Supernova Progenitor Survey (SPY; Napiwotzki *et al.*, 2003), which ob-

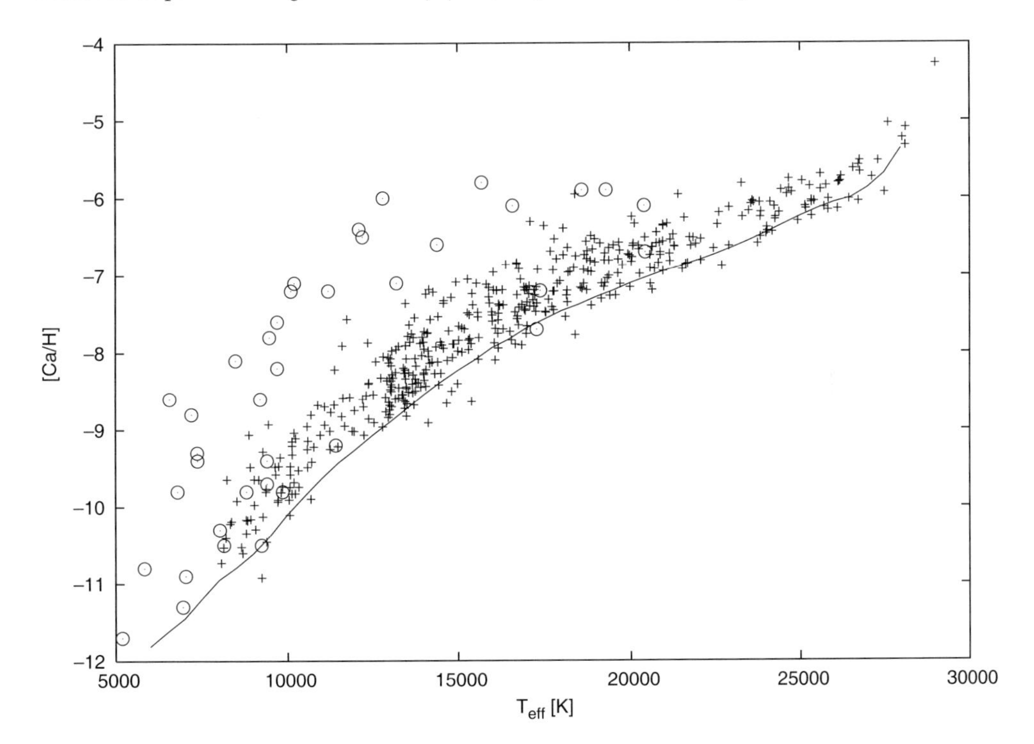

Figure 5.4 Calcium-to-hydrogen abundances (circles) and upper limits (plus symbols) for over 550 DA white dwarfs from the surveys of Zuckerman *et al.* (2003) and Koester *et al.* (2005b). The plot demonstrates the observational bias precluding the detection of modest metal abundances in warmer white dwarfs. The solid line represents an equivalent width detection limit of 15 mÅ. From Koester and Wilken (2006), reproduced with permission. © ESO.

served over one thousand white dwarfs with the UVES spectrograph on the Very Large Telescope (VLT) Unit 2. This extensive survey aimed to identify radial velocity variable, double degenerate binaries and, thus, required high spectral resolution. As a by-product, the search uncovered 18 new DAZ and nine new DBZ stars among warmer, $T_{\rm eff} \gtrsim 10\,000$ K white dwarfs, all displaying weak calcium K lines with equivalent widths less than 0.3 Å (Koester *et al.*, 2005a, 2005b). Importantly, nearly 500 DA stars of various temperatures were searched for calcium K line absorption, and upper limit abundances were determined for null detections (see Figure 5.4). These results provide a visually straightforward demonstration of the observational bias against the optical detection of metals in warmer white dwarfs.

5.3.3
The Spectacular Case of GD 362

The second white dwarf discovered to have circumstellar dust came 18 years (!) after G29-38, and resulted from two groups simultaneously recognizing the significance of its very metal-rich spectrum. First, Kawka and Vennes (2005) and, soon thereafter, Gianninas *et al.* (2004) reported that the optical spectrum of GD 362 had the strongest calcium lines seen in any DA star to date (see Figure 5.5), strong lines of magnesium and iron, and nearly *solar* abundances of these elements. These authors concluded that the cool star was too nearby to have attained its spectacular

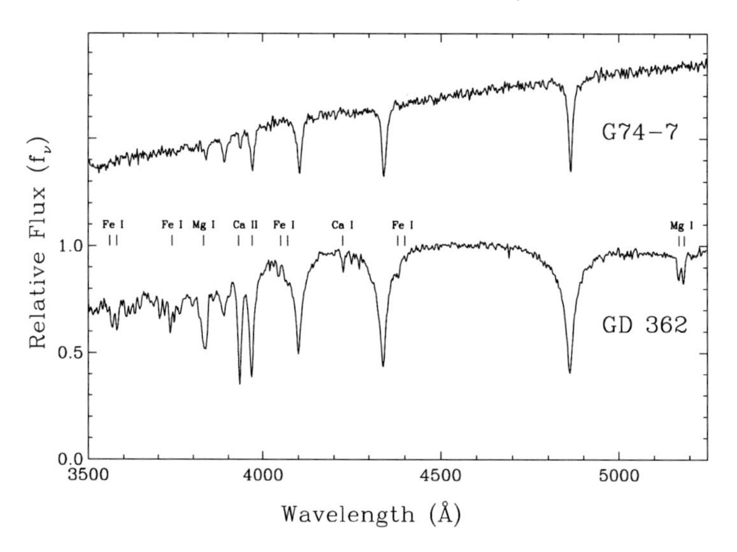

Figure 5.5 The highly metal-rich optical spectrum of GD 362, which exhibits strong lines of calcium, magnesium, and iron in addition to hydrogen Balmer lines, thus appearing as a DAZ-type star. This heavily polluted white dwarf has an atmosphere dominated by helium rather than hydrogen (Zuckerman *et al.*, 2007). Notably, the calcium H line is so sufficiently strong that it overwhelms Hϵ. Shown for comparison is the spectrum of G74-7, the prototype DAZ white dwarf. From Gianninas *et al.* (2004), reproduced with the permission of the AAS.

metal content from an interstellar cloud, but otherwise offered no explanation for their amazing find.

The significance of the spectacular pollution in this star was evident to at least two groups of astronomers, each involved in the early stages of Spitzer observing programs on white dwarfs. Using different observational methods, both teams simultaneously published evidence for a circumstellar disk around GD 362 in the same issue of the *Astrophysical Journal*. Chronologically, Kilic *et al.* (2005) observed the white dwarf first, using low-resolution near-infrared spectroscopy at 0.8–2.5 μm. Their spectrum revealed continuum excess beginning near 2.0 μm, matched well by adding 700 K blackbody radiation to the expected stellar flux. The authors inferred that the infrared emission is due to heated circumstellar dust, but neither modeled nor constrained its physical characteristics or origin.

Becklin *et al.* (2005) performed ground-based, mid-infrared photometric observations of GD 362 and detected the source in the N'-band (11.3 μm) at a level of 1.4 mJy. Because the expected flux from the stellar photosphere at this wavelength is only 0.01 mJy, the detection by itself is strong evidence for warm dust. Becklin *et al.* (2005) also obtained near-infrared $J H K L'$ photometry, constraining the spectral energy distribution of the infrared excess, and revealing that the emitting surface area was too large for a substellar companion. They applied the flat dust ring model of Jura (2003) to the infrared emission from GD 362 and showed that the innermost dust is located within 10 stellar radii and has a temperature around 1200 K, where typical dust grains rapidly sublimate. Thus, the location of the dust was found to be consistent with a tidally disrupted minor planet, and the resulting circumstellar disk was the probable source of the accreted metals.

5.3.4 Spectroscopic Searches for Near-Infrared Excess

The success of the near-infrared spectroscopic detection of excess emission from GD 362 was soon repeated. Kilic *et al.* (2006) detected a very strong continuum excess for the metal-rich white dwarf GD 56, sufficient to produce a photometric excess in the H-band (Farihi, 2009) and stronger overall than the near-infrared excesses of both G29-38 and GD 362 (see Figure 5.6). Though the measured spectra vary somewhat in slope and strength, the inner dust temperature for all of these stars is likely to be essentially the same, as the excess emission is largely a function of the emitting solid angle as seen from Earth. This discovery accompanied the first published survey for near-infrared excess that specifically targeted metal-rich white dwarfs. Among 18 DAZ stars, the search identified two additional targets with marginal K-band excesses: GD 133 and PG 1015+161. In both cases, the potential excess was considered too uncertain due to relatively low signal-to-noise (S/N) or nearby sources of potential confusion (Kilic *et al.*, 2006), and no conclusions were made for these stars. Using the same technique roughly one year later, Kilic and Redfield (2007) identified a continuum K-band excess for one more metal-enriched white dwarf, EC 11507−1519 (= WD 1150−153; also shown in Figure 5.6).

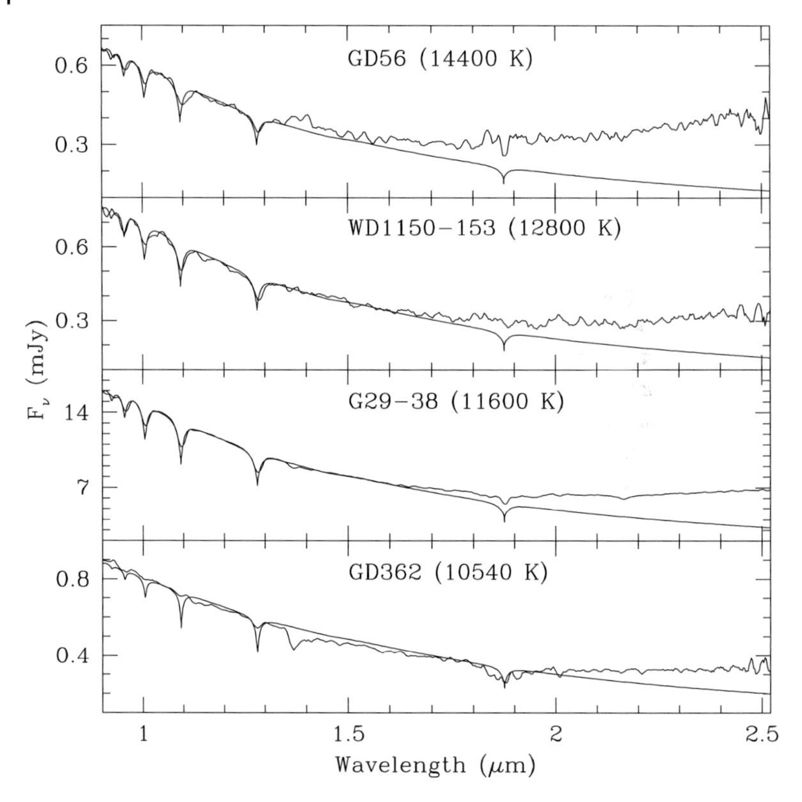

Figure 5.6 Low-resolution, near-infrared spectra of four metal-rich white dwarfs with circumstellar dust obtained with the SpeX spectrograph (Rayner *et al.*, 2003) compared to model stellar atmospheres. This figure was provided by M. Kilic (private communication) and is an updated version of Figure 1 in Kilic and Redfield (2007).

Generally, near-infrared data alone cannot distinguish between various dust emission models, as the excess at these wavelengths only represents "the tip of the iceberg" (as shown later). The near-infrared excesses measured for GD 56 and EC 11507−1519 were attributed to $T \approx 900$ K dust rather than substellar companions, but the distribution of the orbiting material was not modeled (Kilic and Redfield, 2007; Kilic *et al.*, 2006). Hence, the science that emerged from these few years of discoveries made in the near-infrared was generally qualitative and provided some statistical limits on the frequency of DAZ dust disk emission similar to that of G29-38. An identical search of 15 DBZ stars (Kilic *et al.*, 2008) failed to produce any new dust disks, but again provided some limiting statistics on the frequency of metal-rich white dwarfs with warm circumstellar dust. In hindsight, there were as many as six white dwarfs with circumstellar dust that went unidentified in these near-infrared spectroscopic observations.

5.3.5
Spectroscopy at Longer Wavelengths

There are few sufficiently bright, metal-rich white dwarfs that can be observed from the ground in a variety of infrared modes. Tokunaga *et al.* (1990) obtained a noisy, yet pioneering, IRTF low-resolution spectrum of G29-38 between roughly 3–4 μm with an early generation near-infrared spectrometer. The authors concluded that no spectral features were present, specifically not those typically seen in comets. By modern standards, the data quality was insufficient to draw a firm conclusion. Thus, many years later, their experiment was repeated with an 8-m telescope and twenty-first century instrumentation. No spectral features were detected, but the data were of sufficient quality to rule out hydrocarbon emission features associated with comets, the interstellar medium, and a variety of nebulae (Farihi *et al.*, 2008b).

5.4
The Initial Impact of Spitzer

Prior to the launch of Spitzer, there was only one previously published, mid-infrared study of white dwarfs: an ISO search for dust emission from 11 nearby white dwarfs, six of which have metal-enriched photospheres (Chary *et al.*, 1999). The ISO imaging exposures were executed in bandpasses centered near 7 and 15 μm, and included observations of G29-38, G238-44, and vMa 2. Most of the white dwarfs were detected at 7 μm, but only a few were sufficiently bright to be seen at 15 μm. All measured fluxes were consistent with photospheric emission, with the sole exception of G29-38. These data eventually formed part of the basis for the flat dust ring model hypothesized by Jura (2003).

5.4.1
Infrared Capabilities of Spitzer

Spitzer opened up a previously-obscured space to white dwarf researchers, and a few groups were primed to take advantage of its promised, unprecedented sensitivity to substellar companions and circumstellar dust at mid-infrared wavelengths. The entire observatory was cryogenically cooled to 5.5 K, which prevented the facility from being a significant source of thermal background radiation for its instruments. Spitzer was launched in late 2003 with three instruments that had the capability to detect sources at the μJy level (Werner *et al.*, 2004). The Infrared Array Camera (IRAC; Fazio *et al.*, 2004) is a dual-channel, near- and mid-infrared imager with filters centered at 3.6, 4.5, 5.8, and 7.9 μm. The Infrared Spectrograph (IRS; Houck *et al.*, 2004) provided low and moderate-resolution spectroscopy between 5 and 40 μm, plus limited field of view imaging at 16 and 22 μm. The Multi-Band Imaging Photometer for Spitzer (MIPS; Rieke *et al.*, 2004) produced imaging and photometry in three wide bandpasses at 24, 70, and 160 μm. In the middle of May of 2009, Spitzer's cryogen was depleted, and only near-infrared 3.6 and 4.5 μm observations are possible during the remainder of its "warm" mission.

5.4.2 First Results

Results from the first two years of Spitzer white dwarf observations did not emerge chronologically as they were proposed or obtained, and were heavily influenced by a rapidly evolving field. At the time observing proposals from the general community were solicited in late 2003 (and before the deadline arrived in early 2004), only the DAZ study of Zuckerman *et al.* (2003) had been published, and the highly metal-enriched atmosphere of GD 362 had not yet been discovered. The ground-based disk searches and further metal-rich white dwarf discoveries published in 2005 both had an understandable impact on evolving programs and proposals to use Spitzer.

G29-38

It is not surprising that the most highly sought Spitzer white dwarf target was G29-38. Utilizing all three observatory instruments, Reach *et al.* (2005) imaged the prototype dusty white dwarf at 4.5, 7.9, 16, and 24 μm, and obtained a low-resolution spectrum between 5 and 15 μm. These data represented three major advances relative to previous infrared observations of G29-38. First, the 2–5% photometry at 3–8 μm was by far the most accurate data yet obtained for the star. Second, the mid-infrared photometry included novel, longer wavelength coverage and the first observations at $\lambda \geq 10$ μm with total errors better than 20%. Third, and the most scientifically significant, the spectroscopy over a wide range of mid-infrared wavelengths at $S/N > 20$ was unprecedented.

Figure 5.7 displays these Spitzer data for G29-38, revealing both a strong thermal continuum of $T \approx 900$ K, and a remarkable 9–11 μm silicate dust emission feature. A comparison of the shape of its silicate emission to that observed in the interstellar medium, the envelope of the mass-losing giant star Mira, comet Hale-Bopp, and the zodiacal dust cloud was the first concrete evidence that the G29-38 dust had a circumstellar and, hence, planetary origin. Notably, its spectrum does not exhibit signatures of polycyclic aromatic hydrocarbon (PAH) molecules, which often dominate the mid-infrared spectra of the interstellar medium (Allamandola *et al.*, 1989; Draine, 2003). The appearance of the 9–11 μm feature in G29-38 (see Figure 5.8) differs from interstellar silicates and also from silicates forming in the ejecta of Mira (particles that will eventually become part of the interstellar medium). Of the four distinct sources compared to G29-38, its emission most resembles the emission from the rocky particles of the zodiacal dust cloud.

Reach *et al.* (2005) modeled both the warm thermal continuum and the silicate emission with an optically thin, circumstellar cloud (i. e., a spherical shell or flattened disk) of dust with an exponentially decreasing radial profile. The model invoked micron-sized olivine (plus some forsterite) dust to account for the strong

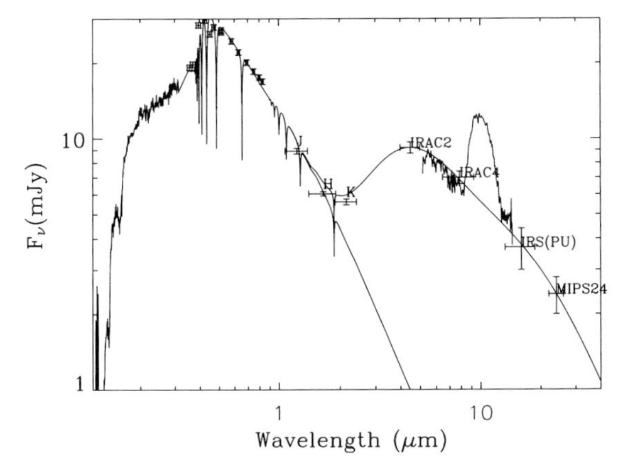

Figure 5.7 Ultraviolet through mid-infrared spectral energy distribution of G29-38, including Spitzer photometric and spectroscopic observations. Photometric data are shown as error bars, while solid lines show an International Ultraviolet Explorer spectrum, a stellar atmosphere model from the ultraviolet through infrared, a model for the thermal continuum, and the measured mid-infrared spectrum. The Spitzer data reveal $T \approx 900$ K dust emission and a strong silicate emission feature at 9–11 μm consistent with micron-sized olivine grains. From Reach *et al.* (2005), reproduced with the permission of the AAS.

emission feature, and similarly small carbon grains to account for the thermal continuum. Silicates are common in the debris of short-period comets and asteroids in the Solar System; hence, the dust in G29-38 is consistent with planetary materials (Lisse *et al.*, 2006). The authors did note that the lack of PAH features in the spectrum was potentially inconsistent with the presence of carbon grains at a ratio of 3 : 1 to the silicate grains, as in their model. It is likely that carbon was utilized in the model primarily for its featureless infrared spectrum, rather than as a likely constituent of planetary debris since externally polluted white dwarfs only rarely show signatures of carbon (Jura, 2006).

The fundamental difference between this model and the disk model of Jura (2003) is the assumed optical depth of the disk material at various wavelengths. In the geometrically flat, vertically optically thick model of Jura (2003), the bulk of material is unseen and effectively shielded from stellar radiation. Such a disk is heated by absorption of ultraviolet radiation – often the major or dominant source of radiant energy in white dwarfs – and the warmest dust grains are located within a few tenths of a solar radius. In order to absorb and re-emit up to 3% of the stellar luminosity, as does G29-38, a flat disk must be seen in a near face-on configuration. For optically thick disk material, it is only possible to infer a minimum mass. In contrast, the optically thin model proposed by Reach *et al.* (2005) yields a total disk mass from the infrared emission; for G29-38, this model predicts on the order 10^{18} g of small dust grains, and potentially more mass contained in larger, inefficiently emitting, particles.

Critically, optically thin dust grains are warmed by the full starlight of the host star and located at a few to several solar radii in the case of G29-38. Poynting–

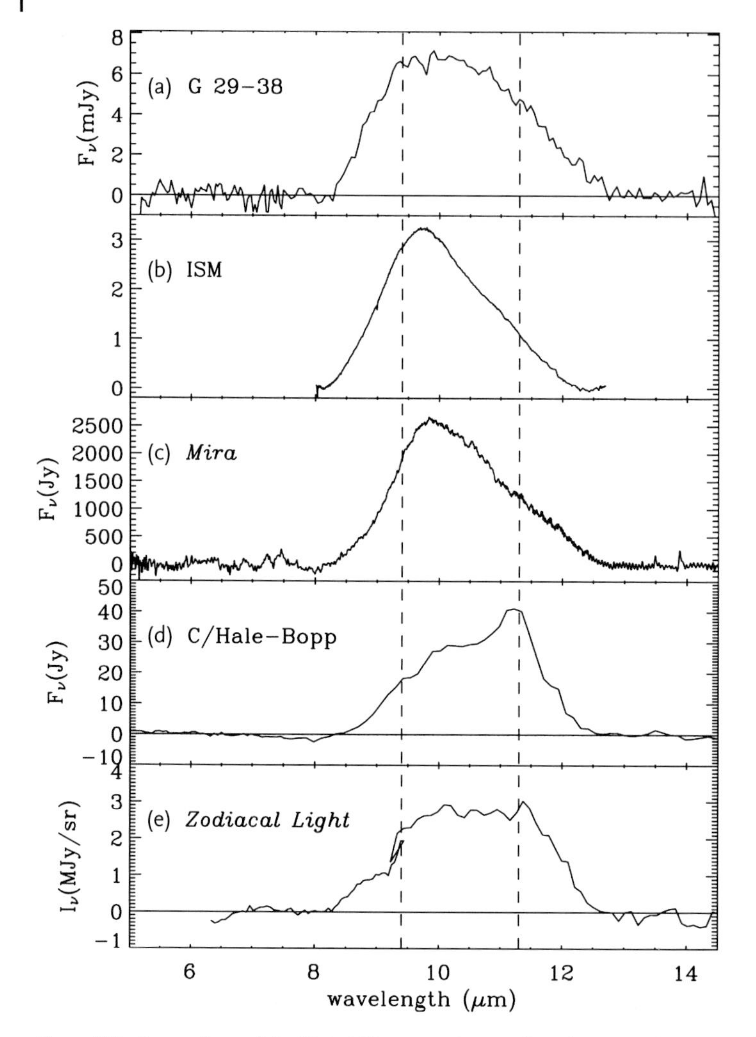

Figure 5.8 Comparison of the high S/N, continuum-subtracted, silicate emission feature observed from G29-38 compared to various astronomical silicates. The G29-38 feature has a red wing that is distinct from interstellar silicates and most similar to the dust in the zodiacal cloud of the Solar System. From Reach *et al.* (2005), reproduced with the permission of the AAS.

Robertson (PR) or radiation drag will cause exposed dust particles to spiral in towards G29-38 on timescales of a few years (Reach *et al.*, 2005). Without a source of replenishment for this closely orbiting dust, it is difficult to reconcile persistent optically thin emission from G29-38 over an 18 year period between its discovery and the first Spitzer observations.

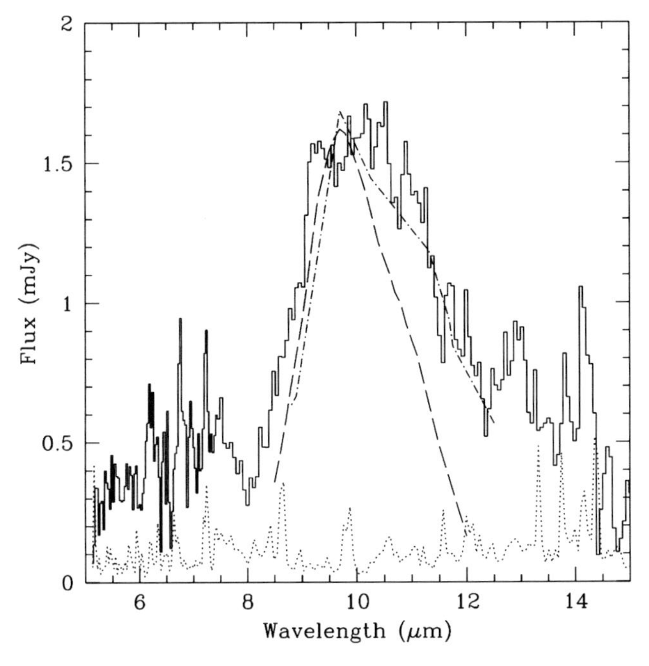

Figure 5.9 Spitzer IRS low-resolution spectrum of GD 362 revealing a very strong silicate feature with a red wing extending to nearly 12 μm. The solid line is the data while the dotted line represents the uncertainties. The dashed and dashed-dotted lines are the emission profiles of interstellar silicates and planetesimal dust of the main sequence star BD +20 307 (Song *et al.*, 2005), respectively. From Jura *et al.* (2007b), reproduced with the permission of the AAS.

GD 362

Naturally, the discovery of infrared excess on top of spectacular metal-enrichment in GD 362 made it an obvious Spitzer target. Jura *et al.* (2007b) conducted observations of GD 362 with IRAC 3–8 μm photometry, IRS 5–15 μm low-resolution spectroscopy, and MIPS 24 μm photometry. These observations detected a striking silicate feature sufficient in strength to influence the 7.9 μm IRAC photometry and by itself re-emit 1% of the stellar luminosity (see Figure 5.9). While the 9–11 μm feature measured for G29-38 is quite strong, the emission detected from GD 362 is extreme; among mature stellar systems, only the very dusty main sequence star BD +20 307 (Song *et al.*, 2005) has a comparably strong silicate emission feature.

Jura *et al.* (2007b) modeled the entire infrared emission with a more sophisticated version of the geometrically thin, vertically optically thick disk model used for G29-38. The model consisted of three radially concentric and distinct regions within a flat disk geometry: two inner, opaque regions and an optically thin outer region. The innermost region was required to be vertically isothermal, and the middle region was modeled to have a temperature gradient between its top and middle layers. The outermost region was then modeled to be the source of the silicate emission, and is generally warmer than the middle region due to the change in optical depth. Importantly, the modeled disk of GD 362 had a *finite* radial extent, and was contained entirely within 1 $R_\odot$ of the white dwarf, where rocky bodies such as large asteroids should be tidally destroyed (Davidsson, 1999). Remarkably, a strictly flat disk model does not have sufficient surface area (i. e., cannot intercept enough starlight) to account for the prodigious, overall infrared emission from GD 362; a warp or slight flaring in the (outer) disk was necessary to reproduce the data.

5.4.3
The First Spitzer Surveys of White Dwarfs

There were four programs approved in the first Spitzer Guest Observer (GO) cycle that aimed to detect infrared excess from white dwarfs. Most of these programs actively sought an excess from substellar or planetary companions, in addition to searching for dust emission similar to that of G29-38. On the one hand, the frequency of brown dwarfs and planets associated with the intermediate-mass, main-sequence progenitors of white dwarfs constrains theories of star and planet formation. The direct or indirect detection of such low-mass objects around A- and early F-type stars ranges from very difficult to impossible compared to their detection around the white dwarf descendants (Farihi *et al.*, 2008b). On the other hand, one possible origin for the photospheric metals in cool white dwarfs is wind capture from an unseen, low-mass stellar or substellar companion (Holberg *et al.*, 1997; Zuckerman and Reid, 1998). Zuckerman *et al.* (2003) reported a 60% fraction of metal-enrichment among DA white dwarfs with close or very close (i. e., spatially unresolved from the ground or known radial velocity variables), low-mass, main-sequence companions. Therefore, Spitzer was primed to detect various astrophysical sources that might account for the metal contamination observed in cool white dwarfs.

An Unbiased Survey

The largest Spitzer survey of white dwarfs to date was carried out in its first GO cycle and observed 124 stars selected for brightness at near-infrared wavelengths from their 2MASS photometry (Mullally *et al.*, 2007). No preference was given to metal-rich white dwarfs, but the sample included G29-38 and 11 other cool stars with photospheric metals. For reasons of efficiency, each of the targets was observed using only one of the wavelength pairs which IRAC obtains simultaneously, 4.5 and 7.9 μm. Of the 12 externally polluted white dwarfs, only G29-38 (the IRAC photometry published by Reach *et al.*, 2005, was part of this survey) and LTT 8452 were found to have infrared excesses consistent with circumstellar dust (von Hippel *et al.*, 2007). Perhaps even more profound is the result that of 112 white dwarfs not considered to be externally polluted, none had an infrared excess attributable to a disk. One must keep in mind that roughly one-quarter of the stars in this survey had effective temperatures above 25 000 K, and that the bulk of white dwarfs in the sample had not been observed with high resolution spectroscopy necessary to detect modest metal abundances. Therefore, some caution is warranted when in-

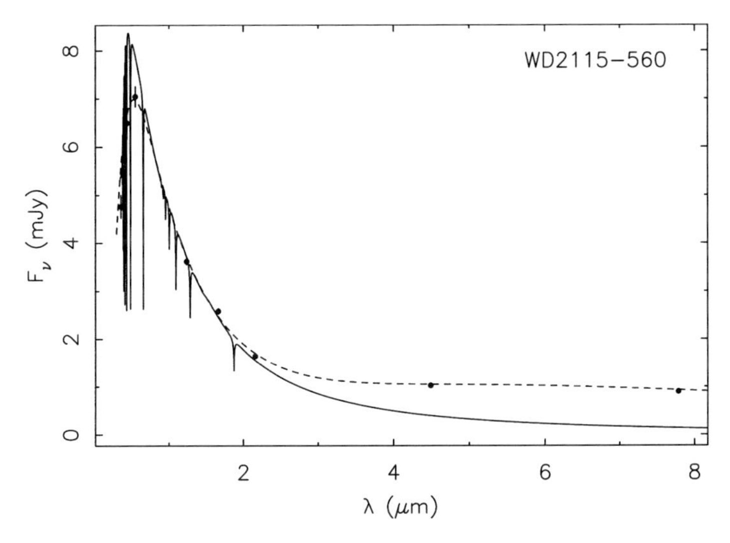

Figure 5.10 Infrared excess from the DAZ white dwarf LTT 8452 as measured by Spitzer IRAC. Short wavelength photometry from the literature and IRAC flux densities are shown as dots with (small) error bars. The solid line is a stellar atmosphere model, while the dashed line represents the addition of an optically thick, flat disk model for the circumstellar dust. From von Hippel *et al.* (2007), reproduced with the permission of the AAS.

terpreting this result, but tentatively speaking, no disks are detected around white dwarfs that are not metal-polluted.

The metal-enriched stars in their sample were rather diverse in effective temperature, calcium abundances, and basic atmospheric composition (i. e., both DAZ and DBZ stars), and the authors did not try to draw any statistical conclusions on the basis of these two disk detections. Nevertheless, they demonstrated that the opaque, flat disk model fit the available photometric data on all known white dwarfs with disks at the time of publication (G29-38, GD 362, GD 56, and LTT 8452; e. g., see Figure 5.10). Importantly, the authors concluded that the analogy with planetary rings suggests viscous spreading lifetimes on par with the Gyr cooling ages of white dwarfs and, therefore, such a disk does not require replenishment as in the optically thin dust model.

Surveys of DAZ White Dwarfs

Together, both Debes *et al.* (2007) and Farihi *et al.* (2008b) targeted 18 cool DAZ stars from Zuckerman *et al.* (2003) with IRAC imaging observations at all four wavelengths, including the prototype DAZ white dwarf G74-7. Given the proliferation of disk discoveries occurring at that time, it was both fairly disappointing and somewhat surprising that neither of the two surveys identified any stars with infrared excess similar to G29-38 and other dusty white dwarfs. Debes *et al.* (2007) suggested that rapid dust depletion due to PR drag within the tidal disruption radius of the white dwarf could be responsible for the absence of disks. The four DAZ stars observed by Debes *et al.* (2007) have temperatures below 9000 K and, hence, metal diffusion timescales longer than 100 yr (Koester and Wilken, 2006); a "recent" absence of dust is conceivable for these stars if the accreted disk material was originally optically thin.

Farihi *et al.* (2008b) surveyed several DAZ stars warmer than 9000 K, and a few warmer than 10 000 K, for which the observed metal abundances essentially require ongoing accretion and would be difficult to reconcile with an absence of orbiting material. They hypothesized that collisions between grains in a developing disk could rapidly destroy dust particles while preserving the reservoir of circumstellar material – primarily in gaseous form – necessary for the inferred, ongoing accretion. A simple calculation showed that optically thin material within the Roche limit of a white dwarf will collide on timescales 10 to 30 times faster than their PR timescales and, thus, a ring of gaseous debris might develop instead of circumstellar dust (Farihi *et al.*, 2008b).

G166-58

An apparent infrared excess was identified for the metal-rich white dwarf G166-58, but only at the two longest IRAC wavelengths of 5.8 and 7.9 μm (see Figure 5.11 and Farihi *et al.*, 2008b). At the time of discovery, the available IRAC mosaicking software was limited to creating images with 1″.2 pixels, and the image of G166-58 overlapped somewhat with a background galaxy 5″ distant. The background source complicated the flux measurements of the star, especially at 7.9 μm, where the galaxy appears brighter than the white dwarf. Both point spread function (PSF) fitting photometry and radial profile analyses supported the conclusion that the

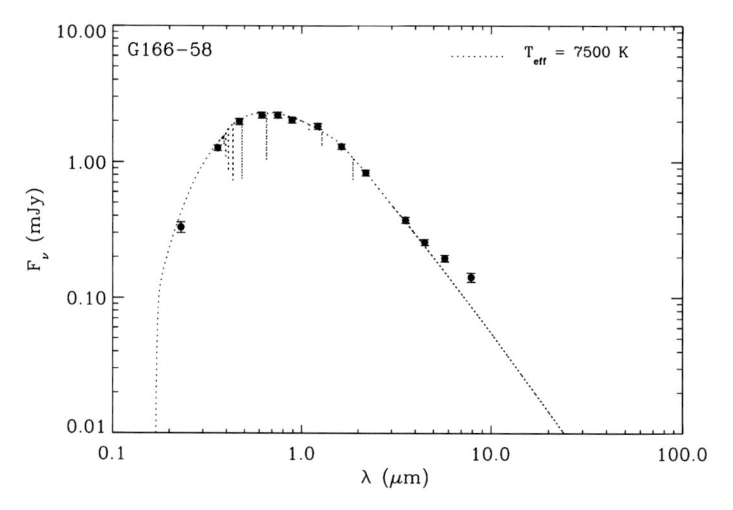

Figure 5.11 Spectral energy distribution of G166-58, the first white dwarf found to have an infrared excess at mid- but not near-infrared wavelengths (Farihi *et al.*, 2008b). Available short wavelength data are shown together with IRAC photometry and a stellar atmosphere model. The warmest dust at G166-58 is only 400–500 K compared to 1000–1200 K in other white dwarf circumstellar disks. From Farihi *et al.* (2010c), reproduced with the permission of the AAS.

measured excess originated in the point-like image of the star. In later software released by the Spitzer Science Center, the ability to construct images with 0″.6 pixels firmly corroborated the point-like nature of the infrared excess (Farihi *et al.*, 2010c).

Due to the diffraction limit of the telescope and the fact that most galaxies have steeply rising (power law) spectral energy distributions that peak at far-infrared wavelengths, Spitzer images of G166-58 at 16 or 24 μm would almost certainly be confused with an even brighter galaxy. Yet, despite the absence of warmer dust, the flux decrease towards 7.9 μm is consistent with a disk contained within the tidal disruption radius. However, it is the inner, dust-poor region that made this star unique upon discovery and still begs for an explanation.

Other Spitzer Disk Searches

Also during the first cycle of Spitzer, two teams independently targeted the most massive white dwarfs to search for infrared excess (Farihi *et al.*, 2008a; Hansen *et al.*, 2006). Two competing hypotheses exist for the origin of white dwarfs with masses larger than roughly $1.0 M_\odot$; remnants of single, high-intermediate mass stars, or mergers of two white dwarfs (Ferrario *et al.*, 2005; Liebert *et al.*, 2005). In support of the first hypothesis, there is at least one, and possibly up to three, massive white dwarfs that descended from single $M \gtrsim 5 M_\odot$ main sequence stars in the 125 Myr old Pleiades open cluster (Dobbie *et al.*, 2006). Such young white dwarfs are excellent targets for planets still warm from formation (Baraffe *et al.*, 2003; Farihi *et al.*, 2008a). Evidence for mergers is rather tenuous, but an example is the comoving visual binary LB 9802. The system consists of two white dwarfs in which the considerably hotter star is also substantially more massive, in stark contrast with expectations from the coeval evolution of two single stars (Barstow *et al.*, 1995). Models of white dwarf mergers suggest massive disks (and possibly second-generation planets) should also form as a repository of shed angular momentum (Hansen *et al.*, 2006).

The Spitzer searches of massive white dwarfs produced no infrared excess candidates, leaving the question of white dwarf mergers still somewhat open. If mergers occur with the simultaneous formation of a massive disk, it is quite possible that the disk would dissipate rapidly as it is expected to be primarily gaseous, at least initially. Although its composition is expected to be unusual, and composed largely of carbon (and oxygen), the disk may evolve similarly to the gas-rich, circumstellar disks observed around young stars and dissipate on similar, Myr timescales. The cooling age of a 50 000 K, $1.2 M_\odot$ white dwarf is around 30 Myr suggesting that any disk could have vanished due to a combination of accretion, early phase radiation pressure, and dust/planetesimal formation. It has been suggested that the class of luminous, post-asymptotic giant R Coronae Borealis stars could be the product of white dwarf mergers (Clayton *et al.*, 2007). These stars are enshrouded in carbon-rich dust (Lambert *et al.*, 2001) and have energetic winds, implying that their circumstellar material will not persist on Myr timescales.

5.5
The Next Wave of Disk Discoveries

A wealth of information about white dwarf disks emerged from the second, third, and fourth Spitzer cycles. Roughly chronologically, the Spitzer observations of GD 362 represented the beginning of something akin to a second generation of white dwarf disk discoveries initially occurring in parallel with the first discoveries. By 2007, all the disk discoveries and searches had been concentrated around DAZ stars, but this was about to change. In 2006, a weak, but definite, helium absorption line was detected in a deep and high resolution optical spectrum of GD 362, demonstrating that the star had an atmosphere dominated by helium (Zuckerman *et al.*, 2007).

5.5.1
The Second Class of Polluted White Dwarfs

The attention paid to DAZ white dwarfs is understandable. Hydrogen-rich atmosphere white dwarfs account for roughly 80% of all white dwarfs at effective temperatures above 12 000 K (Eisenstein *et al.*, 2006). Typical timescales for heavy elements to diffuse below the outer, observable layers of a DA star are a few days for stars between 12 000 and 25 000 K (Koester, 2009). In this temperature range, the convection zone or mixing layer of the star is incredibly thin, on the order of 10^{-15} of its total mass. As a DA star cools below 12 000 K, its convection zone increases in depth rapidly and substantially, growing by five orders of magnitude as it reaches 10 000 K, and another three orders of magnitude by 6500 K. The metal sinking timescales grow commensurately, increasing to 100 yr at 10 000 K and 10^4 yr at 6500 K (Koester, 2009). Hence, the existence of all but the coolest DAZ stars implies the recent or ongoing accretion of heavy elements, and circumstellar disks are an obvious suspect.

In contrast, DBZ white dwarfs above 12 000 K have helium atmospheres with only trace hydrogen abundances, typically at the lower end of the range $10^{-4} - 10^{-6}$ (Voss *et al.*, 2007). It is thought that these stars are the product of very efficient thermal pulses (helium flashes) that expel most of the superficial and primordial hydrogen in the final phases of asymptotic giant mass loss. A white dwarf with a helium-dominated atmosphere is relatively transparent compared to its hydrogen-rich counterparts, facilitating the detection of trace amounts of heavy elements. DB white dwarfs also have significantly larger convection zones than DA stars, roughly four to five orders of magnitude deeper at all but the coolest temperatures. The size of the convection zone determines the timescales for heavy elements to diffuse downward. Hence, metals in DBZ stars can persist for up to 10^6 yr, beginning at temperatures of 12 000 K (Koester, 2009). Therefore, disk searches initially avoided the DBZ class because their photospheric metals could be remnants of long past events.

The potential advantages of searching DBZ white dwarfs for circumstellar dust was highlighted by Jura (2006). He noted that their atmospheric transparency and significantly deep convection zones yielded compelling compositional and mass limits on the polluting material. Based on International Ultraviolet Explorer (IUE) spectroscopic observations of several helium- and metal-rich white dwarfs (Wolff *et al.*, 2002), Jura (2006) identified three stars with measured or upper limit carbon-to-iron ratios that indicated the accretion of refractory-rich and volatile-poor (i. e., rocky) material: GD 40, Ross 640, and HS 2253+8023. Furthermore, the mass of iron alone in the outer, mixing layers of these three stars ranges between 10^{21} and 10^{24} g; these masses are comparable to large Solar System asteroids and Ceres.

The SPY and Hamburg Quasar surveys together uncovered more than one dozen DBZ stars including GD 16, a white dwarf with a distinctive DAZ-type optical spectrum remarkably similar to GD 362 (Friedrich *et al.*, 1999, 2000; Koester *et al.*, 2005a, 2005b). Together with previously known white dwarfs in this class (Dufour *et al.*, 2007; Dupuis *et al.*, 1993b; Wolff *et al.*, 2002), and armed with the knowledge that their atmospheric compositions and total heavy element masses suggested the accretion of rocky material, the DBZ stars made their way into Spitzer and ground-based searches for dust, alongside the DAZ stars.

5.5.2
A Highly Successful Spitzer Search

Recognizing the helium-rich nature and circumstellar disk of GD 362, Jura *et al.* (2007a) disregarded basic atmospheric composition and selected a sample of 11 metal-polluted white dwarfs with potential excess flux in their 2MASS K_s-band photometry. On this basis, GD 56 emerged as the strongest candidate for circumstellar dust, having a large apparent *K*-band excess[2] All of these white dwarfs were observed using both IRAC 3–8 μm and MIPS 24 μm imaging. This was the first use of longer wavelength photometry in a survey of white dwarfs. Four stars with strong infrared excess were identified in the program: GD 40, GD 56, GD 133, and PG 1015+161 (see Figure 5.12).

The detection of circumstellar dust around GD 40 was an excellent confirmation that DBZ white dwarfs held important clues to the nature of their metal-contamination (Jura, 2006) in a different, yet complementary, manner to the DAZ stars. The sizable mixing layers in helium atmospheres provide a strict lower limit to the total mass of any accreted elements. For stars with circumstellar dust, such as GD 40, the minimum total mass of accreted metals places a lower limit on the mass of the asteroid whose debris now orbits the star. Also, the DBZ stars contain only trace or upper limit hydrogen abundances; this is a sensitive diagnostic for a variety of accretion models.

The three DAZ white dwarfs found to have infrared excess were something of a cautionary tale. Perhaps for the first time, it became clear that ground-based observations up to 2.5 μm were sometimes insufficient to confidently identify circumstellar dust. GD 56 displays an unambiguous excess in *K*-band spectroscopy

2) The selection of GD 56, GD 133, and PG 1015+161 as Cycle 2 Spitzer targets was made in early 2005, prior to the publication of their near-infrared spectra (Kilic *et al.*, 2006).

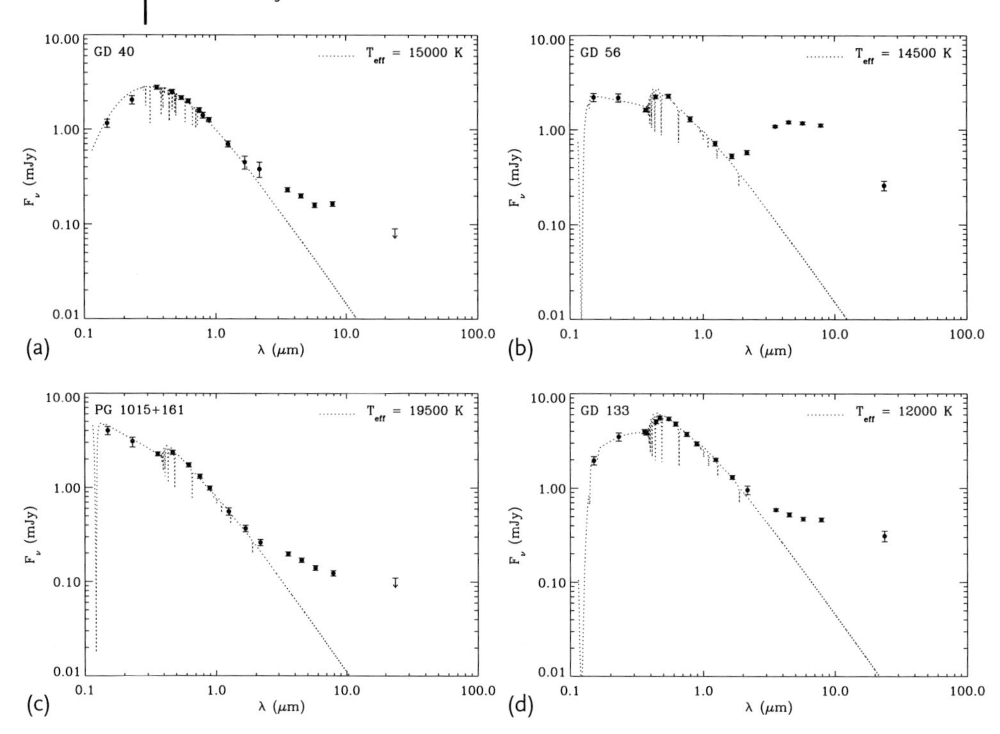

Figure 5.12 Infrared excesses detected for (a) GD 40, (b) GD 56, (c) PG 1015+161 and (d) GD 133, with Spitzer IRAC and MIPS. Short wavelength photometric data from the literature are shown together with appropriate stellar atmosphere models. The downward arrows are 3σ upper limits for nondetections.

and photometry, stronger than both G29-38 and GD 362, and its IRAC photometry reveals particularly strong emission. While the *K*-band spectra of GD 133 and PG 1015+161 were inconclusive (Kilic *et al.*, 2006), their IRAC photometry reveals clear excess emission in each case. It is worth remarking that the reason these two white dwarfs lack notable *K*-band excesses is almost certainly *not* because they lack warm dust, as in the case of G166-58. Rather, the strength of emission from a flat disk depends also on its solid angle with respect to the Sun (Jura, 2003).

In fact, the flat disk model is able to reproduce the infrared data for GD 40, GD 133 and PG 1015+161 very well using inner dust temperatures T = 1000–1200 K, close to the temperature at which grains should rapidly sublimate (Jura *et al.*, 2007a). However, the strong near-infrared emission of GD 56 cannot be duplicated with a flat disk model, even with dust temperatures above 1200 K. The sharp rise requires more emitting surface than is available in the flat disk model, implying some portion of the disk is warped or flared (Jura *et al.*, 2007a).

The importance of the MIPS observations lies in the fact that the longer wavelength fluxes constrain the detected disk material to lie well within the Roche limit of the white dwarf, consistent with a parent body that was tidally destroyed. The publication of these results in 2007 brought the number of white dwarf disks observed at 24 μm to six, with none showing evidence for cool dust. On the contrary, the detections and upper limits at this wavelength implied the coolest grains had temperatures of several hundred K and orbited within $1.2\,R_\odot$ (roughly 100 white dwarf radii). If a circumstellar disk is formed by the gravitational capture of interstellar material at the classical Bondi–Hoyle radius, then one might expect to detect cool dust as it approaches from this initial distance of several AU (Koester and Wilken, 2006). The MIPS results of Jura *et al.* (2007a) demonstrated that emission from dust captured at such distances was strictly ruled out, as the predicted emission would be tens to hundreds of times greater than that observed.

5.5.3
The Detection of Gaseous Debris in a Disk

At roughly the same time, and to the amazement of the white dwarf community, Gänsicke *et al.* (2006) reported the discovery of a single, warm DAZ white dwarf with remarkably strong emission features from both calcium and iron. Like many astronomical discoveries, the identification of metallic emission lines from SDSS J122859.93+104032.9 (hereafter SDSS 1228) was accidental; its spectrum was flagged in a search for weak spectroscopic features due to very low mass (stellar or substellar) companions to apparently single white dwarfs (B. Gänsicke, private communication).

The strongest emission from SDSS 1228 is seen in the calcium triplet centered near 8560 Å (see Figure 5.13) and these lines are rotationally broadened in a manner expected from Keplerian disk rotation. While the detected features are directly

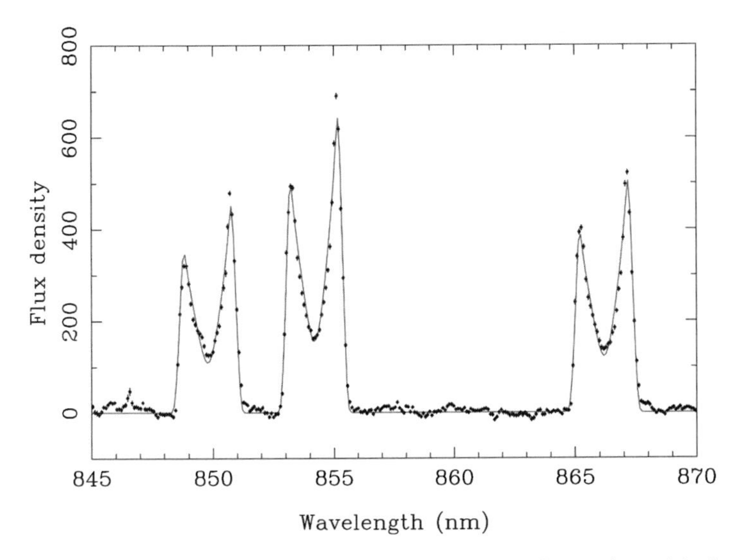

Figure 5.13 Calcium emission in the optical spectrum of SDSS 1228 measured at the William Herschel Telescope with the ISIS spectrograph. The data are shown as points with (small) error bars while the model for the emission is shown as a solid line. From Gänsicke *et al.* (2006), reprinted with permission from AAAS.

analogous to hydrogen and helium emission from accretion disks in cataclysmic variables (Horne and Marsh, 1986), the spectrum of SDSS 1228 has strong hydrogen Balmer lines seen only in absorption, implying the emitting disk is essentially free of light gases. The full optical spectrum of SDSS 1228 is otherwise fairly typical of a warm DAZ star with a high metal abundance, showing strong magnesium absorption, and is similar to the spectrum of G238-44.

Gänsicke *et al.* (2006) showed that the three calcium features were well-modeled by optically thick emission from a highly inclined disk. The peak-to-peak velocity broadening of $\pm 630\,\mathrm{km\,s^{-1}}$, together with the steep outer walls of the feature, limit the gas disk to a maximum radius of $1.2\,R_{\odot}$. This fact shows that the orbiting material is within the Roche limit of the star, consistent with the tidal destruction of a large asteroid. While disk models had been largely successful in reproducing the observed infrared emission from white dwarfs with circumstellar dust, and predicted disk radii generally within $1\,R_{\odot}$, the gaseous metal emission line profiles were the first *empirical* evidence that circumstellar material around metal-enriched white dwarfs orbits within the Roche limit.

At the time of discovery, SDSS 1228 was the first circumstellar disk identified around a metal-lined white dwarf with $T_{\mathrm{eff}} > 15\,000\,\mathrm{K}$. The combined studies of von Hippel *et al.* (2007) and Kilic *et al.* (2006) had targeted 11 DAZ white dwarfs warmer than this, and they speculated that the lack of infrared excesses might be due to dust sublimation within the Roche limit of these higher temperature stars. At that time, this hypothesis was consistent with sublimated debris orbiting the 22 000 K white dwarf SDSS 1228 (Gänsicke *et al.*, 2006), but the pattern would soon be broken. The infrared excess discovered for PG 1015+161 was the first to buck the trend, and more examples would follow, including substantial dust around SDSS 1228 itself. Based on this and additional reasons, it is probable that the gaseous debris around SDSS 1228 is the result of collisions rather than sublimation (Melis *et al.*, 2010).

5.5.4
Dust Deficiency of DAZ Stars – Collisions?

Based on the work of von Hippel *et al.* (2007) and Jura *et al.* (2007a), it became apparent that the DAZ white dwarfs with the highest inferred accretion rates were most likely to harbor circumstellar dust. Previous work had highlighted the correlation of DAZ white dwarf dust disk frequency with higher calcium-to-hydrogen ratios (Kilic and Redfield, 2007; Kilic *et al.*, 2006). However, while metal abundance is correlated with accretion rate, the size of the convection layer also plays a critical role (Koester and Wilken, 2006). It makes physical sense that DAZ stars accreting at the highest rates require the most massive reservoirs, and the more massive the supply of heavy elements, the better the chance of its detection in the infrared. At the same time, several DAZ stars requiring relatively high inferred metal accretion rates failed to show an infrared excess between 3 and 8 μm, the wavelength range accessible with IRAC photometry. An excellent example of this is G238-44, with one of the highest known calcium-to-hydrogen abundances (log[Ca/H] = −6.7) and a diffusion timescale less than a day (Koester and Wilken, 2006).

The idea that collisions within an evolving disk can grind dust grains into gaseous debris was first suggested by Jura *et al.* (2007a) and can, in principle, account for a number of polluted white dwarfs for which no infrared excess is detected by Spitzer out to 8 (and even 24) μm. The basic idea is that Keplerian velocities for particles orbiting a few to several tenths of a solar radius from a white dwarf are roughly between 400 and 800 $km\,s^{-1}$, and that small (i. e., a few percent) deviations from this can lead to collisions at speeds sufficient to "vaporize" dust grains. Warm particles as small as 0.01 μm are inefficient emitters and absorbers of infrared radiation ($2\pi a/\lambda \ll 1$), while smaller particles are essentially the size of gas molecules.

Farihi *et al.* (2008b) further showed that mutual collisions should dominate the initial temporal evolution of optically thin dust particles produced in a tidal disruption event. For dust orbiting cool white dwarfs, collisional timescales for disk particles are at least ten times shorter than their PR timescales, implying gaseous debris produced via collisions is plausible. Furthermore, *avoiding* the self-erosion of dust in this manner likely requires a high disk surface density, so that the particle spacing is comparable to the size of the grains (Farihi *et al.*, 2008b) and collisions are efficiently damped. Such a scenario is consistent with the highest accretion rate stars exhibiting the infrared signature of dust most often, and with the (massive) optically thick disk models (Jura *et al.*, 2007a).

Finally, Jura (2008) extended the idea of disk particle collisions to include multiple, smaller tidal disruption events, under the assumption that the mass distribution of a surviving asteroid belt scales similarly to the Solar System and is, thus, dominated by bodies a few to several km in size. Such a process should more efficiently annihilate solid particles as an infalling, small asteroid impacts a pre-existing, low mass disk at a nonzero inclination. In systems where multiple, smaller asteroids are destroyed on timescales much shorter than the often inferred 10^5 yr disk lifetimes, the resulting debris should be primarily gaseous. While the focus of this chapter is not theoretical, suffice to say that at this point in time, it is thought that some type of collisional scenario (i. e., vaporized debris) is responsible for the lack of observed mid-infrared excess from a number of metal-polluted white dwarfs, especially those for which relatively high metal accretion rates are inferred.

5.5.5 Expanding Searches to the DBZ Stars

At about this time, astronomers turned to the DBZ stars in earnest and began to obtain longer wavelength observations of both classes of polluted white dwarfs. Kilic *et al.* (2008) targeted 20 DBZ white dwarfs with the near-infrared spectrograph that had successfully detected excess emission from several DAZ stars, but yielded no candidates within the helium-rich sample. Surprisingly, GD 40 failed to reveal an infrared excess in these observations, despite a potential K_s-band photometric excess in the 2MASS catalog; this was the reason it was selected as a Spitzer target by Jura *et al.* (2007a). Kilic *et al.* (2008) tentatively identified the relatively long diffusion timescales in these stars as the reason for the lack of dust. They estimated that dust disks had typical lifetimes around an order of magnitude shorter than the 10^6 yr required for metals to begin sinking in the bulk of DBZ stars (Koester, 2009). At the same time, it became clear that ground-based observations had failed to detect 50% of the disks known by the middle of 2008 (G166-58, GD 40 GD 133, PG 1015+161). While LTT 8452 has not yet been observed with *K*-band spectroscopy, its disk signature is not revealed by the $H = 14.00 \pm 0.06$ mag and $K_s = 14.02 \pm 0.06$ mag photometry available from 2MASS.

Farihi *et al.* (2009) undertook a Spitzer Cycle 3 study comprising new observations as well as an analysis of all available archival data on metal-polluted white dwarfs as of the end of 2008. Targets included 20 white dwarfs composed of roughly equal numbers of DAZ and DBZ types, including MIPS 24 μm photometry of several stars previously observed only at shorter wavelengths. New disks were detected around GD 16 and PG 1457−086, while a better characterization of the infrared emission from LTT 8452 was enabled by new photometry at both shorter and longer infrared wavelengths. GD 16 is strikingly similar in optical spectral appearance to GD 362, exhibiting a DAZ-type spectrum while actually having a helium-rich atmosphere, as evidenced by very weak absorption features detected at high resolution with VLT UVES (Koester *et al.*, 2005a). Together with GD 362 and GD 40, the discovery of dust around GD 16 increased the number of known disks around DBZ stars by 50%.

The infrared fluxes of GD 16 and LTT 8452 are fairly typical: strong and well modeled by flat disks with inner and outer radii of roughly one and two dozen stellar radii, respectively (see Figure 5.14 and Farihi *et al.*, 2009). In contrast, the infrared excess of PG 1457−086 is rather mild in comparison to previously detected disks, with a fractional luminosity, $\tau = L_{IR}/L = 0.0007$. For comparison, G29-38 and GD 362 have $\tau = 0.03$, roughly 50 times brighter than the excess of PG 1457−086 (Farihi *et al.*, 2009). The observations and model for this star, along with the scientific implications of its existence, are discussed in Section 5.6.3.

Prior to this point in time, the DBZ stars had been, understandably, treated rather differently than the DAZ stars. Because the metal sinking timescales are relatively rapid in the DAZ white dwarfs, a steady-state balance between accretion and diffusion is a reasonably safe assumption to make (Koester and Wilken, 2006). From this balance and the observed calcium abundance in each star, a total metal accretion rate can be calculated under the assumption that calcium, together with other elements, accretes in solar proportions but without any hydrogen or helium. This is equivalent to assuming a gas-to-dust ratio of 100 : 1 in the interstellar medium but that only dust is accreted (Jura *et al.*, 2007a). The analysis of Farihi *et al.* (2009) attempted to put all metal-contaminated stars on an equal footing by calculating time-averaged accretion rates for the DBZ stars in a similar manner. While physically unmotivated due to the long diffusion timescales in DBZ stars (i. e., a steady state cannot be assumed), a time-averaged accretion rate is still a useful diagnostic.

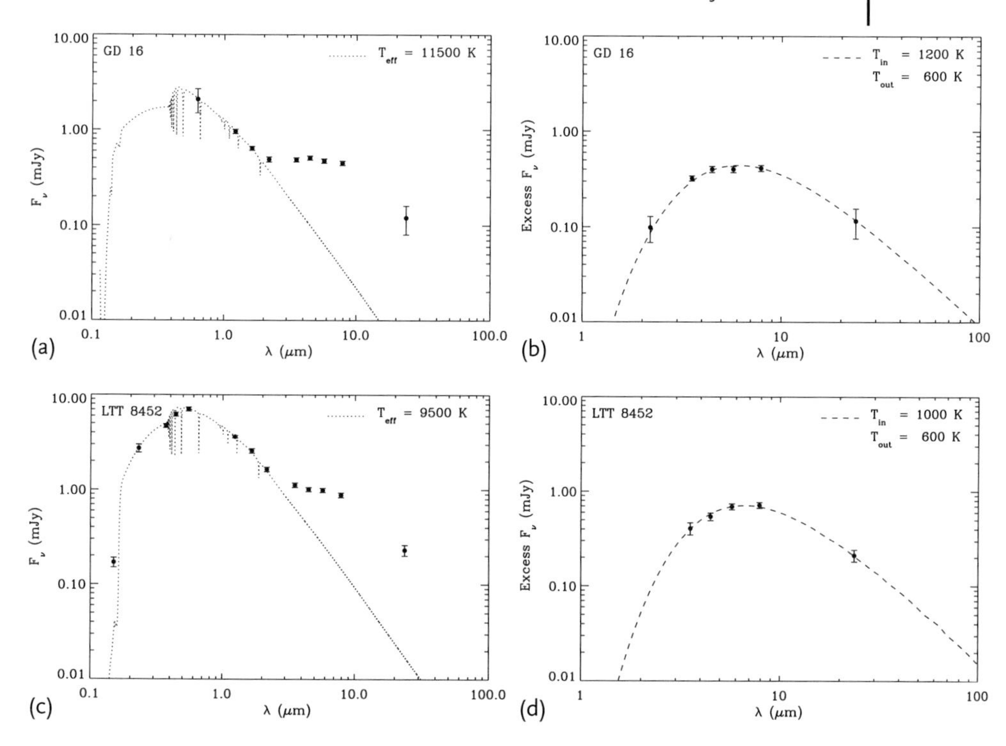

Figure 5.14 Spitzer IRAC and MIPS 3–24 μm photometry for GD 16 and LTT 8452. Panels (a) and (c) show available short wavelength photometry together with the Spitzer data, and reveal strong infrared excesses compared to photospheric models (dotted lines). Panels (b) and (d) show the excesses only with optically thick, flat disk models (dashed lines).

5.5.6 Additional Disks with Gaseous (and Solid) Debris

SDSS 1228 was the first of three metal-enriched white dwarfs found by the end of 2008 to have circumstellar, gaseous debris, via the detection of emission lines from the calcium triplet. The second single DA white dwarf confirmed to manifest metallic emission was SDSS J104341.53+085558.2 (hereafter SDSS 1043; Gänsicke *et al.*, 2007). At 18 300 K, SDSS 1043 is somewhat cooler than SDSS 1228, but its spectrum also displays magnesium absorption indicating atmospheric pollution.

With this second discovery, it was again suggested that the warm temperatures of SDSS 1043 and SDSS 1228 were consistent with solid debris becoming sublimated within their stellar Roche limits. These stars were potentially dust-poor analogs of white dwarfs with infrared excess (Gänsicke *et al.*, 2007). However, the dust disks around both PG 1015+161 (Jura *et al.*, 2007a) and PG 1457−086 (Farihi *et al.*, 2009) orbit stars of 19 300 K and 20 400 K, respectively (Koester and Wilken, 2006). Yet, the optical spectrum of the former shows no evidence for calcium emission (B. Gänsicke, private communication). Gänsicke *et al.* (2007) searched both GD 362 and G238-44 for calcium emission lines, but the former only exhibits absorption lines in the triplet, while the latter reveals only a stellar continuum.

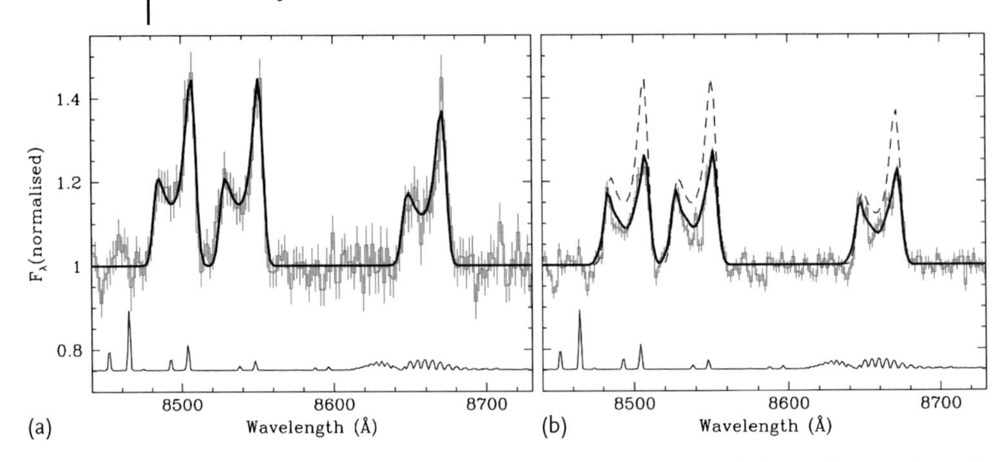

Figure 5.15 Variable, asymmetric line profiles of the calcium emission from Ton 345. The observed changes imply a shift in disk eccentricity from 0.4 (a) (2004 December) to 0.2 (b) (2008 January). Interestingly, no further changes to the line profile were observed in 2008 February and November (Melis *et al.*, 2010). From Gänsicke *et al.* (2008), reproduced with the permission of Wiley-Blackwell.

Importantly, an effort was made to identify additional DA white dwarfs with calcium emission from the vast SDSS DR4 catalogue of Eisenstein *et al.* (2006), and to place some limits on the frequency of these stars. Using an automated routine to select stars with excess flux in the region of the calcium triplet, SDSS 1228 and SDSS 1043 emerged as the only two candidates for emission among over 400 DA white dwarfs with $g < 17.5$ AB mag. Another eight, relatively weak candidates resulted from selecting among 7360 stars with $g < 19.5$ AB mag. The frequency of this phenomenon is less than 0.2% (Gänsicke *et al.*, 2007).

Gänsicke *et al.* (2008) expanded their search for calcium emitting white dwarfs in the SDSS DR6, while removing their restriction to DA stars. This provided over 15 000 likely white dwarfs and nearly 500 candidates for excess calcium flux. However, when visually inspected, the bulk of the spectra were found to suffer from poor night sky line subtraction resulting in an apparent excess in the triplet region. The only stars to pass muster were SDSS 1228, SDSS 1043, and SDSS J084539.17+225728.0; the latter is a star previously catalogued as Ton 345 and classified as an sdO star in the Palomar Green survey (Green *et al.*, 1986). Ton 345 is a DBZ star with calcium triplet emission and photospheric absorption lines of calcium, magnesium, and silicon (Gänsicke *et al.*, 2008). Remarkably, the emission lines of Ton 345 display a marked asymmetry, suggesting that the disk of emitting material is not circular, but has significant nonzero eccentricity in the range 0.2–0.4. Additionally, temporal monitoring of this star has revealed variability in the shape of the asymmetric emission line profiles (see Figure 5.15), strongly suggesting disk evolution on timescales of several years that appears to be episodic rather than ongoing (Gänsicke *et al.*, 2008; Melis *et al.*, 2010).

Overall, the white dwarfs with gaseous debris have provided significant insight into the composition, geometry, and evolution of circumstellar disks. Importantly, all three white dwarfs with gaseous debris also have infrared excess from dust (Brinkworth *et al.*, 2009; Farihi *et al.*, 2010c; Melis *et al.*, 2010) and atmospheres enriched with heavy elements. The solid material is modeled to be spatially coincident with the gas, arguing against sublimation of dust interior to some critical radius. In each case, the calcium triplet line profiles constrain the emitting material to lie within roughly 1 $R_\odot$, while emission from hydrogen or helium is distinctly absent. Therefore, these stars belong to the same class of white dwarfs polluted by debris from tidally disrupted asteroids.

While this is an evolving field, it appears unlikely that the temperature of the white dwarf plays a role in generating the gas, which is a natural result of collisions among solid particles as in the multiple asteroid scenario of Jura (2008). However, stellar effective temperature dictates how efficiently the circumstellar gas can be heated and, thus, detected in the infrared as it emits while cooling (Melis *et al.*, 2010).

5.6 Studies and Statistics

Table 5.1 lists all confirmed and suspected white dwarfs with mid-infrared excess indicating the presence of circumstellar dust as of late 2010. They are ordered chronologically by the publication date for the data revealing each infrared excess. The white dwarf effective temperature, estimated distance from the Earth, apparent K-band magnitude, and the telescope that was used to discover the infrared excess are also listed in the table.

5.6.1 Spectroscopic Confirmation of Rocky Circumstellar Debris

Near the halfway mark of Spitzer Cycle 3, there were ten white dwarfs known to have circumstellar dust, but infrared spectra had only been obtained for G29-38 and GD 362. Jura *et al.* (2009a) used IRS to observe seven of the remaining dusty stars between 5 and 15 μm with the low resolution modules; G166-58 was deemed too problematic for spectroscopy due to its neighboring galaxy (Farihi *et al.*, 2008b). All but one of the IRS targets were detected; owing to a combination of intrinsic faintness and exposure time restrictions in a high background region of the sky, no signal was obtained for PG 1015+161 (Jura *et al.*, 2009a).

Despite the groundbreaking sensitivity of Spitzer, most of the white dwarfs were confidently detected only between 8 and 12 μm, where the silicate emission peak is typically 20 to 60% stronger than the 6–8 μm thermal continuum (see Figure 5.16). Each of these white dwarfs exhibits strong 10 μm emission with a red wing extending to at least 12 μm. The measured features are inconsistent with interstellar silicates (see Figure 5.8), but are instead typical of glassy (amorphous) silicate dust grains, specifically olivines typically found in the inner Solar System, and in evolved solids associated with planet formation (Lisse *et al.*, 2008).

Table 5.1 The 2010 census of white dwarfs with circumstellar dust disks[a].

WD	Name	SpT	T_{eff} (K)	d (pc)	K (mag)	Publication year	Discovery telescope	Reference
2326+049	G29-38	DAZ	11 700	14	12.7	1987	IRTF	1
1729+371	GD 362	DBZ	10 500	57	15.9	2005	IRTF, Gemini	2,3
0408−041	GD 56	DAZ	14 400	72	15.1	2006	IRTF	4
1150−153	EC 11507−1519	DAZ	12 800	76	15.8	2007	IRTF	5
2115−560	LTT 8452	DAZ	9700	22	14.0	2007	Spitzer	6
0300−013	GD 40	DBZ	15 200	74	15.8	2007	Spitzer	7
1015+161	PG	DAZ	19 300	91	16.0	2007	Spitzer	7
1116+026	GD 133	DAZ	12 200	38	14.6	2007	Spitzer	7
1455+298	G166-58	DAZ	7400	29	14.7	2008	Spitzer	8
0146+187	GD 16	DBZ	11 500	48	15.3	2009	Spitzer	9
1457−086	PG	DAZ	20 400	110	16.0	2009	Spitzer	9
1226+109[b]	SDSS 1228	DAZ	22 200	142	16.4	2009	Spitzer	10
0106−328	HE 0106−3253	DAZ	15 700	69	15.9	2010	Spitzer	11
0307+077	HS 0307+0746	DAZ	10 200	77	16.3	2010	Spitzer	11
0842+231[b]	Ton 345	DBZ	18 600	120	15.9	2010	AKARI	11
1225−079	PG	DBZ	10 500	34	14.8	2010	Spitzer	11
2221−165	HE 2221−1630	DAZ	10 100	70	15.6	2010	Spitzer	11
1041+091[b]	SDSS 1043	DAZ	18 300	224	17.7	2010	CFHT, Gemini	12
0735+187[c]	SDSS 0738	DBZ	14 000	136	17.3	2010	Gemini	13
1929+011[c]	GALEX 1931	DAZ	20 900	55	14.7	2010	WISE	14

a White dwarfs without robust infrared excess detections are noted with a colon. The central star of the Helix Nebula (Su *et al.*, 2007) is excluded from this list.

b Confirmation of the presence of a dust disk for SDSS 1228, Ton 345, and SDSS 1043 followed the earlier discovery of a metal-rich *gaseous* disk around each of these white dwarfs (Gänsicke *et al.*, 2006, 2007, 2008).

c Published near-infrared data indicate a likely or potential excess, but confirmation with mid-infrared data is pending. These two white dwarfs are neither discussed nor shown in figures elsewhere in this chapter.

1 Zuckerman and Becklin, (1987), 2 Becklin *et al.*, (2005), 3 Kilic *et al.*, (2005), 4 Kilic *et al.*, (2006), 5 Kilic and Redfield, (2007), 6 von Hippel *et al.*, (2007), 7 Jura *et al.*, (2007a), 8 Farihi *et al.*, (2008b), 9 Farihi *et al.*, (2009), 10 Brinkworth *et al.*, (2009), 11 Farihi *et al.*, (2010c), 12 Melis *et al.*, (2010), 13 Dufour *et al.*, (2010), 14 Debes *et al.*, (2011).

None of the eight circumstellar dust disks showed evidence for emission from polycyclic aromatic hydrocarbons. These (carbon-rich) molecular compounds are found in the infrared spectra of the interstellar medium (Allamandola *et al.*, 1989; Draine, 2003), some circumstellar environments (Jura *et al.*, 2006; Malfait *et al.*, 1998), and in comets (Bockelée-Morvan *et al.*, 1995) with strong features near 6, 8, and 11 μm. Hence, the dust around white dwarfs must be intrinsically carbon-

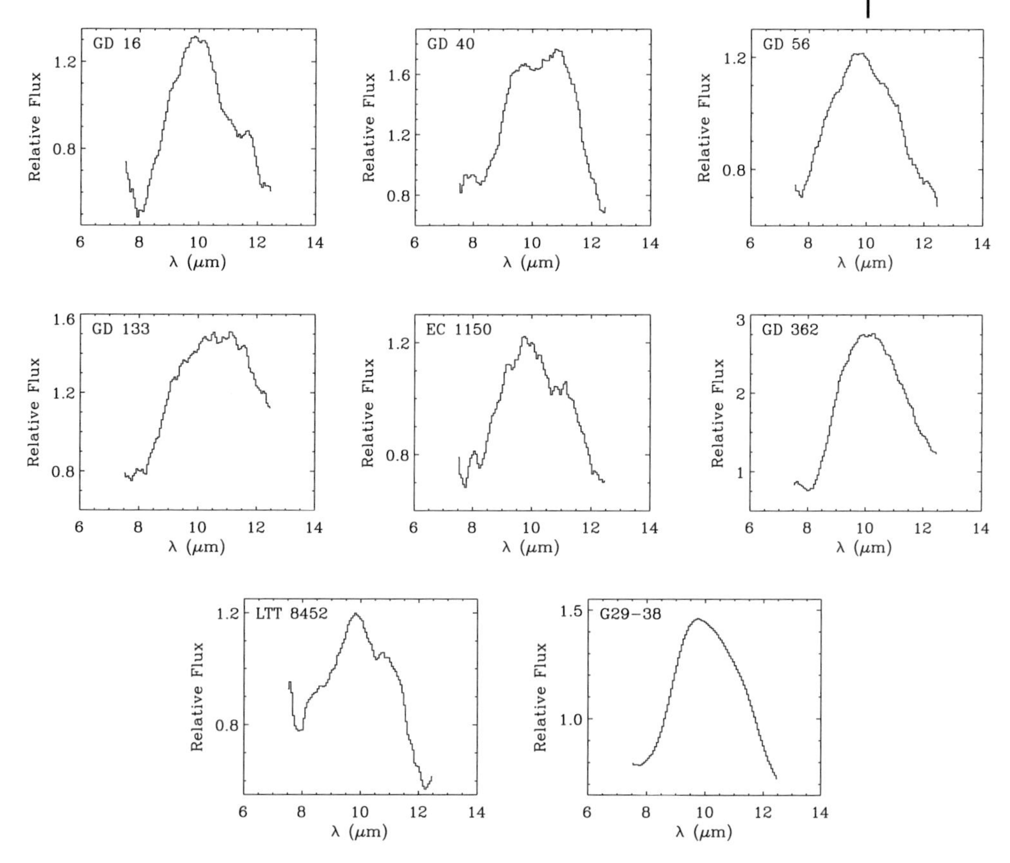

Figure 5.16 Silicate emission features detected in Spitzer IRS low-resolution observations of metal-rich white dwarfs with infrared excess. Each spectrum has been normalized to the average of the 5.7 and 7.9 μm IRAC fluxes, and smoothed by 15 pixels (0.8 μm). The binned data points are highly correlated and structures within the broad silicate feature are probably not real. The detections are modest in most cases, and the data below 8 μm are not shown as this region is typically noise-dominated. Nonetheless, the binned data clearly show that the features are broad, with red wings extending to 12 μm, and are typical of minerals associated with planet formation (Jura *et al.*, 2009a).

deficient (Jura *et al.*, 2009a), and similar to the rocky material of the inner Solar System (Lodders, 2003). The findings from infrared spectroscopy are consistent with, and almost certainly mirrored by, the carbon-poor atmospheres established for several metal-enriched white dwarfs such as GD 40, GD 61, and HS 2253+8023 (Desharnais *et al.*, 2008; Jura, 2008; Zuckerman *et al.*, 2007).

The only white dwarf with circumstellar dust bright enough to be studied spectroscopically with Spitzer at wavelengths longer than 15 μm is G29-38. Reach *et al.* (2009) obtained a second low-resolution IRS spectrum of this prototype dust-polluted white dwarf, this time between 5 and 35 μm. The repeat observations over the short wavelength range revealed no change in the shape of the continuum or silicate emission feature, but the longer wavelength data revealed an additional,

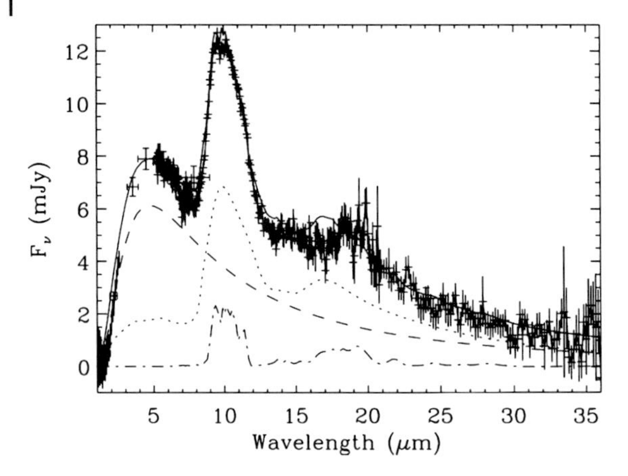

Figure 5.17 Spitzer IRS 5–35 μm spectrum of the circumstellar dust around G29-38. Data are represented by crosses while the best fitting, flat disk plus optically thin layer model is shown as a solid line. The dashed line is the contribution of the optically thick, flat disk, while the emission from silicates is shown as dotted (olivine) and dash-dotted (pyroxene) lines. Multiple model fits to the data produce equally good agreement but the model shown here is the most physically plausible. From Reach *et al.* (2009), reproduced by permission of the AAS.

weaker silicate feature (or combination of features) between 18 and 20 μm (see Figure 5.17). The entire infrared spectrum could be reproduced using any of three physically-distinct models: 1) an optically thin shell, 2) a moderately optically thick and physically thick disk, and 3) an optically thick, physically thin disk with an optically thin layer or outer region.

While the first two models employed by Reach *et al.* (2009) are attractive as they permit the co-identification of various minerals and water ice with the observed infrared emission, they are essentially optically thin models that invoke no more than 10^{19} g of disk mass. In contrast, the third model is physically equivalent to the model of Jura *et al.* (2007b) for GD 362, implying the 10^{19} g of required optically thin material represents only a tiny fraction of the total disk mass. As discussed earlier, the PR timescales for warm, optically thin dust around white dwarfs are very short; 15 yr for 1 μm silicate particles 1 $R_\odot$ distant from G29-38. Thus, for either of the optically thin models proposed by Reach *et al.* (2009), a mechanism is required to replenish 10^{19} g of material around G29-38 roughly every 15 yr, yet its infrared excess has been present for more than two decades (Zuckerman and Becklin, 1987). A disk whose total mass is orders of magnitude greater can accomplish this readily, and the asteroid-sized masses of heavy elements in DBZ stars such as GD 40 argue strongly for commensurate disk masses. Reach *et al.* (2009) found the flat disk model with an optically thin layer or outer region can reproduce the entire infrared emission of G29-38, requiring only silicates.

The importance of these observations cannot be overstated. More telling than the atmospheric composition of the metal-contaminated white dwarfs themselves, this is the strongest evidence that circumstellar dust around white dwarfs is derived from rocky planetary bodies (Jura *et al.*, 2009a).

5.6.2
First Statistics and the Emerging Picture

By the end of 2008, a sufficient number of metal-polluted white dwarfs had been observed with Spitzer to merit statistical analysis: 52 stars with IRAC and 31 with MIPS. Because no white dwarf has been detected at 24 μm without a simultaneous and stronger detection at IRAC wavelengths, the IRAC observational statistics better constrain the frequency of dust disks. From these 52 IRAC observations comprising 11 dust disk detections, it was established that dust disk frequency is correlated with 1) time-averaged accretion rate and 2) cooling age (see Figure 5.18 and Farihi *et al.*, 2009). Based on the cumulative Spitzer IRAC observations of over 200 white dwarfs, infrared excess from circumstellar dust is only detected from those stars with atmospheric metal contamination (Farihi *et al.*, 2008a; Hansen *et al.*, 2006; Mullally *et al.*, 2007).

Applying the time-averaged metal accretion rate analysis to the IRAC dataset, Farihi *et al.* (2009) found over 50% of all cool white dwarfs with metal accretion rates $dM/dt \gtrsim 3 \times 10^8\ \mathrm{g\,s^{-1}}$ have dust disks. Furthermore, when these cool, metal-polluted stars are statistically accounted for as members of larger samples of white dwarfs from which they are drawn, it is found that between 1 and 3% of all white dwarfs with cooling ages less than about 0.5 Gyr have both photospheric metals and circumstellar dust (Farihi *et al.*, 2009). These results signify an underlying population of asteroids that have survived the post-main sequence evolution of their host star, and imply that a commensurate fraction of main sequence A- and F-type stars harbor asteroid belts, and probably build terrestrial planets. Evidence is strong that white dwarfs can be used to study disrupted minor planets.

The MIPS dataset implies that dust is not observed outside the Roche limit of the metal-rich white dwarfs. All white dwarf disks detected at 24 μm have coexisting, strong 3–8 μm IRAC excess fluxes, implying the dust is not drifting inward from the interstellar medium (Farihi *et al.*, 2009). Dust grains captured near the Bondi–Hoyle radius of a typical metal-enriched white dwarf should have temperatures below 100 K, warming as they approach the star under PR drag. Both the MIPS detections of white dwarfs with dust, and the nondetections for the remaining metal-rich stars, argue against an influx of interstellar material (Farihi *et al.*, 2009; Jura *et al.*, 2007a). White dwarfs with infrared excess always display decreasing flux towards 24 μm, indicating a compact arrangement of dust, consistent with disks created via the tidal disruption of rocky planetesimals.

Successful disk models are vertically optically thick at wavelengths up to 20 μm, and geometrically thin (Jura, 2003). There are two reasons such models are likely to be accurate. First, particles in an optically thin disk would not survive PR forces for more than a few days to years (Farihi *et al.*, 2008b). Second, the orbital periods of particles within $1\,R_\odot$ of a white dwarf can be under 1 h, implying that the disk will rapidly relax into a flat configuration. Using this model, the warmest dust has been successfully modeled to lie within the radius at which blackbody grains in an optically thin cloud should sublimate rapidly. Generally, the circumstellar disks have inner edges which approach the sublimation region for silicate dust in an

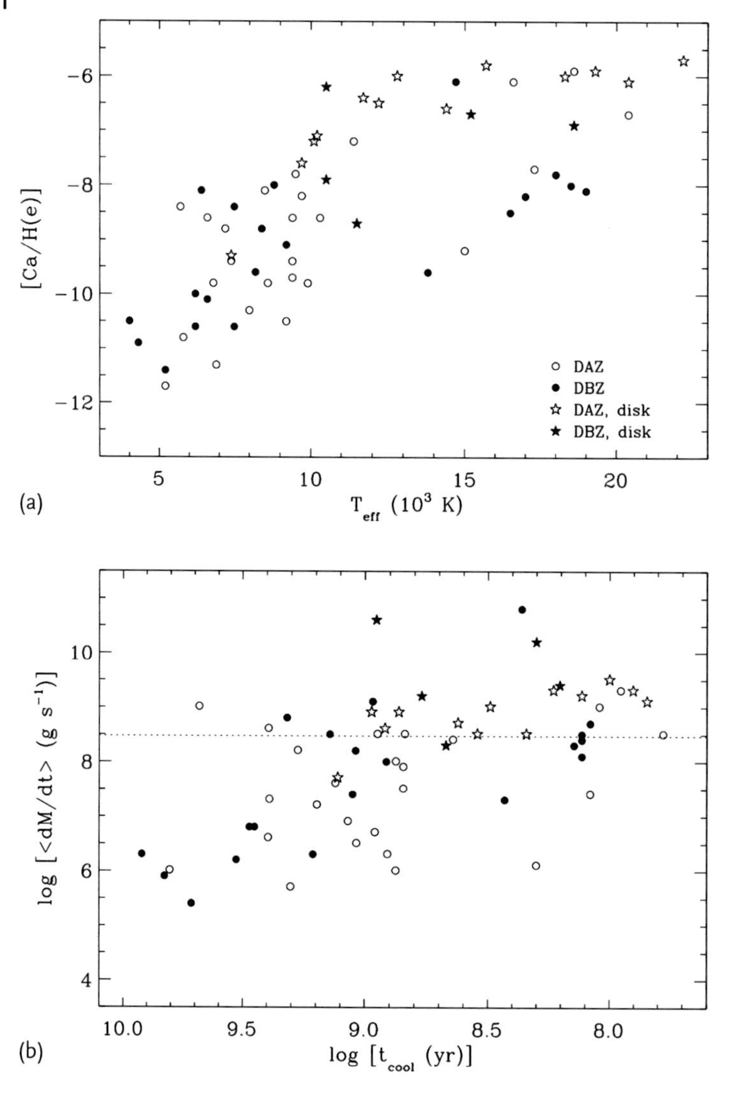

Figure 5.18 Dust disk frequency among all 61 metal-rich white dwarfs observed by Spitzer IRAC up to early 2010. Panel (a) uses a purely observational approach, plotting both disk detections and nondetections versus calcium abundance and effective temperature. Panel (b) employs a more astrophysical approach, by showing the calculated time-averaged metal accretion rate and cooling age for each star. The dotted line in (b) corresponds to 3×10^8 g s^{-1}. G166-58 is the only star with a disk that is located significantly below this accretion rate benchmark, and with a cooling age beyond 1 Gyr. From Farihi *et al.* (2010c), reproduced with the permission of the AAS.

optically thick disk; precisely the behavior expected for a dust disk that is feeding heavy elements to the photosphere of its white dwarf host.

The majority of both DAZ and DBZ white dwarfs do not have infrared excesses when viewed with Spitzer IRAC and MIPS (Farihi *et al.*, 2009). Circumstellar gas disks are a distinct possibility around dust-poor, yet polluted, white dwarfs with metal accretion rates $dM/dt \gtrsim 3 \times 10^8\,\mathrm{g\,s^{-1}}$, while fully accreted disks are a possibility for the DBZ stars with metal diffusion timescales near 10^6 yr. It is possible that a critical mass and density must be reached to prevent the dust disk from rapid, collisional self-annihilation, and when this milestone is not reached, a gas disk results.

For $T_{\mathrm{eff}} \lesssim 20\,000$ K, white dwarfs with younger cooling ages are more likely to be orbited by a dusty disk (Farihi *et al.*, 2009). This observational fact is consistent with a scenario in which a remnant planetary system gradually resettles following the post-main sequence evolution of its parent star (Debes and Sigurdsson, 2002). Because an asteroid must be perturbed into a highly eccentric orbit in order to pass within the stellar Roche limit, planets of conventional size are also expected to persist around white dwarfs. This scenario is strengthened by recent results indicating higher metal accretion for warmer white dwarfs (Zuckerman *et al.*, 2010), and predicts that rocky planetary bodies are destroyed within the Roche limit of hotter white dwarfs more often than for cool white dwarfs. High resolution ultraviolet spectra of hot DA white dwarfs with photospheric metals have revealed velocity-shifted, circumstellar (and distinctly *not* interstellar) heavy element absorption features in several white dwarfs (Bannister *et al.*, 2003). At $\mathrm{T_{eff}} \geq 30\,000$ K, white dwarfs will sublimate even the most refractory materials within their Roche limits, implying that hotter white dwarfs might host disks composed only of gaseous heavy elements (von Hippel *et al.*, 2007). The bulk of white dwarfs with dust have cooling ages less than 0.5 Gyr, with only one older than 1 Gyr, suggesting the possibility that surviving minor planet belts tend to become stable or depleted on these timescales.

5.6.3
Dust-Deficiency in DAZ Stars – Narrow Rings?

During the final cryogenic Spitzer cycle, additional examples of two previously known white dwarfs with unusual infrared excess were found, potentially representing subclasses of narrow circumstellar dust rings, instead of disks, around white dwarfs.

Disks with Subtle Infrared Emission

The infrared excess of PG 1457−086 is 3–6σ above the predicted photospheric flux at the three shortest IRAC wavelengths, but might not have been recognized without supporting near-infrared photometry that suggested a slight *K*-band excess (Farihi, 2009; Farihi *et al.*, 2009). Similarly, the very mild infrared excess discovered for HE 0106−3253 is at the 4–6σ level in the IRAC bandpasses (Farihi *et al.*, 2010c). The spectral energy distributions of these two white dwarfs are shown

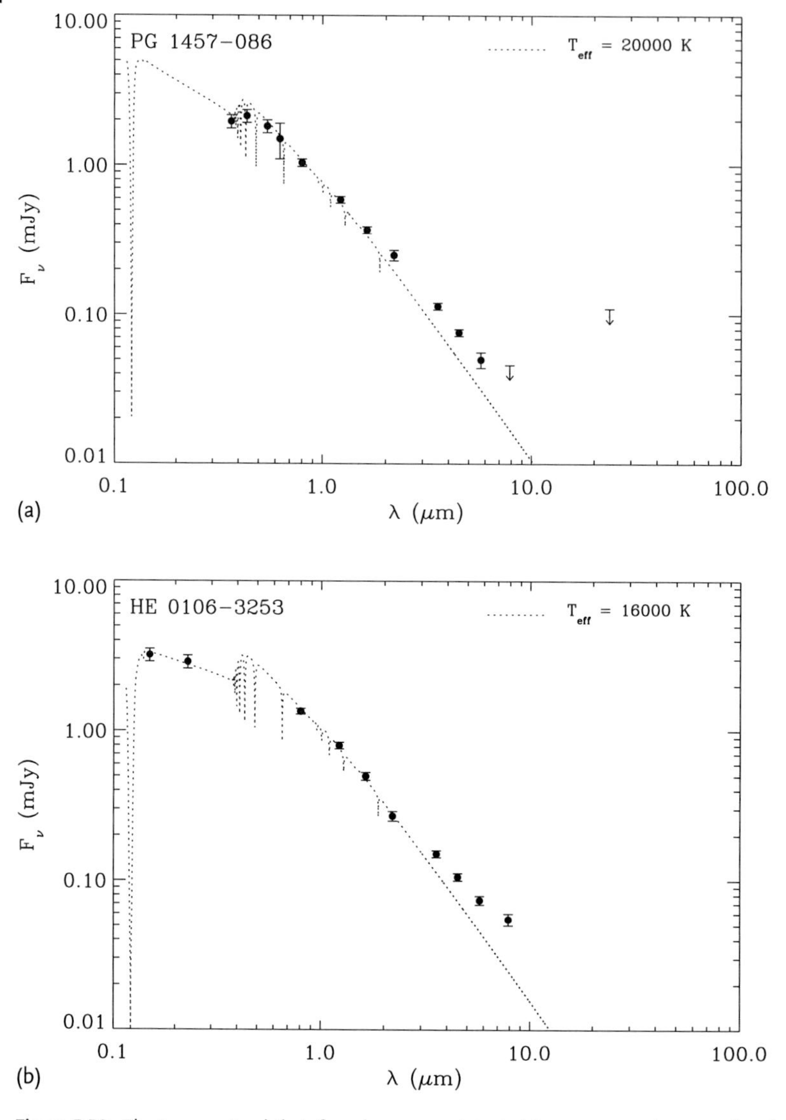

Figure 5.19 The two most subtle infrared excesses detected from narrow, circumstellar dust rings (Farihi *et al.*, 2009, 2010c).

in Figure 5.19. As a benchmark comparison, the IRAC excess of G29-38 is significant at a level of 15–20σ.

In cases like this, establishing just how much of the detected flux is excess and how much is photosphere can be a major problem, as many white dwarfs are not well-constrained by optical and near-infrared photometry (McCook and Sion, 1999). Furthermore, it is probably the case that including IRAC 7.9 μm photometry in model fits without the benefit of a MIPS 24 μm constraint will bias the outer disk radius towards larger values. This is because the 7.9 μm filter bandpass is sufficiently wide to include flux from silicate emission, as clearly occurs for GD 362 and, to a lesser degree, for G29-38 (see Figure 5.20). Farihi *et al.* (2010c) fitted the 2–6 μm spectral energy distributions of all white dwarfs with infrared excess in a

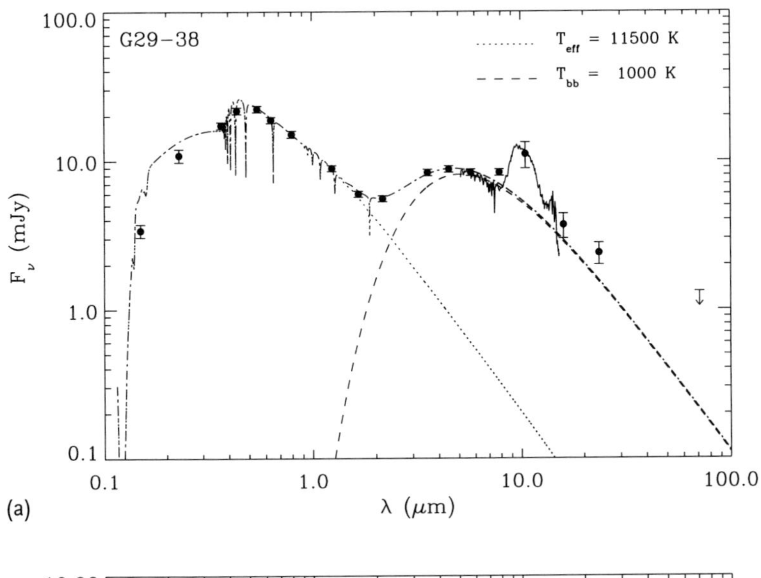

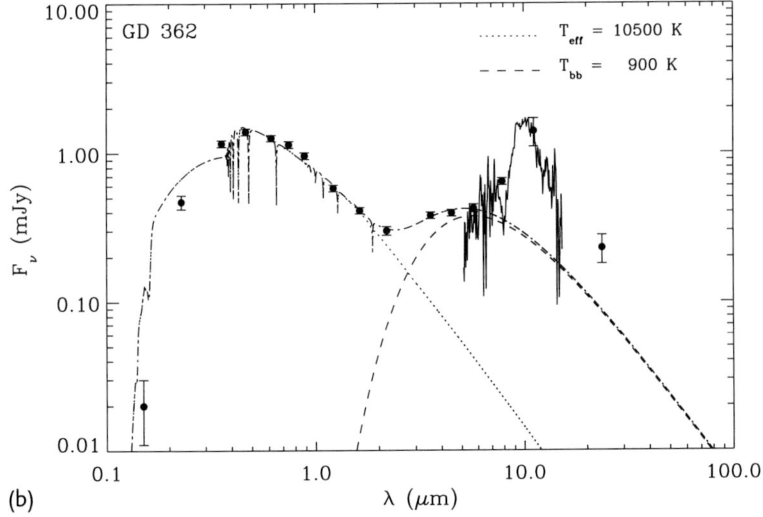

Figure 5.20 A comparison of the full spectral energy distributions of (a) G29–38 and (b) GD 362. Available short wavelength data are shown together with Spitzer photometry and spectra (described in the text) as well as 10 μm (*N*- or *N*′-band) ground-based photometry. Also shown is an upper limit to the flux of G29–38 at 70 μm from Spitzer MIPS imaging. Stellar atmosphere models are plotted as dotted lines, while simple blackbodies fitted to the near-infrared continua are shown as dashed lines.

uniform manner to establish the best fractional infrared luminosity of their disks and compared these values to the cooling ages of the white dwarfs (see Tables 5.1 and 5.2 and Figure 5.21). There are insufficient data to confidently identify a trend, but it appears that the narrowest rings are found in younger systems, while rings comparable to those of Saturn occur in more evolved systems. If the trend is real, then possible explanations include viscous spreading over long timescales (Farihi *et al.*, 2008b; von Hippel *et al.*, 2007) or a decrease over time in the frequency of additional asteroid impacts during disk evolution (Jura, 2008).

There are at least three white dwarfs whose infrared excesses are sufficiently subtle that flat disk model fits to their IRAC data predict dust rings of radial extent $\Delta r < 0.1 R_\odot$: HE 0106−3253, PG 1457−086, and SDSS 1043, which all have $\tau < 10^{-3}$ (Farihi *et al.*, 2009, 2010c). Additional stars that might also have rings this narrow include HE 0307+0746 and PG 1015+161. Since the inclination is unknown, the radial extent of the rings could be even smaller than predicted for nonzero inclination models. For example, if HE 0106−3253 or PG 1457−086 are near face-on ($i = 0°$), then their dust rings would have $\Delta r \approx 0.01 R_\odot$ (or roughly an Earth radius, which is 10 times smaller than the rings of Saturn).

Narrow rings found around a few white dwarfs suggest that asteroid accretion might be relevant to additional, and potentially many, metal-contaminated stars without an obvious infrared excess (Farihi *et al.*, 2010c). A dust ring of radial extent $0.01 R_\odot$ would be difficult to confirm via infrared photometry above an inclination of $i = 50°$ as it would produce an excess under 2σ for typical IRAC data. At the same time, even such a narrow ring has the potential to harbor over 10^{22} g of dust in

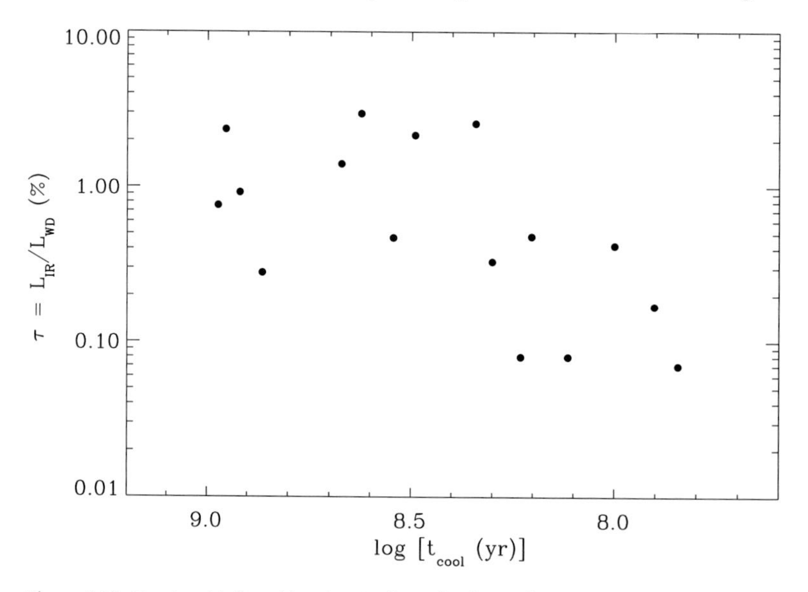

Figure 5.21 Fractional infrared luminosity from the thermal continuum (see Table 5.2) of white dwarf circumstellar dust disks versus cooling age. PG 1225−079 and G166–58 are not plotted due to their unusual excesses.

Table 5.2 Thermal Continuum Excess of White Dwarfs with Dust. Measured infrared excess from thermal continuum emission between 2 and 6 μm, as most stars lack spectroscopic data on any potential silicate emission in the 8–12 μm region.

WD	Name	Stellar model T_{WD} (K)	2–6 μm blackbody[a] T_{IR} (K)	$\tau = L_{IR}/L_{WD}$
0106−328	HE 0106−3253	16 000	1400	0.0008
0146+187	GD 16	11 500	1000	0.0141
0300−013	GD 40	15 000	1200	0.0033
0408−041	GD 56	14 500	1000	0.0257
0307+077	HS 0307+0746	10 500	1200	0.0028
0842+231	Ton 345	18 500	1300	0.0048
1015+161	PG	19 500	1200	0.0017
1041+091	SDSS 1043	18 500	1500	0.0008
1116+026	GD 133	12 000	1000	0.0047
1150−153	EC 11507−1519	12 500	900	0.0216
1226+110	SDSS 1228	22 000	1000	0.0042
1225−079	PG	10 500	300	0.0005
1455+298	G166-58	7500	500	0.0015
1457−086	PG	20 000	1800	0.0007
1729+371	GD 362	10 500	900	0.0235
2115−560	LTT 8452	9500	900	0.0092
2221−165	HE 2221−1630	10 100	1000	0.0076
2326+049	G29-38	11 500	1000	0.0297

a This single temperature is a zeroth order approximation of the true disk spectral energy distribution.

an optically thick, flat disk configuration, and supply metals at 10^9 g s^{-1} for nearly 10^6 yr. Circumstellar disks or rings that produce even more subtle infrared excesses probably await detection, and might apply to the bulk of all metal-polluted white dwarfs.

Disks with Enlarged Inner Holes

G166-58 displays an infrared excess that becomes obvious only at 5.7 μm, while its shorter wavelength IRAC data are consistent with the stellar continuum (see Figure 5.11 and Farihi *et al.*, 2008b, 2010c). As such, it is the only white dwarf with circumstellar dust in which the inner disk edge does not coincide with the region where silicate grains rapidly sublimate at temperatures near 1200 K (Farihi *et al.*, 2009).

However, during Spitzer Cycle 5, PG 1225−079 was found to have a measured excess only at 7.9 μm (see Figure 5.22 and Farihi *et al.*, 2010c). Without repeat observations or data at other wavelengths, it is difficult to assign any certainty to the measured excess of PG 1225−079. If real (and confirmed in the future), then this

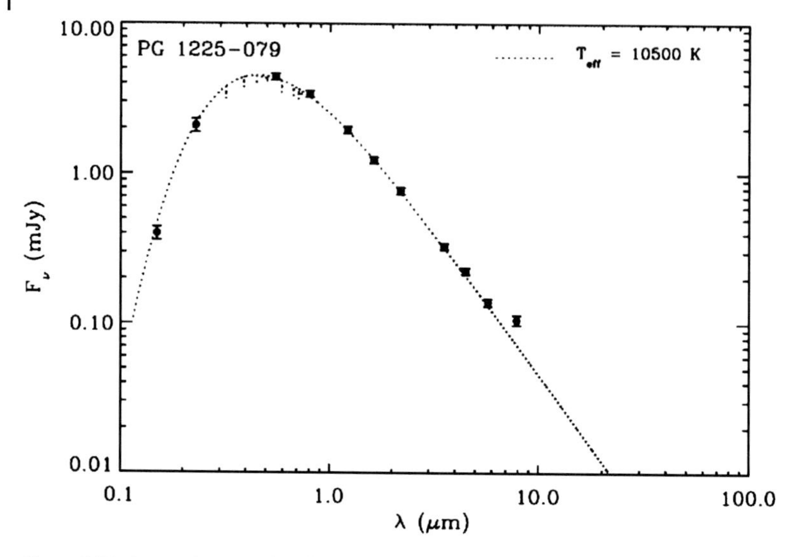

Figure 5.22 Spectral energy distribution of PG 1225−079 with Spitzer IRAC flux measurements. From Farihi *et al.* (2010c), reproduced with the permission of the AAS.

cool, metal-rich white dwarf joins G166-58 in having a dust disk with a relatively large inner region that is dust-poor.

The multiple asteroid model can account for disks with distinct regions dominated by either dust or gas (Jura, 2008). Hence, it is plausible that the inner regions of the dust disks around G166-58 and PG 1225−079 were bombarded by small asteroids that vaporized solids there, but had insufficient mass or orbital energy to destroy all the dust. Additionally, the same model can produce narrow, or otherwise attenuated, dust disks via impacts that annihilate rocky particles in some, but not all, originally dusty regions (Farihi *et al.*, 2010c).

An alternative is that the inner circumstellar regions of these two stars are largely free of matter, having been near- to fully-accreted from the inside out. This possibility suggests that the extant photospheric metals are remnants from prior infall of disk material. G166-58 is by far the coolest white dwarf with an infrared excess, and its 2000 yr timescale for metals to diffuse below the photosphere is relatively long for a DAZ white dwarf (Koester, 2009). PG 1225−079 is a DBZ white dwarf and, therefore, does not require accretion to be ongoing. An accurate and detailed determination of the lighter and heavier elements polluting these stars might be able to constrain their accretion history.

5.6.4
The Composition and Masses of Asteroids around GD 362 and GD 40

To date, the polluted stellar and circumstellar environments of two white dwarfs have been studied in great detail, facilitated by their relatively transparent, helium-rich atmospheres.

GD 362

Using a several hour exposure from HIRES on the Keck telescope, Zuckerman *et al.* (2007) detected 15 elements heavier than helium in the optical spectrum of GD 362, positively shattering the record for number of metals detected in a cool white dwarf photosphere at any wavelength (Friedrich *et al.*, 1999; Sion *et al.*, 1990b; Wolff *et al.*, 2002). This remarkable star is spectacularly polluted and manifests detectable abundances of strontium and scandium, highly refractory elements that comprise about one part in 10^9 of the Sun (Lodders, 2003).

The array of ingredients polluting GD 362 is rich in refractory and transitional elements, while relatively poor in volatiles (Zuckerman *et al.*, 2007). There are detectable abundances of six elements with condensation temperatures above 1400 K present in the star: scandium, aluminum, titanium, calcium, strontium, and vanadium. The transitional elements magnesium, silicon, and iron are the most abundant elements in GD 362 by a substantial margin; together with oxygen, these three elements also comprise 94% of the bulk Earth (Allègre *et al.*, 1995). In fact, overall, the pattern of heavy elements in the white dwarf is most consistent with a combination of the bulk Earth and Moon (see Figure 5.23 and Zuckerman *et al.*, 2007).

Jura *et al.* (2009b) used X-ray observational upper limits to constrain the total mass accretion rate of GD 362, noting that this helium-rich white dwarf has an anomalously high abundance of hydrogen (log[H/He] $= -1.1$; Zuckerman *et al.*, 2007). Together, the mass of heavy elements in the convection zone of GD 362

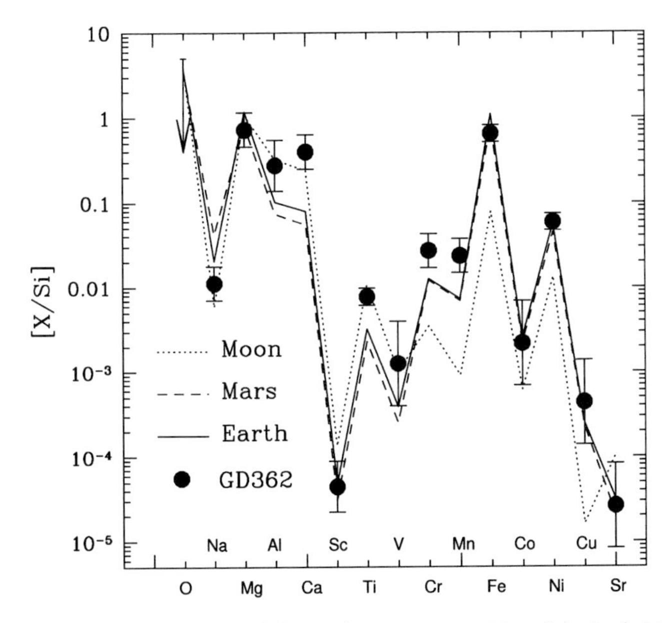

Figure 5.23 The remarkable heavy element abundances in the atmosphere of GD 362, showing the measured abundances for 14 detected metals (and an upper limit for oxygen) relative to silicon, together with the bulk composition of the Earth, Moon, and Mars. The best overall match is achieved with a combination of the Earth and Moon compositions. From Zuckerman *et al.* (2007), reproduced with the permission of the AAS.

is 1.8×10^{22} g, comparable to the mass contained in a 240 km diameter Solar System asteroid (e. g., Themis). On the other hand, the mass of hydrogen in its outer layers is 7.0×10^{24} g (Koester, 2009), orders of magnitude larger than for all known helium-rich white dwarfs of comparable temperature, except GD 16 (which is also polluted with heavy elements from a circumstellar disk; Farihi *et al.*, 2009; Jura *et al.*, 2009b).

One possibility for the atmospheric hydrogen in these two stars is delivery via water-rich planetary bodies. Because surface ices will not survive the giant phases of stellar evolution, any extant water in asteroids around white dwarfs would have to be sufficiently buried. The Solar System objects Ceres and Callisto are thought to have internal water that comprises roughly 25 and 50% of their total mass, respectively (Canup and Ward, 2002; McCord and Sotin, 2005; Thomas *et al.*, 2005). One possibility is that GD 362 accreted its hydrogen in the form of water from a few to several hundred large asteroids (unlikely) or an even larger body with internal water. If one supposes the current accretion event is the result of the latter possibility, then all of the observed properties of GD 362 – disk, atmospheric pollution, large hydrogen abundance, and X-ray upper limit – can be accounted for by the destruction and subsequent accretion of a parent body with a total mass between that of Callisto and Mars (Jura *et al.*, 2009b).

GD 40

In contrast, GD 40 exhibits hydrogen-deficiency typical of DB stars in its temperature range, but is nonetheless interesting. Already known to host magnesium, iron, silicon, and carbon from ultraviolet observations and calcium from optical spectroscopy (Friedrich *et al.*, 1999), the discovery of its disk was in large part motivated by the recognition of its relative carbon-deficiency and the asteroid-sized mass of metals contained in the star (Jura, 2006). Again using HIRES on Keck, Klein *et al.* (2010) detected an additional four elements and better constrained the abundances of all previously detected elements except carbon via multiple strong lines.

From this optical dataset, seven of the eight heavy elements present in the atmosphere of GD 40 form a subset of those detected in GD 362, with the exception of oxygen. Interestingly, Klein *et al.* (2010) found that by assuming that all of the atmospheric metals were delivered in their common oxides, and that all of the hydrogen was delivered in water, there is an excess of oxygen in the outer layers of the star. A solution to this conundrum is that the photospheric abundances differ from the accreted abundances, and GD 40 has been accreting for a minimum of a few diffusion timescales. In this scenario, the disk has a lifetime greater than 10^6 yr, and the apparent excess of oxygen is due to the fact that heavier elements such as iron sink more rapidly than oxygen (Koester, 2009).

The mass of heavy elements currently in the convection zone of GD 40 is 3.6×10^{22} g, already an asteroid-sized mass. However, with the caveat that the star must be in a steady state and has been accreting for at least a few diffusion timescales, the minimum mass of metals in the destroyed parent body grows to 3×10^{23} g (Klein *et al.*, 2010) or about the size of Vesta, the second largest asteroid in the Solar System.

In a steady state accretion mode, Klein *et al.* (2010) found that the water content of the now-destroyed minor planet can be no more than a few percent. Lastly, GD 40 exhibits a silicon-to-magnesium ratio significantly distinct from that found in stars and in the bulk Earth, hinting at the possibility that the parent body polluting GD 40 was at least partially differentiated (Klein *et al.*, 2010).

5.6.5 Evidence for Water in Debris Orbiting White Dwarfs

Starting from a default position that white dwarf atmospheric pollutants are delivered via planetary system remnants, the frequency of this phenomenon likely reflects the frequency of rocky planetary systems around the main sequence, A- and F-type progenitors of the current population of white dwarfs. From those white dwarfs displaying both atmospheric metals and circumstellar dust, a very conservative minimum frequency of remnant planetary systems is around 3% (Farihi *et al.*, 2009). However, evidence is mounting that most, if not all, metal-contaminated white dwarfs are the result of accretion of rocky planetesimals (Farihi *et al.*, 2010a). In this more liberal and, likely, more accurate view, the fraction of intermediate mass main sequence stars with terrestrial planetary systems that survive partly intact into the white dwarf phase is 20–30% (Zuckerman *et al.*, 2003, 2010). In light of this, some well-established observational properties of white dwarfs require re-examination. Perhaps most intriguing is the trace hydrogen seen in helium-rich white dwarfs.

In short, if DB white dwarfs accreted interstellar hydrogen at the fluid rate, then they would quickly become DA white dwarfs. Therefore, this does not occur (Koester, 1976; Wesemael, 1979). Convective mixing can dredge up helium in DA stars with $T_{eff} < 12\,000$ K, resulting in a helium-dominated atmosphere with traces of hydrogen (Tremblay and Bergeron, 2008), but hydrogen is found in DB and DBZ stars at warmer temperatures as well (Voss *et al.*, 2007). Lastly, trace hydrogen in helium-rich white dwarfs can be either primordial or accreted from interstellar space at rates comparable to the Eddington rate (i. e., direct impact). However, if helium atmosphere white dwarfs obtain trace metals and hydrogen from independent sources, then stars exhibiting *both* should be less frequent than stars exhibiting *either*. This is not supported by the observations, suggesting that helium atmosphere white dwarfs with both metals and hydrogen are polluted by water-rich asteroids (Farihi *et al.*, 2010a).

Of the four DBZ white dwarfs with circumstellar dust, GD 16 and GD 362 have more than 10^{24} g of atmospheric hydrogen, while GD 40 and Ton 345 have abundances and upper limits four orders of magnitude lower (Jura *et al.*, 2009b). These stark differences are not due to sensitivity, as trace hydrogen is detectable in DBZ white dwarfs typically down to one part in 10^5 with moderate-resolution optical spectra (Dufour *et al.*, 2007; Voss *et al.*, 2007). The hydrogen-poor atmospheres of the latter two stars are fairly typical of DB white dwarfs, while the hydrogen content of the former two stars is rather remarkable.

One possibility is that these four stars represent extremes of post-main sequence water content in planetary bodies. Oxygen has yet to be detected in either GD 16 or GD 362, but as both stars have $T_{eff} < 12\,000$ K and only optical spectra to date, this is not really surprising. It takes enormous oxygen abundances to produce detectable, optical absorption lines in cool white dwarfs (Gänsicke *et al.*, 2010). Sensitive searches for photospheric oxygen in these two stars via ultraviolet spectroscopy will be able to further constrain scenarios of pollution by parent bodies with internal water.

Interestingly, there are currently two warm DBZ stars with marked oxygen detections made in the far-ultraviolet: GD 61 and GD 378 (Desharnais *et al.*, 2008). Jura and Xu (2010) recognized the significance of these data, noting that both atmospheres had $O/C > 1$ and trace hydrogen in a temperature range where convection could not transform a DA into a DB. Their carbon-deficiency rules out pollution by interstellar matter, and accounting for their multiple atmospheric metals as oxides leaves an excess of oxygen in both stars (Jura and Xu, 2010). Because the heavy element diffusion timescales in these DBZ white dwarfs are of the order 10^5 yr, it remains uncertain what fraction of the excess oxygen can be attributed to mineral oxides common to rocky planetary bodies, but whose metal components have sunk more rapidly than the oxygen. Under some reasonable assumptions about their accretion history, Jura and Xu (2010) found that GD 378 was likely polluted by a water-rich asteroid, while the case for water in GD 61 is less compelling but still possible. It appears likely that some fraction of A- and F-type stars build planetary systems replete with water-rich asteroids, the building blocks of habitable planets and potentially analogous to the parent bodies that supplied the bulk of water in Earth's oceans (Morbidelli *et al.*, 2000). The near future will probably yield confirmation of this water in the circumstellar debris that orbits white dwarfs.

5.6.6
A Last Look at the Interstellar Accretion Hypothesis

The totality of published results and works in progress on disks and metal-pollution of white dwarfs circa 2009 strongly argues against interstellar accretion. Yet, even as the evidence for circumstellar disks and rocky pollutants began accumulating, and subsequently disseminated at international conferences, the scientific community continued to cite the interstellar medium as a source for accreted metals in cool white dwarfs (Desharnais *et al.*, 2008; Dufour *et al.*, 2007; Koester and Wilken, 2006). Given the dearth of circumstellar disk detections for metal-rich white dwarfs cooler than 10 000 K (Farihi *et al.*, 2009, 2010c) and their 10^4–10^6 yr metal diffusion timescales, it is perhaps understandable that some researchers remained skeptical.

The class of stars used most often to argue in favor of interstellar accretion are the coolest DBZ stars. These are spectrally classified as DZ because helium transitions become difficult to excite below 12 000 K (Voss *et al.*, 2007; Wesemael *et al.*, 1993). Because detectable metals can persist in DZ white dwarfs for as long as several Myr, it is conceivable that these stars are now located up to a few hundred pc from a region in which they became polluted by interstellar matter. Classically, the glaring problem with this scenario is the lack of accreted hydrogen in these stars, as the interstellar medium (cosmic or solar abundance) is 91% hydrogen and 0.01% elements heavier than neon, by number (Aannestad *et al.*, 1993; Däppen, 2000). Calcium-to-hydrogen abundances in DZ stars are typically super-solar, despite the fact that calcium continually sinks in their atmospheres (Farihi *et al.*, 2010a).

Farihi *et al.* (2010a) used 146 DZ stars from the SDSS DR4 with stellar parameters modeled by Dufour *et al.* (2007) to evaluate the interstellar accretion hypothesis by calculating several diagnostics. No correlation is found between calcium abundance and tangential speed, as expected if the stars were accreting at Bondi–Hoyle fluid rates necessary to produce the observed pollution (Dupuis *et al.*, 1993a; Koester and Wilken, 2006). More than half the sample white dwarfs are currently situated above the ± 100 pc thick gas and dust layer of the Galaxy, with nearly one-fifth of the stars located over 200 pc above the plane. Furthermore, roughly half the sample are now moving back into the disk of the Galaxy rather than away, implying that they have been out of the interstellar medium for several to tens of Myr (Farihi *et al.*, 2010a).

Comparing a commensurate number of cool, helium-rich SDSS white dwarfs, with and without detected photospheric metals, also represents a problem for the interstellar accretion scenario. The two classes of stars appear to belong to the same population of disk stars in temperature, atmospheric composition, Galactic positions and velocities. Among all these helium-rich white dwarfs, there are pairs within several pc of one another for which: (1) only one star is metal-rich but the pair share similar space velocities, or (2) both stars are polluted but their space velocities differ dramatically.

Perhaps the strongest evidence against interstellar accretion in white dwarfs actually comes from the same mechanism that permits the metals to reside for Myr timescales; namely, their relatively deep convection zones (Koester, 2009; Paquette *et al.*, 1986). Because any atmospheric metals are thoroughly mixed in the convective layers of the star, the measured abundances can be translated into masses by knowing the mass of the convective envelope. Figure 5.24 shows the mass of calcium detected in the 146 SDSS DZ stars as a function of effective temperature, revealing that these masses are typical of large Solar System asteroids. Moreover, because metals in the interstellar medium are locked up in dust grains, they will not be captured by a passing white dwarf at the fluid rate, but at a rate around ten times the Eddington accretion rate (Alcock and Illarionov, 1980). The calcium masses plotted in Figure 5.24 cannot be accounted for by the accretion of interstellar dust grains, even assuming Myr passages within the densest of environments (Farihi *et al.*, 2010a).

At present, there is no observational evidence in favor of interstellar accretion onto white dwarfs, yet plentiful and compelling data supporting the accretion of rocky planetary material. Given the existing observational data on metal-enriched white dwarfs, the interstellar accretion hypothesis is no longer viable.

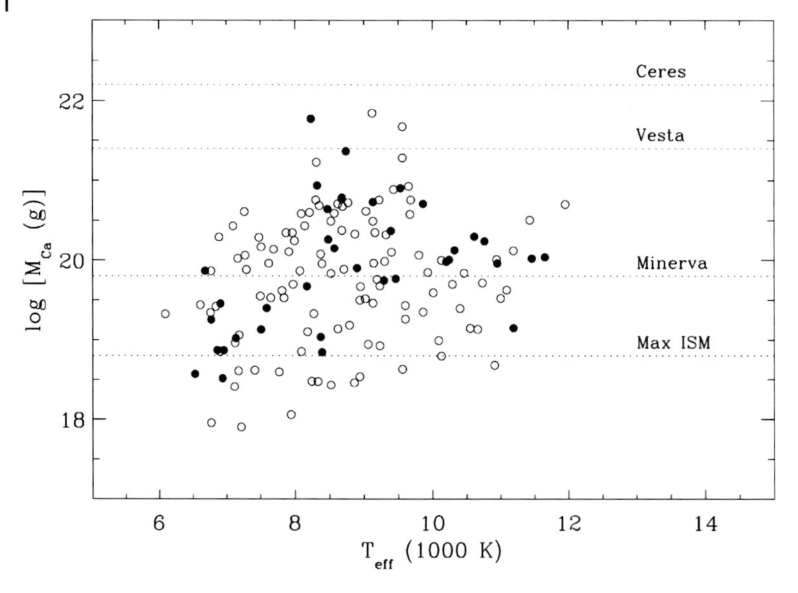

Figure 5.24 Calcium masses in the convective envelopes of 146 cool and metal-polluted white dwarfs with helium atmospheres from the SDSS. The open and filled circles represent stars with trace hydrogen abundance upper limits and detections, respectively. The top three dotted lines represent the mass of calcium contained in the two largest Solar System asteroids Ceres and Vesta, and the 150 km diameter asteroid Minerva, assuming calcium is 1.6% by mass as in the bulk Earth (Allègre *et al.*, 1995). The dotted line at the bottom is the maximum mass of calcium that can be accreted over 10^6 yr by a cool white dwarf moving at a velocity of 50 km s^{-1} through an interstellar cloud with a number density of $\rho = 1000\,\text{cm}^{-3}$. This figure is an updated version of Figure 2 in Farihi *et al.* (2010a).

5.7
Related Objects

The discussion in this chapter has been focused on apparently single, cool and metal-enriched white dwarfs because only these stars are observed to have circumstellar dust. Additionally, all stars in this class require external sources of pollution, for which circumstellar matter is currently the strongest candidate. It has already been mentioned that Spitzer observations have not identified infrared excess in over 150 white dwarfs without photospheric metals, but this needs some qualification. A brief discussion of white dwarfs that are (potentially) related to the stars of this chapter is presented below.

5.7.1
White Dwarfs Polluted by Companions?

Zuckerman *et al.* (2003) identified several DAZ stars among white dwarfs with low mass stellar companions, M dwarfs, in suspected or confirmed short period orbits. The orbital separations of these systems are either established by radial velocity studies or constrained by direct imaging, and consistent with detached, post-common envelope binaries (Schultz *et al.*, 1996; Farihi *et al.*, 2005; Hoard *et al.*, 2007). Although these stars are not interacting in the conventional sense of Roche lobe overflow, the relatively high frequency of photospheric metal detections in their white dwarf components indicates that these DAZ stars are capturing material from the stellar wind of the companion (Debes, 2006; Zuckerman *et al.*, 2003). Recent high resolution imaging with the Hubble Space Telescope (HST) has strengthened this interpretation by finding that all of the DA+dM systems are visual pairs, while the DAZ+dM systems are typically spatially unresolved (Farihi *et al.*, 2010b).

These observational results strengthened the hypothesis that apparently single, metal-enriched white dwarfs might be polluted by unseen, unresolved companions (Dobbie *et al.*, 2005; Holberg *et al.*, 1997). However, any such companions would have to be sufficiently low in both mass and temperature that they are not detected in large near-infrared surveys such as 2MASS (Hoard *et al.*, 2007). Five years of Spitzer IRAC studies have laid this hypothesis firmly to rest; the infrared spectral energy distributions of metal-rich white dwarfs fail to reveal the expected signature of low mass companions down to $25\,M_{\text{Jup}}$, according to substellar cooling models (Baraffe *et al.*, 2003; Burrows *et al.*, 2003; Farihi *et al.*, 2008a, 2009). Therefore, apparently single, metal-rich white dwarfs are not polluted by mass transfer or wind capture from unseen, substellar companions.

Interestingly, two of the five known brown dwarf companions to white dwarfs are in close orbits analogous to the DAZ+dM systems found by Zuckerman *et al.* (2003): GD 1400B and WD 0137−049B (Farihi and Christopher, 2004; Maxted *et al.*, 2006). Neither of the two cool white dwarfs in these systems show signs of atmospheric metal pollution from their close substellar secondaries, despite both stars having a large number of co-added, high resolution spectra obtained with VLT UVES (R. Napiwotzki, private communication). Thus, it appears that the mid- to late-L dwarfs in these systems do not generate winds comparable to their higher mass, M dwarf counterparts.

5.7.2
Dust in the Helix?

Su *et al.* (2007) reported a strong infrared excess at the location of the central star of the Helix Nebula (NGC 7293), detected at 24 and 70 μm with Spitzer MIPS (see Figure 5.25). An analysis of the excess emission using photometry at all IRAC and MIPS wavelengths, together with high resolution IRS spectroscopy between 10 μm and 35 μm, reveals a 100 K continuum. Su *et al.* (2007) attribute this emission to cold dust grains at several tens of AU from the central star, which, itself, may be best described evolutionarily as a pre-white dwarf (Napiwotzki, 1999).

The 24 μm flux source was reported to be partly spatially-resolved, but this is potentially due to imperfect subtraction of the diffuse nebular emission, which is nonuniform and contains both continuum and emission components (Su *et al.*, 2007). Assuming the detected excess is indeed point-like, the 6″ size of the

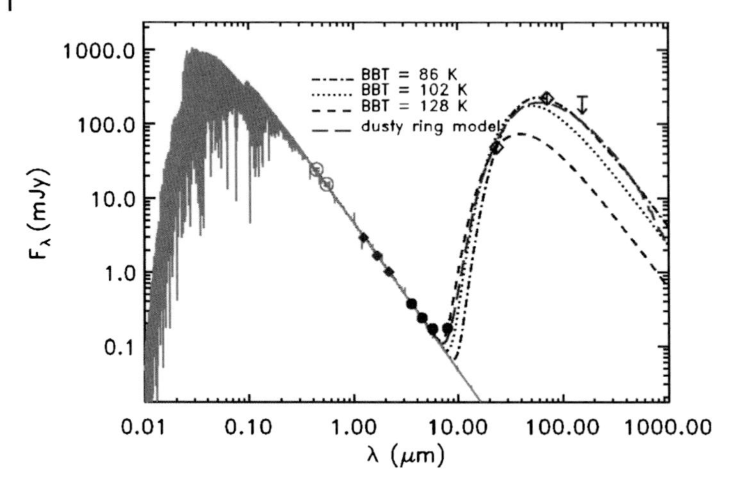

Figure 5.25 Strong infrared excess detected by Spitzer at the location of the central star in the Helix Nebula. The combination of multiple known emission sources within the nebula and the low spatial resolution of the MIPS beam makes the interpretation of the excess ambiguous. A cold debris disk is one possibility, but scenarios involving a companion star are perhaps more plausible (Bilikova *et al.*, 2009). From Su *et al.* (2007), reproduced with the permission of the AAS.

MIPS 24 μm PSF only limits the physical size of the emitting region to 1300 AU. Therein lies one difficulty in the interpretation of the detected excess.

A distinct alternative to cold dust, analogous to that detected in main sequence stars, is dust captured by a companion star and heated by the pre-white dwarf. Bilikova *et al.* (2009) reported the results of a Spitzer survey for additional 24 μm excesses from 72 hot and pre-white dwarfs, about half of which are in planetary nebulae. They found 12 new cases of strong excess, all of which are associated with planetary nebulae, and at least two of which have binary central stars. Given that (asymmetric) planetary nebulae are likely to be intimately connected with binary star evolution (de Marco, 2009), and in light of the findings of Bilikova *et al.* (2009), it seems plausible that the infrared excess in the Helix Nebula is *not* associated with a circumstellar dust disk around the pre-white dwarf.

5.8
Outlook for the Present and Near Future

At the time of writing, Spitzer IRAC is still operating at 3.6 μm and 4.5 μm with Cycle 7 observations beginning in the middle of 2010, and eighth and possibly ninth cycles planned. Observing at mid-infrared wavelengths similar to cryogenic Spitzer, the Wide-field Infrared Survey Explorer (WISE) has just completed its coverage of the entire sky. While not as sensitive as pointed observations with Spitzer IRAC, WISE has the capability to detect bright ($\gtrsim$ 100 μJy), warm dust disks around white dwarfs not yet known to be metal-polluted and, hence, not previously targeted by Spitzer or ground-based searches for infrared excess.

The Herschel Space Observatory is currently observing in the far-infrared and solicited its first open round of proposals from the scientific community, but may not be sufficiently sensitive for white dwarf dust studies. The mid-infrared excesses observed by Spitzer are typically tens to hundreds of μJy and falling towards longer wavelengths, while Herschel reports 5–10 μJy detection limits. Unless white dwarfs harbor especially prominent excesses from cold dust, around one to ten times their peak optical fluxes, Herschel will not detect dust in many white dwarf systems. At somewhat longer wavelengths, the Atacama Large Millimeter Array (ALMA) is set to begin science operations in 2011. ALMA should have both μJy sensitivity and subarcsecond spatial resolution, making it a promising tool for direct imaging of spatially-resolved, cold debris around white dwarfs.

Further into the future, the James Webb Space Telescope (JWST) will operate in the near- and mid-infrared with sensitivity superior to Spitzer. There are two leaps forward that can be envisioned with such a capability. First, better quality mid-infrared spectra of warm dust around white dwarfs will enable a more detailed analysis of temperature, location, geometry, and composition. Second, infrared spectroscopy might reveal very subtle infrared excesses for metal-rich white dwarfs where Spitzer IRAC photometry could not. This technique was successfully employed for main sequence stars with Spitzer IRS, where high S/N spectroscopy revealed at least two cases of spectacular, yet subtle, warm dust emission (Beichman *et al.*, 2005, 2006; Lisse *et al.*, 2007).

On the other end of the electromagnetic spectrum, the recently-installed ultraviolet Cosmic Origins Spectrograph (COS) on HST promises to be a useful instrument with which to study the elemental abundances of polluted white dwarfs. Optimized for the far-ultraviolet, COS will function best for metal-rich white dwarfs warmer than 16 000 K, but near-ultraviolet observations are possible with lower efficiency for all but the coolest white dwarfs with metals. The science enabled by instruments like COS is really at the core of white dwarf circumstellar dust studies. The stellar atmosphere distills the accreted debris, providing the bulk chemical composition of the destroyed parent body. In this manner, the investigation of circumstellar dust around white dwarfs has the potential to reveal more about the anatomy of extrasolar, terrestrial planetary bodies than any other technique.

Acknowledgments

I would like to thank my colleagues and collaborators for many years of helpful discussions and interesting projects, and the editor, D.W. Hoard, for allowing me the opportunity to contribute to this book. The published and online[3] versions of the white dwarf catalogue assembled by G. McCook and E.M. Sion have been an invaluable resource over the years and the white dwarf community owes the authors and caretakers an enormous debt of gratitude for their effort. I would also

3) http://www.astronomy.villanova.edu/WDCatalog/index.html (15 June 2011)

like to thank D. Koester for making the white dwarf atmosphere models used in this chapter available.

References

Aannestad, P.A., Kenyon, S.J., Hammond, G.L., and Sion, E.M. (1993) *Astron. J.*, **105**, 1033.

Alcock, C. and Illarionov, A. (1980) *Astrophys. J.*, **235**, 534.

Alcock, C., Fristrom, C.C., and Siegelman, R. (1986) *Astrophys. J.*, **302**, 462.

Allamandola, L.J., Tielens, A.G.G.M., and Barker, J.R. (1989) *Astrophys. J. (Suppl.)*, **71**, 733.

Allègre, C.J., Poirier, J.P., Humler, E., and Hofmann, A.W. (1995) *Earth Planet. Sci. Lett.*, **4**, 515.

Bannister, N.P., Barstow, M.A., Holberg, J.A., and Bruhweiler, F.C. (2003) *Mon. Not. R. Astron. Soc.*, **341**, 477.

Baraffe, I., Chabrier, G., Barman, T.S., Allard, F., and Hauschildt, P.H. (2003) *Astron. Astrophys.*, **402**, 701.

Barstow, M.A., Jordan, S., O'Donoghue, Burleigh, M.R., Napiwotzki, R., and Harrop-Allin, M.K. (1995) *Mon. Not. R. Astron. Soc.*, **277**, 971.

Becklin, E.E., Farihi, J., Jura, M., Song, I., Weinberger, A.J., and Zuckerman, B. (2005) *Astrophys. J. Lett.*, **632**, L119.

Beichman, C.A. *et al.* (2005) *Astrophys. J.*, **626**, 1061.

Beichman, C.A. *et al.* (2006) *Astrophys. J.*, **639**, 1166.

Bilikova, J., Chu, Y.H., Su, K., Gruendl, R., Rauch, T., De Marco, O., and Volk, K. (2009) *J. Phys. Conf. Ser.*, **172**, 012055.

Bockelée-Morvan, D., Brooke, T.Y., and Crovisier, J. (1995) *Icarus*, **116**, 18.

Brinkworth, C.S., Gänsicke, B.T., Marsh, T.R., Hoard, D.W., and Tappert, C. (2009) *Astrophys. J.*, **696**, 1402.

Barnbaum, C. and Zuckerman, B. (1992) *Astrophys. J. Lett.*, **396**, L31.

Burrows, A., Sudarsky, D., and Lunine, J.I. (2003) *Astrophys. J.*, **596**, 587.

Canup, R.M. and Ward, W.R. (2002) *Astron. J.*, **124**, 3404.

Chary, R., Zuckerman, B., and Becklin, E.E. (1999) *The Universe as Seen by ISO* (eds P. Cox and M.F. Kessler), ESA-SP, **427**, 289.

Chayer, P., Fontaine, G., and Wesemael, F. (1995) *Astrophys. J. (Suppl.)*, **99**, 189.

Clayton, G.C., Geballe, T.R., Herwig, F., Fryer, C., and Asplund, M. (2007) *Astrophys. J.*, **662**, 1220.

Däppen, W. (2000) *Allen's Astrophysical Quantities*, vol. 4 (ed. A.N. Cox), AIP Press Springer, New York, p. 27.

Davidsson, B.J.R. (1999) *Icarus*, **142**, 525.

Debes, J.H. (2006) *Astrophys. J.*, **652**, 636.

Debes, J.H. and Sigurdsson, S. (2002) *Astrophys. J.*, **572**, 556.

Debes, J.H., Sigurdsson, S. and Hansen, B. (2007) *Astron. J.*, **134**, 1662.

Debes, J.H., Hoard, D.W., Kilic, M., Wachter, S., Leisawitz, D.T., Cohen, M., Kirkpatrick, J.D., and Griffith, R.L. (2011) *Astrophys. J.*, in press.

de Marco, O. (2009) *Publ. ASP*, **121**, 316.

Desharnais, S., Wesemael, F., Chayer, P., Kruk, J.W., and Saffer, R.A. (2008) *Astrophys. J.*, **672**, 540.

Dobbie, P.D., Burleigh, M.R., Levan, A.J., Barstow, M.A., Napiwotzki, R., Holberg, J.B., Hubeny, I., and Howell, S.B. (2005) *Mon. Not. R. Astron. Soc.*, **357**, 1049.

Dobbie, P.D., Napiwotzki, R., Lodieu, N., Burleigh, M.R., Barstow, M.A., and Jameson, R.F. (2006) *Mon. Not. R. Astron. Soc.*, **373**, L45.

Draine, B.T. (2003) *Annu. Rev. Astron. Astrophys.*, **41**, 241.

Dufour P. *et al.* (2007) *Astrophys. J.*, **663**, 1291.

Dufour, P., Kilic, M., Fontaine, G., Bergeron, P., Lachapelle, F.-R., Kleinman, S.J., and Leggett, S.K. (2010) *Astrophys. J.*, **719**, 803.

Dupuis, J., Fontaine, G., Pelletier, C., and Wesemael, F. (1992) *Astrophys. J. (Suppl.)*, **82**, 505.

Dupuis, J., Fontaine, G., Pelletier, C., and Wesemael, F. (1993a) *Astrophys. J. (Suppl.)*, **84**, 73.

Dupuis, J., Fontaine, G., and Wesemael, F. (1993b) *Astrophys. J. (Suppl.)*, **87**, 345.

Eisenstein, D.J. *et al.* (2006) *Astron. J.*, **132**, 676.

Farihi, J. (2009) *Mon. Not. R. Astron. Soc.*, **398**, 2091.

Farihi, J. and Christopher, M. (2004) *Astron. J.*, **128**, 1868.

Farihi, J., Becklin, E.E., and Zuckerman, B. (2005) *Astrophys. J. (Suppl.)*, **161**, 394.

Farihi, J., Becklin, E.E., and Zuckerman, B. (2008a) *Astrophys. J.*, **681**, 1470.

Farihi, J., Zuckerman, B., and Becklin, E.E. (2008b) *Astrophys. J.*, **674**, 431.

Farihi, J., Jura, M., and Zuckerman, B. (2009) *Astrophys. J.*, **694**, 805.

Farihi, J., Barstow, M.A., Redfield, S., Dufour, P., and Hambly, N.C. (2010a) *Mon. Not. R. Astron. Soc.*, **404**, 2123.

Farihi, J., Hoard, D.W., and Wachter, S. (2010b) *Astrophys. J. (Suppl.)*, **190**, 275.

Farihi, J., Jura, M., Lee, J.E., and Zuckerman, B. (2010c) *Astrophys. J.*, **714**, 1386.

Fazio, G.G. *et al.* (2004) *Astrophys. J. (Suppl.)*, **154**, 10.

Ferrario, L., Wickramasinghe, D.T., Liebert, J., and Williams, K.A. (2005) *Mon. Not. R. Astron. Soc.*, **361**, 1131.

Fontaine, G. and Michaud, G. (1979) *Astrophys. J.*, **231**, 826.

Friedrich, S., Koester, D., Heber, U., Jeffery, C.S., and Reimers, D. (1999) *Astron. Astrophys.*, **350**, 865.

Friedrich, S., Koester, D., Christlieb, N., Reimers, D., and Wisotzki, L. (2000) *Astron. Astrophys.*, **363**, 1040.

Gänsicke, B.T., Marsh, T.R., Southworth, J., and Rebassa-Mansergas, A. (2006) Science, **314**, 1908.

Gänsicke, B.T., Marsh, T.R., and Southworth, J. (2007) *Mon. Not. R. Astron. Soc.*, **380**, L35.

Gänsicke, B.T., Koester, D., Marsh, T.R., Rebassa-Mansergas, A., and Southworth J. (2008) *Mon. Not. R. Astron. Soc.*, **391**, L103.

Gänsicke, B.T., Koester, D., Girven, J., Marsh, T.R., and Steeghs, D. (2010) Science, **327**, 188.

Gianninas, A., Dufour, P., and Bergeron, P. (2004) *Astrophys. J. Lett.*, **617**, L57.

Graham, J.R., Matthews, K., Neugebauer, G., and Soifer, B.T. (1990) *Astrophys. J.*, **357**, 216.

Green, P.J., Ali, B., and Napiwotzki, R. (2000) *Astron. Astrophys.*, **540**, 992.

Green, R.F., Schmidt, M., and Liebert, J. (1986) *Astrophys. J. (Suppl.)*, **61**, 305.

Hansen, B.M.S., Kulkarni, S., and Wiktorowicz, S. (2006) *Astron. J.*, **131**, 1106.

Hoard, D.W., Wachter, S., Sturch, L.K., Widhalm, A.M., Weiler, K.P., Pretorius, M.L., Wellhouse, J.W., and Gibiansky, M. (2007) *Astron. J.*, **134**, 26.

Holberg, J.B., Barstow, M.A., and Green, E.M. (1997) *Astrophys. J. Lett.*, **474**, L127.

Horne, K. and Marsh, T.R. (1986) *Mon. Not. R. Astron. Soc.*, **218**, 761.

Houck, J.R. *et al.* (2004) *Astrophys. J. (Suppl.)*, **154**, 18.

Jura, M. (2003) *Astrophys. J. Lett.*, **584**, L91.

Jura, M. (2006) *Astrophys. J.*, **653**, 613.

Jura, M. (2008) *Astron. J.*, **135**, 1785.

Jura, M. *et al.* (2006) *Astrophys. J. Lett.*, **637**, L45.

Jura, M., Farihi, J., and Zuckerman, B. (2007a) *Astrophys. J.*, **663**, 1285.

Jura, M., Farihi, J., Zuckerman, B., and Becklin, E.E. (2007b) *Astron. J.*, **133**, 1927.

Jura, M., Farihi, J., and Zuckerman, B. (2009a) *Astron. J.*, **137**, 3191.

Jura, M., Muno, M., Farihi, J., and Zuckerman, B. (2009b) *Astrophys. J.*, 699 1473.

Jura, M. and Xu, S. (2010) *Astron. J.*, **140**, 1129.

Kawka, A. and Vennes, S. (2005) *ASP Conf. Ser.* Proc. of the 14th European Workshop on White Dwarfs, vol. 334 (eds D. Koester and S. Moehler), ASP, San Francisco, p. 101.

Kilic, M. and Redfield, S. (2007) *Astrophys. J.*, **660**, 641.

Kilic, M., von Hippel, T., Leggett, S.K., and Winget, D.E. (2005) *Astrophys. J. Lett.*, **632**, L115.

Kilic, M., von Hippel, T., Leggett, S.K., and Winget, D.E. (2006) *Astrophys. J.*, **646**, 474.

Kilic, M., Farihi, J., Nitta, A., and Leggett, S.K. (2008) *Astron. J.*, **136**, 111.

Klein, B., Jura, M., Koester, D., Zuckerman, B., and Melis, C. (2010) *Astrophys. J.*, **709**, 950.

Kleinman, S.J. (1994) *Astrophys. J.*, **436**, 875.

Koester, D. (1976) *Astron. Astrophys.*, **52**, 415.

Koester, D. (2009) *Astron. Astrophys.*, **498**, 517.

Koester, D. and Wilken, D. (2006) *Astron. Astrophys.*, **453**, 1051.

Koester, D., Provencal, J., and Shipman, H.L. (1997) *Astron. Astrophys.*, **230**, L57.

Koester, D., Napiwotzki, R., Voss, B., Homeier, D., and Reimers, D. (2005a) *Astron. Astrophys.*, **439**, 317.

Koester, D., Rollenhagen, K., Napiwotzki, R., Voss, B., Christlieb, N., Homeier, D., and Reimers, D. (2005b) *Astron. Astrophys.*, **432**, 1025.

Lacombe, P., Wesemael, F., Fontaine, G., and Liebert, J. (1983) *Astrophys. J.*, **272**, 660.

Lambert, D.L., Rao, N.K., Pandey, G., and Ivans, I.I. (2001) *Astrophys. J.*, **555**, 925.

Liebert, J., Bergeron, P., and Holberg, J.B. (2005) *Astrophys. J. (Suppl.)*, **156**, 47.

Lisse, C.M. *et al.* (2006) Science, **313**, 635.

Lisse, C.M., Beichman, C.A., Bryden, G., and Wyatt, M.C. (2007) *Astrophys. J.*, **658**, 584.

Lisse, C.M., Chen, C.H., Wyatt, M.C., and Morlok, A. (2008) *Astrophys. J.*, **673**, 1106.

Lodders, K. (2003) *Astrophys. J.*, **591**, 1220.

Malfait, K., Waelkens, C., Waters, L.B.F.M., Vandenbussche, B., Huygen, E., and de Graauw, M.S. (1998) *Astron. Astrophys.*, **332**, L25.

Maxted, P.F.L., Napiwotzki, R., Dobbie, P.D., and Burleigh, M.R. (2006) *Nature*, **442**, 543.

McCook, G.P. and Sion, E.M. (1999) *Astrophys. J. (Suppl.)*, **121**, 1.

McCord, T.B. and Sotin, C. (2005) *J. Geophys. Res. (Planet.)*, **110**, 5009.

McLean, I.S. (1997) Electronic Imaging in Astronomy, in *Detectors and Instrumentation*, John Wiley & Sons, Ltd, Chichester.

Melis, C., Jura, M., Albert, L., Klein, B., and Zuckerman, B. (2010) *Astrophys. J.*, **722**, 1078.

Morbidelli, A., Chambers, J., Lunine, J.I., Petit, J.M., Robert, F., Valsecchi, G.B., and Cyr, K.E. (2000) *Meteoritics and Planetary Science*, **35**, 1309.

Mullally, F., Kilic, M., Reach, W.T., Kuchner, M., von Hippel, T., Burrows, A., and Winget, D.E. (2007) *Astrophys. J. (Suppl.)*, **171**, 206.

Napiwotzki, R. (1999) *Astron. Astrophys.*, **350**, 101.

Napiwotzki, R. *et al.* (2003) *The Messenger*, **112**, 25.

Paquette, C., Pelletier, C., Fontaine, G., and Michaud, G. (1986) *Astrophys. J. (Suppl.)* 61, 197

Patterson, J., Zuckerman, B., Becklin, E.E., Tholen, D.J., and Hawarden, T. (1991) *Astrophys. J.*, **374**, 330.

Probst, R. (1981) PhD Thesis, University of Virginia.

Probst, R. (1983) *Astrophys. J. (Suppl.)*, **53**, 335.

Probst, R.G. and O'Connell, R.W. (1982) *Astrophys. J. Lett.*, **252**, L69.

Rayner, J.T., Toomey, D.W., Onaka, P.M., Denault, A.J., Stahlberger, W.E., Vacca, W.D., Cushing, M.C., and Wang S. (2003) *Publ. ASP*, **115**, 362.

Reach, W.T., Kuchner, M.J., von Hippel, T., Burrows, A., Mulally, F., Kilic, M., and Winget, D.E. (2005) *Astrophys. J. Lett.*, **635**, L161.

Reach, W.T., Lisse, C., von Hippel, T., and Mullally, F. (2009) *Astrophys. J.*, **693**, 697.

Rieke, G. *et al.* (2004) *Astrophys. J. (Suppl.)*, **154**, 25.

Schultz, G., Zuckerman, B., and Becklin, E.E. (1996) *Astrophys. J.*, **460**, 402.

Shipman, H.L. (1986) *Astrophysics of Brown Dwarfs* (eds M.C. Kafatos, R.S. Harrington, and S.P. Maran), Cambridge University Press, New York, p. 71.

Sion, E.M., Hammond, G.L., Wagner, R.M., Starrfield, S.G., and Liebert, J. (1990a) *Astrophys. J.*, **362**, 691.

Sion, E.M., Kenyon, S.J., and Aannestad, P.A. (1990b) *Astrophys. J. (Suppl.)*, **72**, 707.

Skrutskie, M.F. *et al.* (2006) *Astron. J.*, **131**, 1163.

Song, I., Zuckerman, B., Weinberger, A.J., and Becklin, E.E. (2005) *Nature*, **436**, 363.

Su, K.Y.L. *et al.* (2007) *Astrophys. J. Lett.*, **657**, L41.

Telesco, C.M., Joy, M., and Sisk, C. (1990) *Astrophys. J. Lett.*, **358**, L21.

Thomas, P.C., Parker, J.W., McFadden, L.A., Russell, C.T., Stern, S.A., Sykes, M.V., and Young, E.F. (2005) *Nature*, **437**, 224.

Tokunaga, A.T., Becklin, E.E., and Zuckerman, B. (1990) *Astrophys. J. Lett.*, **358**, L17.

Tokunaga, A.T., Hodapp, K.W., Becklin, E.E., Cruikshank, D.P., Rigler, M., Toomey, D.W., Brown, R.H., and Zuckerman, B. (1988) *Astrophys. J. Lett.*, **332**, L71.

Tremblay, P.E. and Bergeron, P. (2008) *Astrophys. J.*, **672**, 1144.

van Maanen, A. (1917) *Publ. ASP*, **29**, 258.

van Maanen, A. (1919) *Publ. ASP*, **31**, 42.

von Hippel, T., Kuchner, M.J., Kilic, M., Mullaly, F., and Reach, W.T. (2007) *Astrophys. J.*, **662**, 544.

Vauclair, G., Vauclair, S., and Greenstein, J.L. (1979) *Astron. Astrophys.*, **80**, 79.
Voss, B., Koester, D., Napiwotzki, R., Christlieb, N., and Reimers, D. (2007) *Astron. Astrophys.*, **470**, 1079.
Wachter, S., Hoard, D.W., Hansen, K.H., Wilcox, R.E., Taylor, H.M., and Finkelstein, S.L. (2003) *Astrophys. J.*, **586**, 1356.
Wegner, G. (1972) *Astrophys. J.*, **172**, 451.
Wehrse, R. (1975) *Astron. Astrophys.*, **39**, 169.
Weidemann, V. (1960) *Astrophys. J.*, **131**, 638.
Welsh B.Y., Craig N., Vedder P.W., and Vallerga J.V. (1994) *Astrophys. J.*, **437**, 638.
Welsh B.Y., Sfeir D.M., Sirk M.M., and Lallement R. (1999) *Astron. Astrophys.*, **352**, 308.
Werner, M.W. *et al.* (2004) *Astrophys. J. (Suppl.)*, **154**, 1.
Wesemael, F. (1979) *Astron. Astrophys.*, **72**, 104.
Wesemael, F., Greenstein, J.L., Liebert, J., Lamontagne, R., Fontaine, G., Bergeron, P., and Glaspey, J.W. (1993) *Publ. ASP*, **105**, 761.
Wolff, B., Koester, D., and Liebert, J. (2002) *Astron. Astrophys.*, **385**, 995.
York, D.G. *et al.* (2000) *Astron. J.*, **120**, 1579.
Zuckerman, B. (2001) *Annu. Rev. Astron. Astrophys.*, **39**, 549.
Zuckerman, B. and Becklin, E.E. (1987) *Nature*, **330**, 138.
Zuckerman, B. and Becklin, E.E. (1992) *Astrophys. J.*, **386**, 260.
Zuckerman, B. and Reid, I.N. (1998) *Astrophys. J. Lett.*, **505**, L143.
Zuckerman, B., Koester, D., Reid, I.N., and Hünsch, M. (2003) *Astrophys. J.*, **596**, 477.
Zuckerman, B., Koester, D., Melis, C., Hansen, B.M.S., and Jura, M. (2007) *Astrophys. J.*, **671**, 872.
Zuckerman, B., Melis, C., Klein, B., Koester, D., and Jura, M. (2010) *Astrophys. J.*, **722**, 725.

6
The Origin and Evolution of White Dwarf Dust Disks

John H. Debes

6.1
Introduction

Despite a growing number of white dwarf dust disks that have been observed, little about the origin or evolution of these systems is understood. Often, these disks are pointed to as the observational endpoint for a planetary system similar to the Solar System. However, it is hard to judge the merit of that hypothesis without a better understanding of the relation of these disks to planetary systems. Are they a smoking gun for the presence of planetary systems? The answer lies in a physical understanding of the disks themselves as well as mechanisms for their origin and evolution.

Both the bulk atomic compositions and the location of the dust are consistent with the scenario that an asteroid or other minor body in a possibly unstable planetary system has been perturbed into a highly eccentric orbit that results in tidal disruption by the white dwarf (Debes *et al.*, 2002; Jura, 2003; Zuckerman *et al.*, 2007). Subsequently, the fragments settle into a geometrically thin disk that can persist long enough to be observed in $\sim$ 1–3% of white dwarfs. In some cases, the flux of white dwarf crossing orbits may be small enough that a dust disk does not form, but a gaseous disk resides in a tight orbit around the white dwarf, providing a steady stream of metal rich gas onto the white dwarf's surface (Jura, 2008). These systems would then present as metal rich DAZs or DZs, which comprise a rather significant fraction of the white dwarf population (Jura, 2008; Koester *et al.*, 2005; Zuckerman *et al.*, 2003).

While this is a compelling scenario, a coherent physical model that starts with a particular planetary architecture and successfully reproduces the observed distribution of dusty white dwarfs or explains the frequency or lifetime of these disks does not exist today. The prime works cited by most authors for the above scenario deal with (1) a dynamical investigation of the stability of planetary systems in the presence of mass loss from a central star (Debes *et al.*, 2002) and (2) the recognition that the disks tend to reside within the tidal disruption radius of the white

White Dwarf Atmospheres and Circumstellar Environments. First Edition. Edited by D. W. Hoard.

dwarfs (Jura, 2003). Are there other explanations that match the observations? How accurate are the assumptions in the general scenario laid out above?

This chapter is designed to stimulate an equal growth to the theoretical side of white dwarf dust disks that has been seen on the observational side. As the white dwarf dust disk community moves beyond the discovery phase and into a statistical investigation of the white dwarf disk population, it will require a more sophisticated understanding of the underlying physical processes that shape the evolution of such disks, as well as the types of planetary systems (or other processes) that create such disks.

6.2
Orders of Magnitude around a White Dwarf

In each of the known dusty disks observed around white dwarfs, it is commonly assumed that the inner radius of the disk is truncated at the sublimation radius of the dust, while the outer radius should correspond to the tidal disruption radius of the white dwarf. The exact sublimation temperature depends on the material within the disk and its typical sublimation lifetime. This value can be determined observationally by noting at which wavelength the excess infrared emission begins. Typically, emission from the dust appears in the *K* band, implying interior temperatures of $\sim$ 1300 K. If olivine is assumed to be the primary constituent of the disk, then the lifetime of a dust grain before sublimation, t_{sub}, is given by

$$t_{\text{sub}} = \frac{r_{\text{dust}}\,\rho_{\text{dust}}}{\dot{\sigma}(T)} = \left(\frac{1}{1.2\times 10^{7}}\right)\left(\frac{r_{\text{dust}}}{1\,\mu\text{m}}\right)\left(\frac{\rho_{\text{dust}}}{3\,\text{g}\,\text{cm}^{-3}}\right)\dot{\sigma}(T)^{-1}\,\text{h}\,, \tag{6.1}$$

where r_{dust} and ρ_{dust} are the radius and density of a dust grain, respectively, and $\dot{\sigma}(T)$ is the mass production rate per unit area from the dust ($\text{g}\,\text{cm}^{-2}\,\text{s}^{-1}$) as a function of the dust grain temperature, T, given by

$$\dot{\sigma}(T) = \dot{\sigma}_0\sqrt{\frac{T_0}{T}}\,\text{e}^{-T_0/T}\,, \tag{6.2}$$

(Jura, 2008; Kimura *et al.*, 2002).

For olivine, $\dot{\sigma}_0 = 1.5\times 10^{9}\,\text{g}\,\text{cm}^{-2}\text{s}^{-1}$ and $T_0 = 65\,300$ K. Assuming a temperature of 1500 K, grains that are $\sim$ 1 μm in size have $t_{\text{sub}} = 68$ h. Intuitively, one would expect that the timescale for sublimation should be shorter than for other mechanisms if it dominates the rate of grain removal in the disk. Assuming that is the case, then the radius at which grains reach the sublimation temperature can be defined as a characteristic inner radius,

$$R_{\text{in}} = \left(\frac{L_\star}{16\pi\sigma_{\text{SB}}T_{\text{sub}}^4}\right)^{1/2} \approx 3.7\times 10^{10}\left(\frac{L_\star}{0.001\,L_\odot}\right)^{1/2}\left(\frac{1000\,\text{K}}{T_{\text{sub}}}\right)^{2}\,\text{cm}\,, \tag{6.3}$$

where $L_\star$ is the white dwarf luminosity, σ_{SB} is the Stefan–Boltzmann constant, and T_{sub} is the dust grain sublimation temperature (Jura, 2004, 2008).

The blackbody grain approximation breaks down if the grains are in an optically thick ring, as proposed initially for G29-38 (Jura, 2003). In this case, the temperature is lower than in a strictly optically thin case, as will be shown in Section 6.3.

The tidal disruption radius depends on the binding energy of the smaller bodies that encounter the gravitational field of the white dwarf as well as their rotation. For simplicity, however, one can assume that the following prescription for the tidal disruption radius gives an order of magnitude estimate for the effective outermost edge of the disk,

$$R_{out} = C_{tide}\left(\frac{\rho_\star}{\rho_a}\right)^{1/3} R_\star \approx 69\left[\left(\frac{\rho_\star}{10^6\,\mathrm{g\,cm^{-3}}}\right)\left(\frac{3\,\mathrm{g\,cm^{-3}}}{\rho_a}\right)\right]^{1/3} R_\star\,, \quad (6.4)$$

where C_{tide} is a constant of order unity, $R_\star$ and $\rho_\star$ are the radius and average density of the white dwarf, respectively, and ρ_a is the density of the planetesimal (Davidsson, 1999). Figure 6.1 shows these two important radii, R_{in} and R_{out}, as a function of effective temperature assuming olivine grains and $\rho_a = 3\,\mathrm{g\,cm^{-3}}$ for various hydrogen white dwarf models (Bergeron *et al.*, 1995; Fontaine *et al.*, 2001; Holberg and Bergeron, 2006). There is a critical point at which the disruption radius lies interior to the sublimation radius – this transition might mark where hot gaseous disks around white dwarfs arise; for example, those discovered by Ca triplet emission (Gänsicke *et al.*, 2006; Gänsicke *et al.*, 2007; Gänsicke *et al.*, 2008).

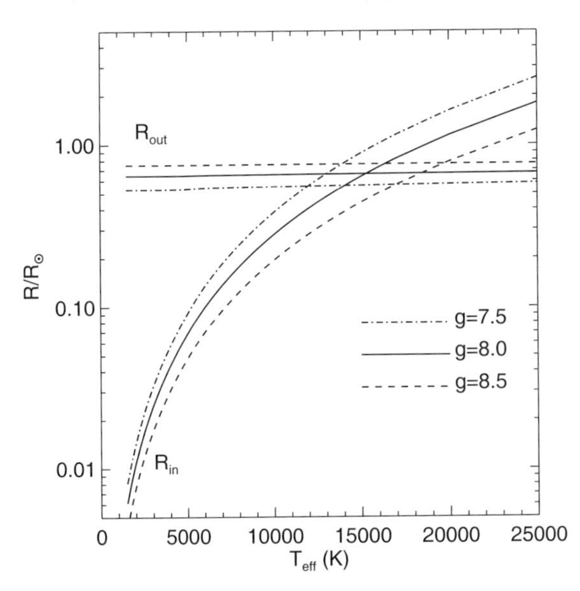

Figure 6.1 Dust disk inner and outer radii, R_{in} and R_{out}, versus white dwarf effective temperature, T_{eff}, for $\log g = 7.5$ (dashed-dotted line), 8.0 (solid line), and 8.5 (dashed line). In regions where $R_{in} > R_{out}$, all debris is sublimated within the tidal disruption radius and the resultant disks are expected to show evidence of strong gaseous emission.

Armed with these interior and exterior radii, one can derive order of magnitude conditions in white dwarf disks to provide a physical intuition for these systems. In particular, any forces acting on the dust, as well as the collisional evolution of dust in the disk, can be investigated.

Many of the forces acting on dust in the Solar System work around white dwarfs as well. In most cases, however, the only force of any consequence acting on grains this close to the white dwarf is Poynting–Robertson (PR) drag. For low luminosity white dwarfs, the drag timescale can be short compared to both human lifetimes and the white dwarf cooling time for the typical R_{in} and R_{out} values calculated above using (6.3) and (6.4). The PR drag timescale is given by

$$t_{PR} = \frac{c a_0^2}{4 G M_\star \beta} \approx 24 \left(\frac{0.6 M_\odot}{M_\star}\right) \left(\frac{a_0}{0.01\,\mathrm{AU}}\right)^2 \beta^{-1} \mathrm{days}\,, \tag{6.5}$$

where c is the speed of light, a_0 is the semimajor axis at which the grain resides, G is the gravitational constant, $M_\star$ is the mass of the white dwarf, and β is the ratio of radiation pressure to the gravitational force, which is given by

$$\begin{aligned} \beta &= \frac{3\, L_\star\, Q_{PR}}{16\, \pi\, G\, c\, M_\star\, r_{dust}\, \rho_{dust}} \\ &\approx 3.2 \times 10^{-4}\, Q_{PR} \left(\frac{L_\star}{0.001 L_\odot}\right) \left(\frac{0.6 M_\odot}{M_\star}\right) \left(\frac{1\,\mu\mathrm{m}}{r_{dust}}\right) \left(\frac{3\,\mathrm{g\,cm^{-3}}}{\rho_{dust}}\right)\,, \end{aligned} \tag{6.6}$$

where Q_{PR} is the efficiency with which the grain absorbs the momentum from light (Artymowicz, 1988; Chen and Jura, 2001; Farihi *et al.*, 2008; Jura, 2004).

In general, β is small for most white dwarfs and, therefore, t_{PR} is relatively long. Unlike debris disks around early type stars, radiation pressure generally cannot rapidly remove dust around a white dwarf. For example, assuming astronomical silicate composition dust (Draine and Lee, 1984; Laor and Draine, 1993), no grains reach $\beta = 0.5$ for white dwarfs with $T_{eff} < 25\,000$ K, the criterion for a hyperbolic orbit of a dust grain and, thus, ejection from the system. For white dwarfs, the PR drag timescale for 1 μm astronomical silicate grains at R_{in} is relatively constant. For 1 μm grains around white dwarfs with $\log g = 8.0$, $t_{PR} \sim 2$ yr (compare this to $t_{PR} \sim 10^{-3}$ yr for 1 μm grains around the Sun at a distance of $\sim 0.3 R_\odot$). Above $T_{eff} = 12\,000$–$15\,000$ K, where R_{in} extends beyond R_{out}, the shortest t_{PR} is determined by R_{out} and drops as T_{eff}^{-4}, becoming ~ 30 days at 25 000 K.

Even though these timescales are long compared to conditions around a solar type star, they have remarkable implications for a dust disk that resides between R_{in} and R_{out}. Grains that feel PR drag (i. e., those in an optically thin distribution) will be removed on very short timescales compared to both evolutionary changes (i. e., cooling) of the white dwarf as well as the cadence of human observations. This provides a strong argument for the presence of an optically thick distribution of dust, unlike more optically thin distributions (Reach *et al.*, 2005, 2009). With an optically thick disk, the upper layers of the disk provide the optically thin region where emission from silicate lines can be seen in the infrared and, no doubt, provide a steady stream of material onto the surface of the white dwarf through PR drag, collisions, and the viscous evolution of the disk itself (which will be discussed further in Section 6.3). PR drag alone, however, cannot account for the observed accretion rates onto the photospheres of white dwarfs (Farihi *et al.*, 2010).

Another important process is the collision of dust grains within the disk. For a $0.6 M_\odot$ white dwarf with $T_{\text{eff}} = 10\,000$ K and $R_{\text{in}} = 2 \times 10^{10}$ cm, the ratio of the PR drag timescale to the collisional timescale, γ, can be calculated as

$$\begin{aligned}\gamma &= \frac{c\, a_0^{1/2}\, \tau(r)}{8\pi\, (G M_\star)^{1/2}\, \beta} \approx 52 \left(\frac{\tau}{\beta}\right) \left[\left(\frac{a_0}{0.01\,\text{AU}}\right)\left(\frac{0.6 M_\odot}{M_\star}\right)\right]^{1/2} \\ &\approx 1.6 \times 10^5 \left(\frac{\tau}{Q_{\text{PR}}}\right)\left(\frac{0.001 L_\odot}{L_\star}\right)\left(\frac{r_{\text{dust}}}{1\,\mu\text{m}}\right)\left(\frac{\rho_{\text{dust}}}{3\,\text{g}\,\text{cm}^{-3}}\right) \\ &\quad \times \left[\left(\frac{a_0}{0.01\,\text{AU}}\right)\left(\frac{M_\star}{0.6 M_\odot}\right)\right]^{1/2}, \end{aligned} \tag{6.7}$$

where τ is the optical depth of the (optically thin) dust distribution (Farihi *et al.*, 2008; Rafikov, 2011). Once again, assuming 1 μm silicate grains, Figure 6.2 shows γ at R_{in} and R_{out} as a function of T_{eff} assuming $\tau = 0.01$.

Given the very short timescales for dust sublimation, collisions, and PR drag at the interior of the disks, it is interesting to consider just what structures can survive for long periods of time around white dwarfs. The atmospheric settling times for cool hydrogen and helium white dwarfs with no infrared excess that nonetheless show metal pollution can be as high as 10^4–10^5 yr (Koester, 2009; Pa-

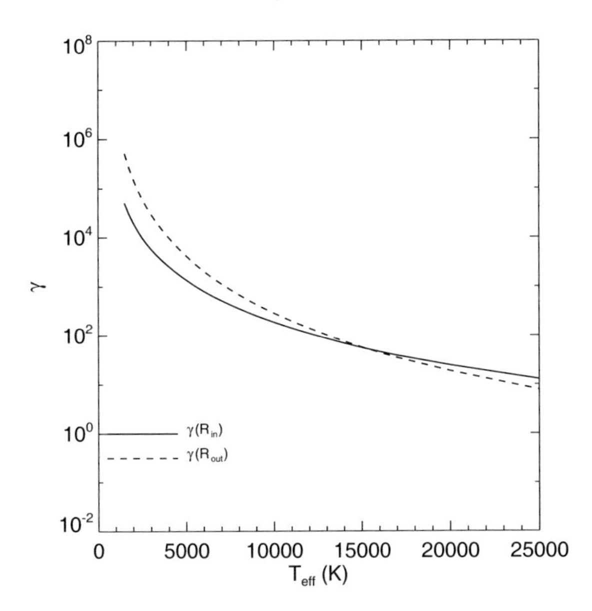

Figure 6.2 Ratio of the PR drag timescale compared to the collisional timescale for dust grains, γ (see (6.7)), as a function of effective temperature for log g = 8.0 white dwarfs. The calculation was performed for 1 μm grains located at the inner and outer radii of the disk (see Figure 6.1).

quette *et al.*, 1986), implying that material from disks must be accreted within Myr timescales (Jura, 2008; Kilic *et al.*, 2008).

6.3
Structure and Evolution of a White Dwarf Dust Disk

What is the correct structure for a white dwarf dust disk? Several general configurations have been proposed, although each one is fairly simple given the relative lack of observational constraints. A few disks have been successfully observed with the Spitzer Space Telescope Infrared Spectrograph (e. g., Jura *et al.*, 2009), and usually display a strong silicate emission feature. The presence of the silicate emission feature implies a population of optically thin, small dust grains, similar to those seen around primordial and transitional disks in young stars. However, the underlying dust continuum is well matched with an optically thick disk at narrow radii, confined to $R_{\text{in}} < R < R_{\text{out}}$. The optically thick disk is well motivated theoretically since it is the only structure that can realistically survive for longer than a few decades.

The optically thick, geometrically thin disk model was first proposed by Jura (2003). In this model, designated here as the "ring model", the grains passively reradiate the incoming white dwarf light back into the infrared and vertical structure in the disk is neglected. The temperature of the dust, T_{ring}, as a function of distance R from a white dwarf with radius $R_\star$ and effective temperature $T_\star$, is then given by

$$T_{\text{ring}} \simeq \left(\frac{2}{3\pi}\right)^{1/4} \left(\frac{R_\star}{R}\right)^{3/4} T_\star \,. \tag{6.8}$$

(Chiang and Goldreich, 1997).

Again, in analogy to young stars, the inner sublimation zone may be optically thin where dust temperatures are higher than the approximation of (6.8). This could also puff up the inner region of the disk with a hot, optically thin inner region (Dullemond *et al.*, 2001). However, for simplicity, these effects are ignored and the expected flux is calculated from the disk emission as

$$F_{\nu,\text{disk}} = 12\pi^{1/3} \frac{R_\star^2 \cos i}{D^2} \left(\frac{2k_{\text{B}} T_\star}{3h\nu}\right)^{8/3} \frac{h\nu^3}{c^2} \int_{x_{\min}}^{x_{\max}} \frac{x^{5/3}}{\text{e}^x - 1} \text{d}x \,, \tag{6.9}$$

where i is the inclination of the disk, k_{B} is Boltzmann's constant, h is Planck's constant, D is the distance to the white dwarf, and $x = h\nu / k_{\text{B}} T_{\text{ring}}$ (Jura, 2003). Figure 6.3 shows $R(T_{\text{ring}} = 1500\,\text{K})$ from (6.8) with R_{out} plotted as a function of T_{eff}. The figure also shows the inferred inner and outer ring radius under the optically thick model for a few known dusty white dwarfs. In general, the dust disks have inner radii consistent with being at the dust sublimation radius.

Figure 6.4 shows both the power and limitation of this model when compared to a full spectrum of G29-38, the canonical dusty white dwarf. In terms of matching

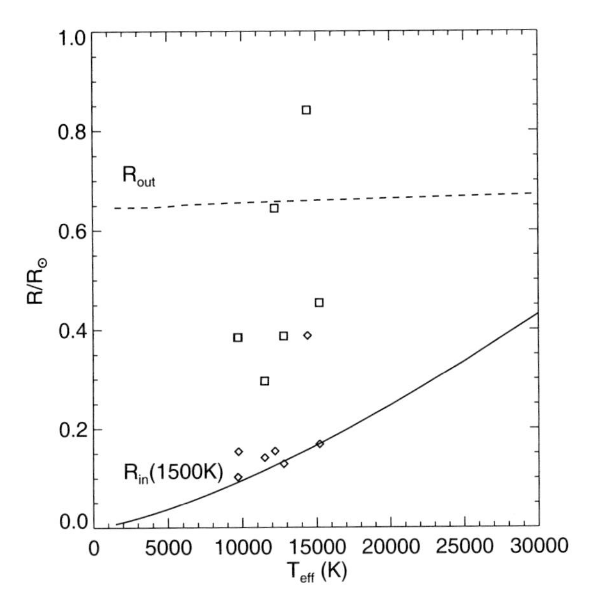

Figure 6.3 Inferred inner and outer disk radii for a subset of dusty white dwarfs taken from Jura *et al.* (2007a, 2009). Diamonds are the inferred inner disk radius, while squares are the outer optically thick disk radius, assumed to roughly correspond to the pericenter of the progenitor planetesimal that was tidally disrupted. The solid curve corresponds to the theoretical inner edge radius for an optically thick disk where the grain temperature reaches 1500 K.

the overall continuum of the infrared excess, the model does remarkably well for only three free parameters, implying an inner temperature of 1600 K, and an outer temperature of 800 K, corresponding to $R_{\text{in}} = 12 R_\star$, and $R_{\text{out}} = 30 R_\star$. Compare this result to a tidal radius of $\sim 60 R_\star$, and it is clear that G29-38's disk fits within the expected maximum radius. However, this model fails to reproduce the emission from the silicate line, primarily because no composition for the dust is assumed; that is, the model utilizes blackbody thermal continuum emission only.

To accurately explain both the silicate feature and the overall thermal continuum of the disk, Jura *et al.* (2007b) calculated a "warped ring" model, where the outer portions of the disk are puffed up due to the irradiation from the central star. In the "warped ring" model, there are three zones. In the inner zone, the disk is optically thick, thermal conduction between gas in the disk and dust is efficient and, thus, the disk is vertically isothermal. The temperature distribution and, hence, the emission follow the "ring model". In a second zone, thermal conduction may be inefficient and temperature gradients can form in the vertical structure of the disk. In the final zone, the disk becomes optically thin to infrared radiation and potentially warps due to the irradiation from the white dwarf (further enhancing the amount of light intercepted from the white dwarf). Using this prescription, Jura *et al.* (2007b) calculated the temperature of the grains in the disk and from that,

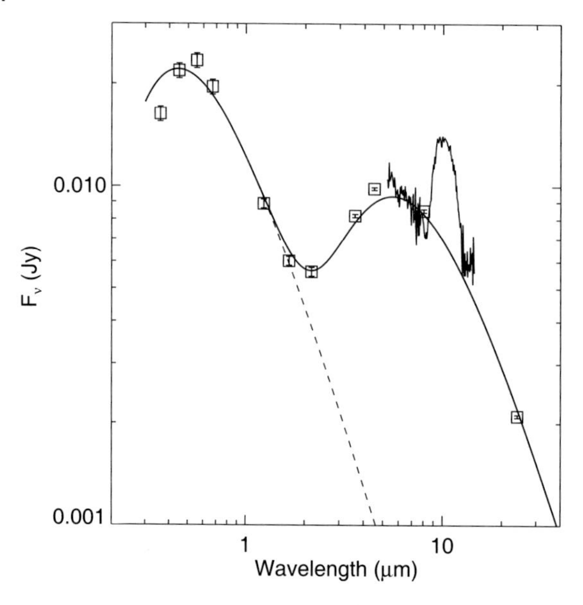

Figure 6.4 Spectral energy distribution of G29-38 and its dusty disk (data from Reach *et al.*, 2005, 2009), fit with a model of a geometrically flat, optically thick ring. The white dwarf is approximated with a black body curve, and the ring model has inner and outer radii of 10 and $30R_\star$, respectively.

the infrared emission from GD 362. The resulting emission fit the overall thermal continuum as well as the silicate emission line (see Figure 6.5).

For G29-38 and six other white dwarfs with observed silicate lines, Jura *et al.* (2009) and Reach *et al.* (2009) introduced similar models that effectively had two zones, an optically thick disk (physically thick in Reach *et al.*, 2009) and then an outer optically thin region (with warps in Jura *et al.*, 2009). This structure is equivalent to an optically thick disk with an optically thin outer region. Figure 6.6 shows cartoons of the possible disk configurations.

One could proceed to more complicated models; for example, those calculated for young primordial and transitional disks, which also have inner sublimation regions, puffed up inner walls, and shadowing (e. g., Calvet *et al.*, 2002; Dullemond *et al.*, 2001). To date, no one has implemented a full version of these specific prescriptions for a white dwarf dust disk to determine whether such a model can successfully reproduce the observed emission.

Ultimately, the major problem amongst all of these models is a lack of "ground truth". In all cases of spectral energy distribution modeling, a three-dimensional structure of unknown composition is folded into a one dimensional spectrum, one which might include at most three or four photometric points. The implicit degeneracies and systematic problems of such a situation are well illustrated in testing of young circumstellar disk modeling (Juhász *et al.*, 2009). The more simple models may be preferred until other observations become available that can provide additional constraints. Typically, such degeneracies are broken by spatially resolving

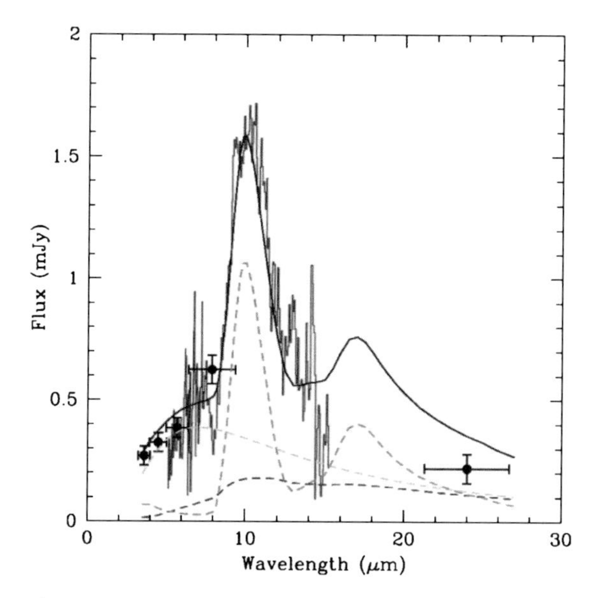

Figure 6.5 Comparison between photosphere-subtracted observations of GD 362 and predicted flux densities for a three-zone model dust disk. The data consist of mid-infrared photometric points and spectrum (thin line) from Spitzer. The dark line shows the total model, which is the sum of the contributions from the three zones (dashed lines). From Jura *et al.* (2007b), reproduced with the permission of the AAS.

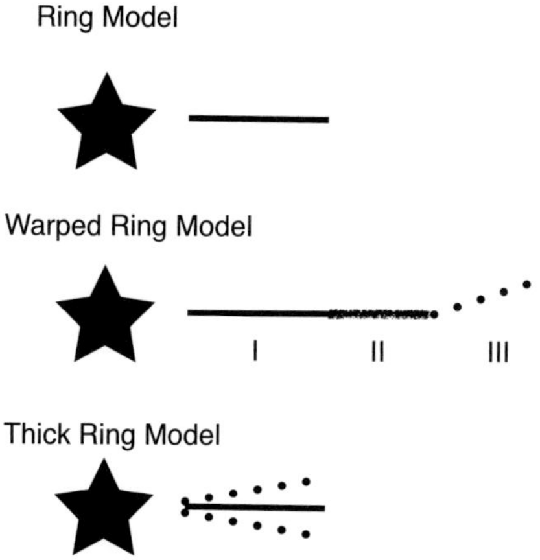

Ring Model

Warped Ring Model

I II III

Thick Ring Model

Figure 6.6 Schematic diagram of the different proposed structures of dusty white dwarf disks. In the "warped ring model", there are three distinct zones: an isothermal optically thick zone (I), an optically thick zone with vertical thermal gradients (II), and an optically thin zone with a warp (III) (Jura *et al.*, 2007b). The thick ring model depicts the possibility that an optically thin upper layer may be present in white dwarf disks. This model produces emission roughly equivalent to a warped ring model with no zone II. Also, see Figure 6 in Reach *et al.* (2009).

the disk – in scattered light, thermal emission, or interferometric measurements. In the case of white dwarfs, however, the prospects for such observations are limited. All of these techniques face immense challenges due to the compact nature of these systems – $1 R_{\odot}$ at a distance of 20 pc corresponds to 4 μas.

Conversely, the fact that G29-38 is a pulsating white dwarf provides unique constraints on its disk, which can in turn inform spectral energy distribution modeling of white dwarfs. This is demonstrated in the work by Graham *et al.* (1990), in which observations of pulsations in the visible and near-infrared were used to determine pulsational modes in the disk thermal emission. In that work, modes at 181 and 243 s were discovered, compared to the visible mode of 614 s. The 181 s mode is confirmed and its relative stability demonstrated by a similar mode detected in Spitzer Infrared Array Camera observations of G29-38 obtained nearly 20 years later (Reach *et al.*, 2009). The detection of particular modes in the nonradial pulsations of G29-38 provide some constraints on the inclination and geometry of the disk. Clearly, however, more work is needed in this area.

6.3.1 Optically Thin Dust Disks?

As a thought experiment, simple steady state models of optically thin, collisional disks can be constructed to determine whether they are observable. A simple one-dimensional model of a collisional disk with a single grain size can be given analytically in terms of the effective optical depth, τ_{eff} (Wyatt *et al.*, 1999). Under the assumption of completely destructive collisions, the distribution of $\tau_{\text{eff}}(r)$ is then

$$\tau_{\text{eff}}(r) = \frac{\tau_{\text{eff}}(R_{\text{out}})}{1 + 4\eta_{\text{o}}\left(1 - \sqrt{r/R_{\text{out}}}\right)} , \tag{6.10}$$

where

$$\eta_{\text{o}} = \frac{5000\tau_{\text{eff}}(R_{\text{out}})}{\beta}\left[\left(\frac{R_{\text{out}}}{1\,\text{AU}}\right)\left(\frac{1 M_{\odot}}{M_{\star}}\right)\right]^{1/2} , \tag{6.11}$$

(Wyatt, 2005). Assuming that R_{out} is the location where dust is being produced by a large reservoir of tidally disrupted material, one can then calculate the expected infrared excess for given effective optical depths. Figure 6.7 shows the rather hopeless case for observing optically thin dust around a $T_{\text{eff}} = 15\,000$ K white dwarf.

6.3.2 Subsequent Evolution of the Dust Disk

Under the assumption that most white dwarf dust disks must be optically thick to survive up to 10^4–10^6 yr, once a disk is formed, then it will behave primarily as a viscous fluid, not unlike Saturn's rings; the viscous lifetime, t_{visc}, is given by

$$t_{\text{visc}} = \frac{R^2}{\nu} , \tag{6.12}$$

where ν is the viscosity of the disk. For a dusty ring with no gas like Saturn's rings, $\nu = \Omega \tau r_{\nu}$, where Ω is the angular speed at a distance R from the white dwarf,

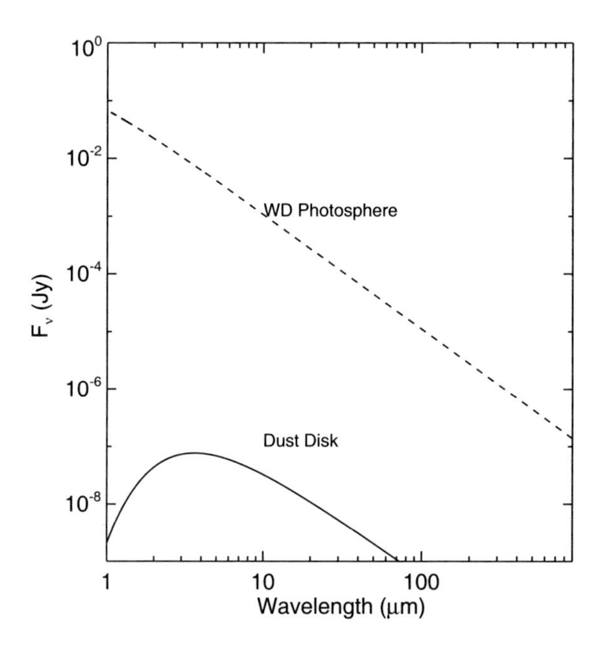

Figure 6.7 Infrared emission from an optically thin dust disk (solid line) around a 15 000 K white dwarf (dashed line), following the collisional models of Wyatt (2005). Optically thin dust disks will be exceedingly difficult to observe around white dwarfs under the assumed collisional and PR drag conditions.

τ is the dust optical depth, and r_ν is a characteristic particle size (Goldreich and Tremaine, 1982). If Saturn's ring viscosity is assumed for a white dwarf disk, then it is, in effect, the same as assuming that the optical depth and characteristic particle size are similar to those of Saturn (von Hippel *et al.*, 2007). From density wave measurements on the surface of Saturn's rings, estimates of the local viscosity can be made. The viscosity varies over three orders of magnitude, but clusters around $100\,\text{cm}^2\,\text{s}^{-1}$ (Tiscareno *et al.*, 2007). Taking that value for $R = 0.5R_\odot$ in (6.12) produces viscous lifetimes in excess of 10^{10} yr, much longer than the typical cooling times with which dusty white dwarfs are observed. Using the longest cooling lifetime as an upper limit on the lifetime of a dust disk around a white dwarf, ~ 1 Gyr, then the viscosity must be about two orders of magnitude higher than in Saturn's rings. This, of course, assumes that the lifetime of the disk is based solely on the viscous evolution of the dust. Assuming pure viscous evolution of the dust ignores the removal of grains through PR drag on the surface, sublimation at the inner edge, and whatever collisions may occur. These effects will hasten a dust disk's demise. The only observational evidence for the geometry and nature of the accretion comes from recent work with the Keck and Whole Earth Telescope observations of the Ca II line in G29-38 that suggests an accretion geometry mainly at the poles of the white dwarf, implying magnetic accretion of ionized gas onto the surface (Thompson *et al.*, 2010).

Once the dust has been either destroyed or accreted, a gaseous disk may persist. Conversely, in the case where $R_{in} > R_{out}$, a gaseous disk will form from the flash sublimation of all dust within the tidal disruption radius of the white dwarf. In these cases, the viscous lifetime of the disk, t_{gas}, can be approximated by assuming an α-disk model and setting $\nu = \alpha v_{orb}/c_s^2 R$, where v_{orb} is the orbital velocity of the gas and $c_s \sim k_B T/m_{gas}$ is the local sound speed in the gas with mean atomic weight m_{gas} (King *et al.*, 2007). The lifetime for a gaseous disk is then given by

$$t_{visc,gas} \sim R \frac{v_{orb}}{\alpha c_s^2} , \tag{6.13}$$

(Jura, 2008). The viscosity parameter α is highly uncertain and could vary by up to two orders of magnitude, but $\alpha \sim 0.001$ provides a timescale for a gaseous disk that is plausible (Jura, 2008).

The implications of assuming purely dusty or purely gaseous disk evolution can be used to back out a surface density of each kind of disk by means of the observed mass accretion of metals onto the white dwarfs. The viscous mass accretion of dust or gas onto the photosphere is

$$\dot{M} = 3\pi\Sigma\nu , \tag{6.14}$$

where Σ is the surface density of the disk (Alibert *et al.*, 2004). An estimate of Σ can be obtained by using observed metal accretion rates of white dwarfs, assuming radii comparable to R_{out}, and combining (6.12) and (6.14). Dusty white dwarfs show accretion rates of $\sim 10^{8.5}\,\mathrm{g\,s^{-1}}$ (Koester and Wilken, 2006), implying surface densities of $3 \times 10^3\,\mathrm{g\,cm^{-2}}$ for $\nu = 10^4\,\mathrm{cm^2\,s^{-1}}$ and $3 \times 10^5\,\mathrm{g\,cm^{-2}}$ for $\nu = 100\,\mathrm{cm^2\,s^{-1}}$. This surface density is two to three orders of magnitude greater than that observed for Saturn's rings (Tiscareno *et al.*, 2007). The implied mass is $\pi\Sigma R^2 \sim 10^{25}$–$10^{27}$ g, in the range of Lunar to Martian masses. Perturbing a single large body of this size inside the white dwarf tidal disruption radius is highly unlikely, unless the cores of hot Jupiters can survive the post-main sequence evolution of the white dwarf progenitor and remain in situ (Reach *et al.*, 2009). It is not realistic to approximate the dust viscosity in this scenario as that of Saturn's rings. In the case of white dwarfs, there is most likely a gaseous region that dominates the accretion region from R_{in} down to the white dwarf photosphere, mediated by the white dwarf's magnetic field. The accretion rate onto the photosphere is then regulated by the viscosity in this gaseous region.

The disks around white dwarfs that do not display an infrared excess are presumed to be optically thin and dominated by gas, that is, their accretion rates are in the range 10^6–$10^8\,\mathrm{g\,s^{-1}}$. The parameter ν is inferred by equating (6.13) to (6.12). Taking $R \sim 3 \times 10^{10}$ cm as a characteristic size for a gaseous disk and assuming a temperature of 300 K results in a gaseous disk lifetime of 5×10^4 yr, $\nu = 5 \times 10^8\,\mathrm{cm^2\,s^{-1}}$, and $\Sigma = 2 \times 10^{-4}$–$2 \times 10^{-2}\,\mathrm{g\,cm^{-2}}$ for accretion rates of 10^6–$10^8\,\mathrm{g\,s^{-1}}$. The mass in gas implied by these surface densities is 10^{18}–10^{20} g, equivalent to planetesimals with radii of 5–20 km (assuming a density similar to Ceres).

Eventually, all of the gas and grains from a collision will accrete onto the white dwarf surface until another planetesimal is perturbed into a disrupting orbit. Over a longer time, enough planetesimals are depleted from the system that observable disruption events are essentially terminated. The approximate timescale for this to occur can be linked to the oldest metal polluted white dwarfs observed, at effective temperatures of $\sim$ 5000 K, which, for a $\log g = 8$ white dwarf, corresponds to a cooling age of $\sim$ 5.7 Gyr.

6.4
Origins of White Dwarf Dust Disks

As stated in Section 6.1, the accepted scenario for the origin of white dwarf dust disks is that a surviving planet or planets perturbs an asteroidal body to within the tidal disruption radius, creating the observed dust, and potentially explaining the presence of metal polluted white dwarfs. Hereafter, this is referred to as the "Unstable Planetary Perturbation" (UPP) model. The works of Debes *et al.* (2002) and Jura (2003) form the cornerstone of this model, augmented by calculations for the survival of asteroids by Jura (2008).

6.4.1
The Unstable Planetary Perturbation Model

The survival of Jovian planets in orbits $>$ 5 AU during post-main sequence evolution is well documented (Duncan and Lissauer, 1998; Villaver and Livio, 2007). If a large population of such planets exist, then they will be the dominant perturbers of relic planetesimals. The UPP model requires that there be at least two planets in the system that become unstable to close approaches, which in turn leads to dynamical rearrangements, outright ejection, and in a small subset of cases, collisions between planets (Ford *et al.*, 2001). The mass loss of the central star during post-main sequence evolution primes the planetary systems for chaotic behavior later in the white dwarf's cooling lifetime.

Two Planet Systems

The critical measure of stability in a two planet system is formally defined within the framework of stability against close approaches. In other words, the orbits of the two objects are considered stable if they remain well separated. One can measure the relative separation of semimajor axes, $\Delta = (a_2 - a_1)/a_1$, where a_1 and a_2 are the inner and outer semimajor axes, respectively. A critical Hill separation, Δ_c, is then the minimum separation between two planets that ensures a lack of close approaches over all time (Hill, 1886). A full treatment of the Hill stability of two planets in the case of static masses can be found in Gladman (1993). Several approximations can be made that simplify the full treatment; for example, equal planetary masses and small eccentricities. The criterion is then given by

$$\Delta_c \simeq \sqrt{\frac{8}{3}\left(e_1^2 + e_2^2\right) + 9\mu^{2/3}}\,, \tag{6.15}$$

where μ is the ratio of the planets' combined mass to that of the central star, e_1 and e_2 are the orbital eccentricities of planets one and two, and Δ_c is in units of the inner planet's semimajor axis, a_1.

If either the mass of the planets or the mass of the star changes, the critical Hill radius will change as well. An increase in planet mass or a decrease in stellar mass will cause μ to become larger, increasing the width of the zone in which orbits are unstable to close approaches. During post-main sequence mass loss, the orbits of planets will widen as the central star loses mass. As long as this process is adiabatic, the planets will simply conserve their angular momentum and widen their orbits in proportion to the mass lost, $a_{\text{final}} = a_{\text{initial}}(M_{\star,\text{initial}}/M_{\star,\text{final}})$. However, since the orbits widen together by the same factor, Δ_c remains the same. Thus, while the critical separation at which the two planets will become unstable widens, their relative separation remains unchanged (Debes *et al.*, 2002). Orbits that are initially marginally stable, or close to being unstable, will become unstable to close planet–planet approaches as a consequence of the mass loss from the central star. This scenario is the physical basis for understanding how formerly stable planetary systems can become unstable.

If the two planet scenario is the origin of dusty white dwarfs, then the problem becomes the timescale for the dynamical erosion of asteroidal material into the tidal disruption radius of the white dwarf after a close approach occurs. Close approaches usually occur within a few to a few thousand orbital periods, which for a planetary system with a Jupiter analogue at some initial semimajor axis a_{initial} would mean

$$\tau \sim 10^3 \left(a_{\text{initial}} \frac{M_{\star,\text{initial}}}{M^2_{\star,\text{final}}}\right)^{3/2} \text{yr}\,. \tag{6.16}$$

If the progenitor is assumed to have lost a factor of ~ 2 in mass, then dusty white dwarfs with cooling ages of ~ 1 Gyr require two planet systems in which the initial semimajor axis is $\sim 10^3$ AU – clearly an unlikely situation. Two planet systems that become unstable will only be important for the youngest and, therefore, hottest white dwarfs.

Multiple Planet Systems

As of mid-2010, roughly 16 of the 395 known planetary systems have more than two planets and, thus, it is useful to develop an idea of how these systems do or do not remain stable under the influence of mass loss of the central star. Chambers *et al.* (1996) found a relation between the separation of a system of planets and the time it would take for the system to suffer a close encounter,

$$\log t = b\delta + c\,, \tag{6.17}$$

where b and c are constants derived through numerical simulations. The parameter δ is related to Δ (from (6.15)), but is defined in a slightly different way. Here, δ is the separation between two planets $(a_{i+1} - a_i)$ in units of mutual Hill radii (R_i) defined as

$$R_i = \left(\frac{\mu_{i+1} + \mu_i}{3}\right)^{1/3} \frac{(a_{i+1} + a_i)}{2} , \tag{6.18}$$

where i can be from one to $(N_{\text{planets}} - 1)$, and planets are assumed to have equal masses and initially circular orbits. If the parameter δ is the same for each pair of adjacent planets, then the separations in units of AU will be different. For example, in the case of three Jovian mass planets ($\mu \sim 10^{-3}$) with $\delta = 6.5$, and the innermost planet located 5.2 AU from the central star, then the next two planets would be located at 9.4 and 16.7 AU. This compares favorably to the actual orbital radii of 9.6 and 19.1 AU for Saturn and Uranus, respectively, in the Solar System, which has presumably been stable for several billion years. The obvious caveat to this comparison is that Saturn and Uranus are significantly less massive than Jupiter and have correspondingly weaker mutual interactions.

Adiabatic mass evolution will have the effect of shortening the time it takes for orbits to suffer close approaches. The knowledge of this has long been used to speed up numerical calculations (Duncan and Lissauer, 1997, and references therein). However, this fact also leads to the hypothesis that planetary systems on the edge of stability for 10^{10} yr will be affected by mass loss. In general, the new time to close approaches for an initial δ with a change in mass is given by

$$\log\left(\frac{t_{\text{final}}}{t_{\text{initial}}}\right) = (b' - b)\delta + (c' - c) . \tag{6.19}$$

The value of c should be expected to have little or no change with a change of mass since it represents the timescale for two planets at $\delta \sim 0$ to suffer a close approach. Mass loss will increase the mutual Hill radii of the planets, which, in turn, will change b to a new value, b', defined here as

$$b' = \left(\frac{\mu_{\text{final}}}{\mu_{\text{initial}}}\right)^{1/3} b , \tag{6.20}$$

where μ_{final} and μ_{initial} are the final and initial mass ratios, respectively. Assuming that (6.17) holds for large values of δ, this suggests that planets that remain stable for the duration of the main sequence lifetime of their parent star will become unstable on a timescale several orders of magnitude shorter after the parent star becomes a white dwarf. This relation holds when orbital separations deviate from exact δ separations and when masses vary by factors of five, suggesting that this type of relation should also hold for systems in which the planetary masses are not equal (Chambers *et al.*, 1996).

This scenario is most powerful because it can account for the presence of instability at any time during a white dwarf's cooling age and, thus, allows for the existence of dusty systems up to Gyr after the change in the central star's mass. However, the large frequency of known white dwarf systems may be at odds with the observed frequency of planetary systems. Furthermore, resonances are often observed around mature planetary systems that enhance their stability; for example, for the planets around HR 8799 (Fabrycky and Murray-Clay, 2010). This scenario also does not work for the Solar System, which is stable against major close approaches between the gas giants for Gyr after the Sun will leave the main sequence, even under the acceleration of mass loss (Batygin and Laughlin, 2008; Duncan and Lissauer, 1998). Unfortunately, despite the large sample of known exoplanets, the completeness of these systems to multiple companions is still not enough to make a direct comparison or to determine how resonances or the general architecture of planetary systems might complicate this mechanism.

Survival of Asteroids and Comets

For the UPP model, or any model that invokes asteroidal or cometary material, to work, a significant reservoir of asteroids and comets must remain after the main sequence evolution of the white dwarf progenitor. The main mechanisms that remove both comets and asteroids occur during post-main sequence evolution, when the central star has evolved into an extremely luminous giant, and is losing large amounts of mass in the form of a slow wind.

Both evaporation and drag from the giant's slow wind act to remove the closest and smallest asteroids. The typical assumption for gas drag is that when the asteroid hydrodynamically captures its mass in gas, it will have lost enough angular momentum to move inward a large enough distance to be removed. This condition is given by

$$R_{\text{ast,min}} = \frac{3G^{1/2} M_{\star,\text{initial}}^{3/2}}{64\pi a_{\text{initial}}^{5/2} \rho_{\text{ast}} V_{\text{wind}}}$$
$$\approx 1.3 \left(\frac{M_{\star,\text{initial}}}{1 M_\odot}\right)^{3/2} \left(\frac{1\,\text{AU}}{a_{\text{initial}}}\right)^{5/2} \left(\frac{3\,\text{g}\,\text{cm}^{-3}}{\rho_{\text{ast}}}\right) \left(\frac{10\,\text{km}\,\text{s}^{-1}}{V_{\text{wind}}}\right) \text{km}\,, \tag{6.21}$$

where $R_{\text{ast,min}}$ is the radius of the smallest surviving asteroid with density ρ_{ast}, $M_{\star,\text{initial}}$ is the initial mass of the white dwarf progenitor star, a_{initial} is the initial orbital radius of the asteroid around the star, and V_{wind} is the speed of the wind from the star (Duncan and Lissauer, 1998; Jura, 2008).

The sublimation timescale for an asteroid, assuming an olivine composition, is given in (6.1). Thus, a minimum asteroid size can be determined by rearranging terms and solving for the radius given a particular red giant branch (RGB) or asymptotic giant branch (AGB) lifetime and luminosity for the progenitor star. Figure 6.8 shows curves of minimum radii of surviving planetesimals at various orbital semimajor axes for both gas drag and sublimation.

The specific orbital evolution of a planetesimal in the wind is more complex since the central stellar mass is decreasing at the same time. The force acting on a sublimating and gas accreting planetesimal experiencing gas drag within the changing

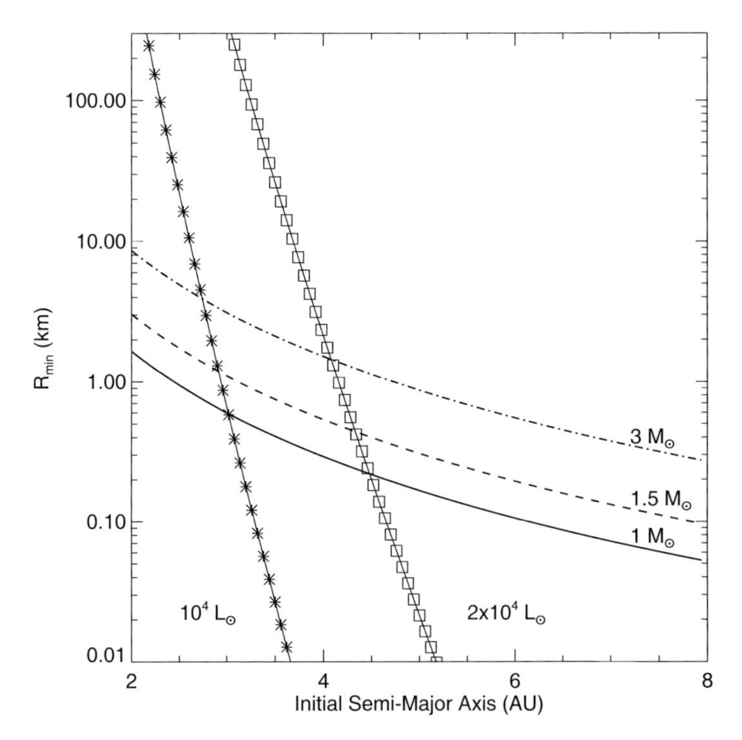

Figure 6.8 Radius of the smallest surviving planetesimals during the post-main sequence evolution of the parent star due to gas drag for stars with initial masses of $1 M_\odot$ (solid line), $1.5 M_\odot$ (dashed line), and $3 M_\odot$ (dash-dot line), and sublimation during the AGB and RGB phases for stars with luminosities of $10^4 L_\odot$ (asterisks) and $2 \times 10^4 L_\odot$ (squares).

potential of a mass losing central star is

$$\boldsymbol{a} = -\frac{G M_\star(t)}{r^2}\hat{r} - \frac{1}{2}\pi C_\mathrm{d} r_\mathrm{ast}(t)^2 \rho_\mathrm{wind}(\boldsymbol{v}_\mathrm{wind} - \boldsymbol{v}_\mathrm{ast}) \,, \tag{6.22}$$

where $M_\star(t)$ is the mass of the central star at time t, r and $\hat{r}$ denote the radial position of the asteroid relative to the central star, C_d is a drag term of order unity, $r_\mathrm{ast}(t)$ is the radius of the sublimating planetesimal at time t, ρ_wind is the density of the wind from the star, $\boldsymbol{v}_\mathrm{wind}$ is the velocity vector of the wind, and $\boldsymbol{v}_\mathrm{ast}$ is the velocity vector of the planetesimal (e. g., see Jura, 2008 and Dong *et al.*, 2010).

Even if an asteroid survives the entire post-main sequence evolution of its star, its final orbit may be quite different from its main sequence location. This has the effect of temporarily enhancing the perturbation rate of planetesimals early in the post-main sequence lifetime of the parent star, as they are shuffled around and either captured or ejected by resonances with any surviving planetary systems. This perhaps explains the presence of dusty excesses around the central stars of planetary nebulae (Dong *et al.*, 2010). This phenomenon will, however, be short-lived as planetesimals are quickly removed from these resonances. Regardless, the results of (6.1) and (6.21) point to the survival of relatively large ($R_\mathrm{ast,min} \gtrsim 1$ km) planetesimals at orbital distances comparable to the asteroid belt in the Solar System.

For icy planetesimals with orbits in the outer regions of planetary systems, sublimation dominates their evolution due to the high luminosity of the RGB and AGB evolution of their central star. Stern *et al.* (1990) showed that pure ice planetesimals will survive only beyond a few hundred AU from their central star. In reality, objects like the Solar System's Kuiper Belt Objects are more complex, with densities that range from highly porous to values consistent with rocky cores. Large planetesimals might survive post-main sequence evolution with only their rocky cores remaining, providing an additional reservoir for tidally disrupted material.

For icy planetesimals at very large separations (i. e., in Oort cloud-like orbits), removal can occur due to asymmetric mass loss during the RGB and AGB phases. "Kicks" from this process can be significant (Heyl, 2007) and can remove up to half or all of a relic Oort cloud reservoir (Parriott and Alcock, 1998).

Assessing the UPP model

The UPP model provides a framework for thinking about how dusty white dwarfs arise: planetary systems somehow perturb surviving planetesimals close to the white dwarf, creating dust and metal accretion within a solar radius of the central star. The UPP model does not rigorously explain how planetesimals are disrupted, how successfully and over what timescale debris from the disruption settles into a mature disk, and how long is the lifetime for that material. Another potential weakness is the reliance on unstable planetary systems. If other planetary systems are generally similar to the Solar System, then few systems will be unstable enough to create significant perturbations for the observed white dwarfs. At least 2% of all white dwarfs show dusty disks (Farihi *et al.*, 2008), and $\sim$ 25% of white dwarfs show metal accretion. Interestingly, the integrated probability of a planetary system around 1–2$M_\odot$ stars, the primary progenitor population of observed nearby white dwarfs, is $\sim$ 24%, strikingly similar to the fraction of metal polluted white dwarfs. Yet, given the stability that the Solar System will maintain beyond the post-main sequence evolution of the Sun, it is difficult to accept that $\sim$ 100% of planetary systems will become unstable.

6.4.2
Dust and Accretion, or Just Accretion?

There is a large population of directly accreting white dwarfs, whose members have metal absorption lines in their atmospheres yet show no observational evidence for a dust disk (e. g., infrared excess). Their position in the sky and space velocity do not agree with an interstellar medium (ISM) accretion scenario, nor do they have companions down to substellar masses (Aannestad *et al.*, 1993; Debes, 2006; Farihi *et al.*, 2010; Kilic and Redfield, 2007; Zuckerman *et al.*, 2003). Despite their apparent lack of disks, the majority of these white dwarfs may well also be accreting material from disrupted asteroids. An elegant explanation for why some dust disks don't arise comes from Jura (2008). In this scenario, multiple smaller asteroids are perturbed into tidal disruption.

The time-averaged metal accretion rate onto the white dwarf surface, $\langle \dot{M}_{\text{metal}} \rangle$, from dust or gas in a disk created via tidal disruption of an asteroid is be given by

$$\langle \dot{M}_{\text{metal}} \rangle = \frac{M_{\text{belt}}(t)}{t_{\text{orbit}}} , \quad (6.23)$$

where $M_{\text{belt}}(t)$ is the total mass of the asteroidal material in the white dwarf system as a function of time t, and t_{orbit} is the typical lifetime of a single asteroid before it is perturbed into tidal disruption (Jura, 2008). The initial tidal disruption event will result in a dusty disk that eventually evolves through grain-grain collisions into a gaseous disk, while the remainder of the surviving dust is accreted onto the white dwarf. Subsequent disruptions will encounter the gas and any remaining dust, which will quickly sputter away the incoming dust, provided that the mass of incoming material is less than that of the gaseous disk. In this case, the mass of the gaseous disk becomes

$$M_{\text{disk}} = \langle \dot{M}_{\text{metal}} \rangle \, t_{\text{gas}} , \quad (6.24)$$

where t_{gas} is the characteristic lifetime of the gaseous disk. The value of t_{gas} is highly uncertain, but Jura (2008) argued for a timescale of $\sim 10^4$ yr based on the fact that gas disks must be long lived to explain the fact that $> 20\%$ of white dwarfs show evidence of metal accretion. Therefore, for the white dwarf to show an appreciable infrared excess from dust, a disrupted asteroid must have a mass of $\sim M_{\text{disk}}$ in order to dominate the evolution of the disk material.

For an asteroid belt with a size distribution, one can estimate the relative fraction of infrared excess versus non-excess sources in the population, f_{IR}. This fraction depends on the lifetime of dusty disks relative to gaseous disks ($t_{\text{dust}}/t_{\text{gas}}$) and the minimum ($M_{\text{min}}$) and maximum ($M_{\text{max}}$) masses of the asteroids in the hypothetical belt,

$$f_{\text{IR}} = \left(\frac{t_{\text{dust}}}{t_{\text{gas}}} \right) \left(\ln \frac{M_{\text{max}}}{M_{\text{min}}} \right)^{-1} . \quad (6.25)$$

(Jura, 2008).

If the minimum mass of the asteroid belt is fixed at the smallest radius that survives post-main sequence evolution (~ 3 km), $M_{\text{max}} \sim 0.2 M_{\text{belt}}(t = 0)$, and assuming that $(t_{\text{dust}}/t_{\text{gas}}) = 3$ and the asteroid belt is not severely depleted, then $f_{\text{IR}} \sim 0.19$. This fraction is about the same as the current empirical ratio of observed dusty disks (18 systems) relative to metal-polluted white dwarfs (64 systems; Farihi *et al.*, 2009, 2010). If it is assumed that

$$M_{\text{belt}}(t) = M_{\text{belt}}(0) \, \mathrm{e}^{-t/t_{\text{orbit}}} , \quad (6.26)$$

then one can determine the expected accretion rate beyond which a dust disk should form and persist for t_{dust}. Figure 6.9 shows a comparison of the data from Farihi *et al.* (2010) with the expected division line as a function of cooling age. Most disk systems lie above the critical accretion rate line, but some non-disk systems

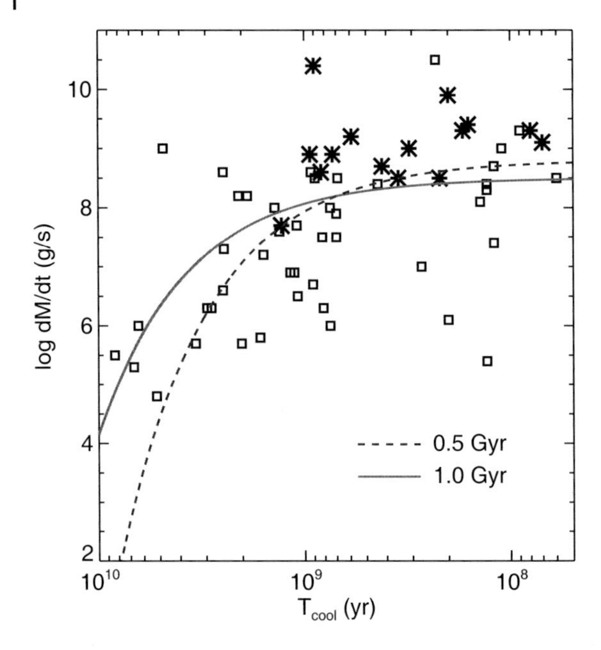

Figure 6.9 Accretion rate in metal enriched white dwarfs as a function of inferred cooling age (data taken from all metal rich white dwarfs observed with Spitzer; Farihi *et al.*, 2010 and references therein). Squares represent white dwarfs without infrared excess, while asterisks represent dusty white dwarfs. Two models for the predicted dividing line between non-disk and disk systems are shown, assuming characteristic times for perturbation of planetesimals of $10^{8.5}$ yr (dashed line) and 10^9 yr (solid line), and assuming a total planetesimal belt mass of 10^{25} g (see (6.23)).

do as well. These systems could either have more massive asteroid belts and, thus, require larger asteroids for the white dwarf to display a dusty excess, or have dust disks that have evolved much quicker than the metal settling time in the white dwarf atmospheres.

The metal polluted white dwarf WD 1257+278, in particular, seems discrepant – it has a large accretion rate of $10^{8.5}\,\mathrm{g\,s^{-1}}$, yet has a relatively cool $T_{\mathrm{eff}} = 8500\,\mathrm{K}$, placing it well beyond the predictions. The settling timescale for WD 1257+278 is $\sim$ 2000 yr (Koester and Wilken, 2006), making it plausible that the dusty disk may have recently dissipated. If that is the case, then its accretion may be out of equilibrium and detectable through the relative abundances of different atomic species (Koester, 2009).

6.4.3
The Mean Motion Perturbation Model

In the years since the UPP model was first proposed, few mechanisms have been proposed for explaining how planetesimals can be perturbed. One possibility is that a single planet perturbs asteroids through mean motion resonances and some fraction of the asteroids that are removed impinge within the tidal disruption radius. This scenario is designated as the Mean Motion Perturbation (MMP) model.

The Kirkwood gaps of the Solar System's asteroid belt are regions where asteroids are quickly removed due to gravitational interactions with Jupiter. Within mean motion resonances, an asteroid's eccentricity random walks until it interacts with a planet (or planets) and is ejected, collides with a planet, or collides with the central star (Gladman *et al.*, 1997; Morbidelli, 1996). The width of such gaps, $\delta a_{\max}$, can be roughly approximated by the maximum libration width for a particular resonance, $p + q{:}q$, in the restricted three body case and assuming low eccentricities,

$$\delta a_{\max} = \pm \left(\frac{16}{3} \frac{|C_r|}{n} e \right)^{1/2} \left(1 + \frac{1}{27 j_2^2 e^3} \frac{|C_r|}{n} \right)^{1/2} - \frac{2}{9 j_2 e} \frac{|C_r|}{n} a \,, \tag{6.27}$$

where C_r is a constant from the resonant part of a disturbing function in the restricted three-body case, n is the mean motion of the planetesimal, e is the eccentricity of its orbit, and j_2 is a direct term in the expansion of the restricted three-body case (Murray and Dermott, 2000). The value of C_r/n is given by

$$C_r/n = \mu \alpha_o |f_d(\alpha_o)| \,, \tag{6.28}$$

where α_o is the ratio of semimajor axes, $a_{\text{ast}}/a_{\text{planet}}$, and $f_d(\alpha_o)$ is another term of the restricted three-body case expansion.

For the 2 : 1 resonance, $j_2 = -1$ and $\alpha_o f_d(\alpha_o) \simeq -0.75$ (Murray and Dermott, 2000). As can be seen in (6.27), the maximum libration width implicitly depends on the mass ratio μ. Just as in Hill stability and the stability of multiplanet systems, the loss of mass from the central star throws previously nonresonant asteroids into resonance as the influence of the perturbing planet grows and the maximum libration width widens. The maximum libration width for any resonances associated with a planet will increase as μ increases. Figure 6.10 demonstrates the growth of $\delta a_{\max}$ with a corresponding change in μ (see additional description of this figure below).

This effect can be observed in numerical simulations of large Solar System asteroids when mass loss from the central star is included. Figures 6.10, 6.11, and 6.12 show the results of ten MERCURY simulations using the Burlirsch Stoer integrator (Chambers, 1999) of 736 Solar System asteroids integrated with Jupiter over 100 Myr to determine how many are perturbed to within $1 R_\odot$ (where they are effectively tidally disrupted). This sample of asteroids was chosen to include all objects with radii $R > 50$ km at any orbital semimajor axis and $R > 10$ km for objects with perihelia > 3 AU based on the latest known Solar System asteroid data (Bowell, 2009). It is assumed that the Sun will sublimate any asteroids smaller and/or closer than this during its post-main sequence evolution (Jura, 2008; Schröder and Connon Smith, 2008). The mass of the Sun was decreased to $0.54 M_\odot$ (Schröder and Connon Smith, 2008) and asteroids were removed if they strayed within $1 R_\odot$.

Four additional 200 Myr simulations were performed to test whether the perturbation rate declines over time. For comparison, a $\log g = 8$ white dwarf has an

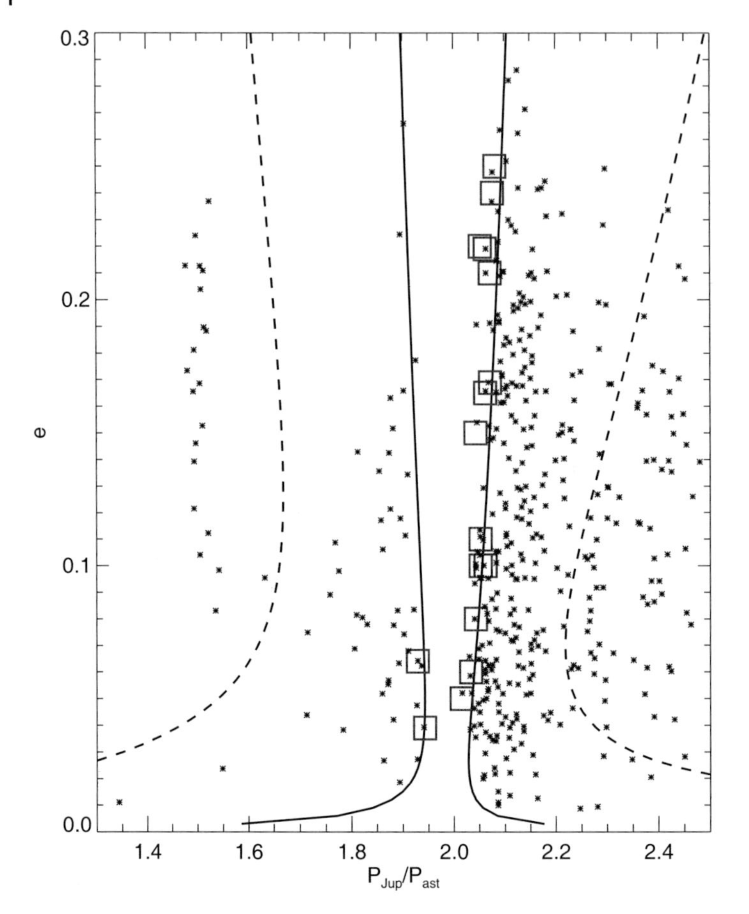

Figure 6.10 The relationship between orbital period and eccentricity near the 2 : 1 resonance in the Solar System, obtained from 14 simulations of the orbital evolution of 736 asteroids through the post-main sequence evolution of the Sun. The initial conditions of the simulation are shown as small asterisks, while large squares represent the asteroids that were caught in the 2 : 1 resonance and perturbed into tidally disruptive orbits. Many asteroids were perturbed in multiple simulations despite slightly differing initial conditions. The maximum libration width ($\delta a_{\max}$) for the 2 : 1 mean motion resonance in the restricted three body problem with $\mu = 9.54 \times 10^{-4}$ (solid line) and $\mu = 1.77 \times 10^{-3}$ (dashed line) are shown. The latter value is representative of post-main sequence evolution for a Sun-like star.

effective temperature of $\sim$ 18 500 K at 100 Myr and $\sim$ 15 000 K at 200 Myr. These simulations match the expected characteristics for some of the known dusty white dwarf systems. Most importantly, these simulations represent a dynamically mature population – the giant planets have had 4.5 Gyr to remove unstable asteroids. The results of the simulations show significant depletion of asteroids within the expanded maximum libration width, $\delta a_{\max}$, following post-main sequence mass loss of the central star (see Figure 6.11).

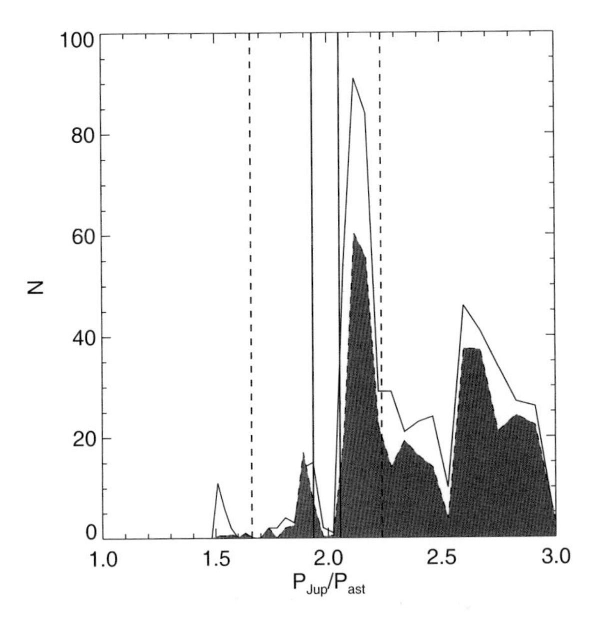

Figure 6.11 Orbital period distributions of asteroids at the start (solid line histogram) and end (shaded area) of the four 200 Myr simulations. The initial maximum libration width ($\delta a_{\max}$) for orbital eccentricity $e = 0.1$ is shown (solid vertical lines), as well as the expanded value at the end of the simulation, following the post-main sequence mass loss of the central star (dashed vertical lines). Asteroids that fall within the post-main sequence $\delta a_{\max}$ are strongly depleted. Depletion around other prominent resonances is also noticeable.

Roughly 1% of the asteroids passed within the tidal disruption radius in the first 100 Myr of the simulations, at a rate of one asteroid approximately every 10 Myr. From 100 to 200 Myr, the average time between asteroid disruption events increased to 17 Myr, implying a roughly linear decrease in the impact frequency on the white dwarf. Figure 6.12 shows the time of disruption as well as the radius of the asteroids. In many cases, the same asteroids were perturbed multiple times, but they are only counted once in the radius distribution. There is a broad peak in the distribution starting at $\sim$ 20 Myr, which is consistent with a white dwarf with $T_{\text{eff}} \sim$ 25 000 K. The hottest dusty white dwarfs are slightly cooler than this value and also represent some of the largest accretors of material, implying dust rich disks occur for white dwarfs with cooling ages consistent with the highest point in the peak of the disruption timescale distribution.

Figure 6.11 shows the orbital semimajor axis (a) versus orbital ellipticity (e) distribution of the asteroids in the simulations at the beginning of the evolution, with δa overplotted for $\mu = 9.54 \times 10^{-4}$. These results can be compared to Minton and Malhotra (2010), who calculated the diffusion of Solar System asteroids (with $D >$ 30 km) out of the asteroid belt due to perturbations from the giant planets. They used second-order mixed variable symplectic mapping (Saha and Tremaine, 1992; Wisdom and Holman, 1991) to integrate test particles between Mars and Jupiter to

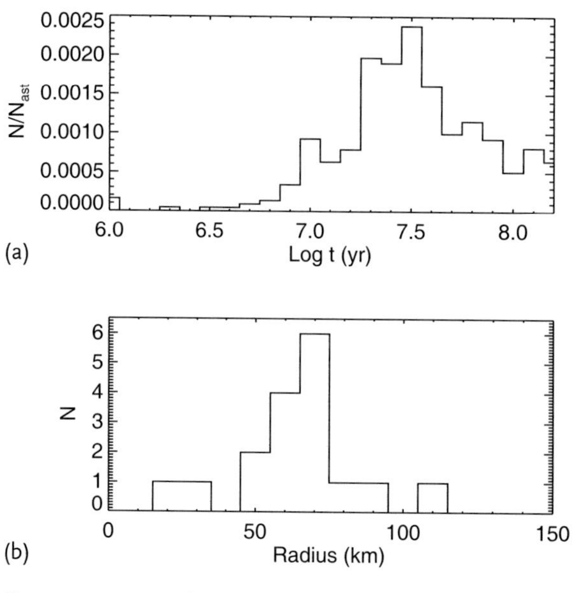

Figure 6.12 (a) Distribution of the times for all disruption events from both the 100 and 200 Myr simulations of the Solar System's asteroid belt during post-main sequence evolution of the Sun (a). The number in each histogram bin is the fraction of the total number of asteroids in the simulation. (b) Distribution of the radii of unique asteroids that suffered tidal disruption events in the simulations.

distances of 1 AU, and then MERCURY (with its hybrid integrator) to perform their integrations within 1 AU. They found that $\sim$ 0.2% of all asteroids in their simulation impacted the Sun over 4 Gyr, a much lower incidence despite the dynamically pristine nature of their sample. Part of this discrepancy comes from the selection of the asteroid samples. Minton and Malhotra (2010) utilized a wider range of initial semimajor axes, while the majority of the asteroids chosen for the MMP simulations described here clustered near the 2 : 1 resonance. Despite this, Minton and Malhotra (2010) found that, of the asteroids perturbed to within 1 AU, 17% went within one solar radius. For the MMP simulations described above, 17% of the asteroids that were perturbed out of the asteroid belt also went within one solar radius. Further work needs to be done to demonstrate whether the mass loss of the system accelerates asteroid perturbation or whether this difference is due to the specific choice of integration schemes or asteroidal populations.

6.4.4
Observability of Ring Progenitors

In the late stages of post-main sequence evolution, when white dwarf progenitors first shed their exterior gaseous layers in the planetary nebula phase, they are luminous enough to light up their relic debris disks as new planetesimals are swept up into planetary resonances (Dong *et al.*, 2010). However, can these planetesimal belts be observed at later times? This would provide direct evidence for the connection between planetary systems and disrupted asteroids as well as reveal where planetary companions may be lurking.

Because of the low luminosity of white dwarfs, there are regions of dust production where the timescale of inward migration due to PR drag is longer than the cooling lifetime of the system. Dust essentially stalls at a particular orbital radius because radiation pressure cannot remove the grains, nor is there a strong stellar wind, so PR drag is inefficient. Currently, the grain size distribution in this situation is unknown. However, the radius at which the PR drag timescale is greater than the cooling age of the white dwarf (t_{cool}) can be estimated from

$$R_{\text{limit}} = \left[\left(\frac{\beta_{\text{max}}}{400} \right) \left(\frac{t_{\text{cool}}}{1\,\text{yr}} \right) \left(\frac{M_\star}{1 M_\odot} \right) \right]^{1/2} \text{AU} . \tag{6.29}$$

Figure 6.13 shows R_{limit} as a function of T_{eff} for $\log g = 8.0$ hydrogen white dwarfs, where β_{max} is the maximum calculated value of β over a range of grain sizes, assuming astronomical silicates. Dust produced in widely separated planetesimal belts will accumulate until all material is ground into small grains. A major unanswered question is what is the minimum grain size? The ISM can remove small grains in debris disks around younger stars both by gas drag and grain-grain collisions (Artymowicz and Clampin, 1997; Debes *et al.*, 2009; Maness *et al.*, 2009).

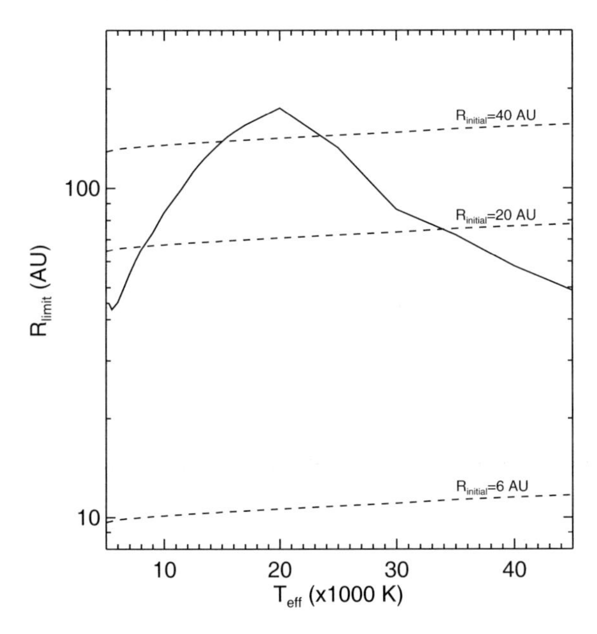

Figure 6.13 Orbital radius R_{limit} at which the PR drag timescale for dust dissipation exceeds the cooling age of white dwarfs with $\log g = 8.0$ over a wide range of effective temperature. Initial semimajor orbital axes (R_{initial}; dashed lines) of 6 AU (bottom), 20 AU (middle), and 40 AU (top) are shown. For $R_{\text{initial}} \gtrsim 40$ AU, dust will not be efficiently removed except through vaporizing collisions.

External ultraviolet flux, as well as that from the white dwarf, can sputter grains (Grigorieva *et al.*, 2007). Further work is needed to explore how the grain size distribution is affected by more subtle grain removal effects and collisional grinding around white dwarfs. However, optically thin disks at distances larger than a few tens of solar radii most likely will be almost impossible to detect against the photosphere of the white dwarf, especially if one extrapolates the amount of planetesimals from those inferred from the Spitzer debris disk results for nearby A-type stars (Bonsor and Wyatt, 2011).

6.5 Conclusion

The presence of dust within a few stellar radii of a white dwarf is nothing short of amazing, but the physical processes that seem to be responsible for this phenomenon also defy easy intuition. When metal rich white dwarfs were first discovered, the possibility that planetary systems may be responsible seemed beyond possibility. However, the main physical picture of these disks requires the delicate perturbation of planetesimals into orbits that are "just so" over hundreds of millions of years after surviving a maelstrom of heat, stellar winds, and dynamical shake-ups.

As demonstrated in this chapter, dusty white dwarfs and metal enriched white dwarfs seem to share a common history, but differ in the amount of material that orbits the central star. For low mass circumstellar disks, collisions and sputtering with many smaller incoming planetesimals quickly turn dusty disks into gaseous disks that only show up as metal enrichment in a white dwarf's photosphere. More massive dusty disks are caused by large planetesimals being perturbed into disrupting orbits, where the tidally disrupted material quickly evolves from an eccentric distribution of tidal debris into a circular, optically thick dust disk. They start out at distances comparable to the Solar System's asteroid belt, and achieve eccentricities in excess of 0.99 until they are tidally disrupted.

The planetesimals are perturbed either through mean motion resonances with giant planets or from the dynamical rearrangement of unstable planetary systems. These perturbations have typical timescales of 10^7–10^9 yr, eventually diminishing as the planetesimal population is fully depleted. These timescales, along with the overall mass of planetesimals, seem to match roughly what would be expected from the dynamical evolution of the Solar System's asteroid belt, implying that well separated, low eccentricity gas giants that are approximately the mass of Jupiter are common amongst stars with masses $> 1.5\,M_\odot$ (i. e., the progenitors of the current Galactic population of white dwarfs).

Yet, some parts of the scenario presented in this chapter remain sketchy or mysterious. They provide the path forward for a more detailed look at how planetary systems evolve after the death of their central star. In turn, these studies can provide limitless opportunities for studying the composition and atomic abundances of exo-planetesimals that are otherwise hopelessly invisible to modern astronomy.

The path forward will require drawing on the experience and techniques that have been used in the study of the Solar System. The roadmap for such work might follow these mileposts:

1. Tidal disruption: Simulations of rubble piles are routinely used to study the tidal disruption of comets and asteroids within the Solar System. These tools can be used to produce a detailed study of how debris from a disruption evolves. What is the timescale for the circularization of debris? What is the maximum separation at which an asteroid will disrupt? Is this maximum separation affected by the bulk physical properties of the planetesimal; for example, composition or porosity? Can anything be learned about planetesimals from the final structure of the disk?
2. Dynamical interactions: Are there other mechanisms that perturb planetesimals; for example, exterior mean motion resonances? (Bonsor *et al.*, 2011) Are there preferred resonances? How do interactions and the characteristic timescales change with planetary/stellar mass? What perturbation contributes the most?
3. Planetary stability: The phase space of stability is poorly studied with regards to mass loss during post-main sequence evolution. Furthermore, the rate of unstable systems is poorly known.
4. Planetesimal dynamics during post-main sequence evolution: What nongravitational effects come into play during post-main sequence evolution? Will sublimation/outgassing from the planetesimals greatly affect their orbits? What is the final architecture of a planetesimal belt at the end of the RGB and AGB phases?
5. Disk evolution: The structures presented in this chapter are all lacking in some way, and the overall evolution of these systems is poorly known. Again, expertise obtained from studying Saturn's rings can be applied to this problem, modified to take into account the special circumstances around a white dwarf. Can observations of dusty disks provide enough constraints for such models to be useful?

Observations in the ultraviolet and mid-infrared provide the most exciting avenues for new discoveries. The Wide-field Infrared Survey Explorer and AKARI mid-infrared all sky surveys have the potential of discovering dozens of new dusty white dwarfs. The Cosmic Origins Spectrograph and Space Telescope Imaging Spectrograph on board the Hubble Space Telescope have both the resolution and sensitivity to detect gas from these disks in addition to the accretion products in the white dwarf photospheres. The wealth of information possible from these observations can potentially open up new avenues beyond those listed above. The opportunity for theoretical work in concert with these observations is rich, and desperately needed given the quick pace of new information.

References

Aannestad, P.A., Kenyon, S.J., Hammond, G.L., and Sion, E.M. (1993) *Astron. J.*, **105**, 1033.

Alibert, Y., Mordasini, C., and Benz, W. (2004) *Astron. Astrophys.*, **417**, L25.

Artymowicz, P. (1988) *Astrophys. J. Lett.*, **335**, L79.

Artymowicz, P. and Clampin, M. (1997) *Astrophys. J.*, **490**, 863.

Batygin, K. and Laughlin, G. (2008) *Astrophys. J.*, **683**, 1207.

Bergeron, P., Wesemael, F., and Beauchamp, A. (1995) *Publicat. ASP*, **107**, 1047.

Bonsor, A., Mustill, A.J., and Wyatt, M. (2011) Dynamical effects of stellar mass-loss on a Kuiper-like belt, *Mon. Not. R. Astron. Soc.*, **594**
Provided by the SAO/NASA Astrophysics Data System

Bowell, E. (2009) *The Asteroid Orbital Elements Database*, version 2010-Aug-29, VizieR Online Data Catalog, **1**, 2001.

Calvet, N., D'Alessio, P., Hartmann, L., Wilner, D., Walsh, A., and Sitko, M. (2002) *Astrophys. J.*, **568**, 1008.

Chambers, J.E. (1999) *Mon. Not. R. Astron. Soc.*, **304**, 793.

Chambers, J.E., Wetherill, G.W., and Boss, A.P. (1996) *Icarus*, **119**, 261.

Chen, C.H. and Jura, M. (2001) *Astrophys. J. Lett.*, **560**, L171.

Chiang, E.I. and Goldreich, P. (1997) *Astrophys. J.*, **490**, 368.

Davidsson, B.J.R. (1999) *Icarus*, **142**, 525.

Debes, J.H. (2006) *Astrophys. J.*, **652**, 636.

Debes, J.H. and Sigurdsson, S. (2002) *Astrophys. J.*, **572**, 556.

Debes, J.H., Weinberger, A.J., and Kuchner, M.J. (2009) *Astrophys. J.*, **702**, 318.

Dong, R., Wang, Y., Lin, D.N.C., and Liu, X.-W. (2010) *Astrophys. J.*, **715**, 1036.

Draine, B.T. and Lee, H.M. (1984) *Astrophys. J.*, **285**, 89.

Dullemond, C.P., Dominik, C., and Natta, A. (2001) *Astrophys. J.*, **560**, 957.

Duncan, M.J. and Lissauer, J.J. (1997) *Icarus*, **125**, 1.

Duncan, M.J. and Lissauer, J.J. (1998) *Icarus*, **134**, 303.

Fabrycky, D.C. and Murray-Clay, R.A. (2010) *Astrophys. J.*, **710**, 1408.

Farihi, J., Jura, M., and Zuckerman, B. (2009) *Astrophys. J.*, **694**, 805.

Farihi, J., Zuckerman, B., and Becklin, E.E. (2008) *Astrophys. J.*, **674**, 431.

Farihi, J., Jura, M., Lee, J.-E., and Zuckerman, B. (2010) *Astrophys. J.*, **714**, 1386.

Fontaine, G., Brassard, P., and Bergeron, P. (2001) *Publ. ASP*, **113**, 409.

Ford, E.B., Havlickova, M., and Rasio, F.A. (2001) *Icarus*, **150**, 303.

Gänsicke, B.T., Marsh, T.R., Southworth, J., and Rebassa-Mansergas, A. (2006) *Science*, **314**, 1908.

Gänsicke, B.T., Marsh, T.R., and Southworth, J. (2007) *Mon. Not. R. Astron. Soc.*, **380**, L35.

Gänsicke, B.T., Koester, D., Marsh, T.R., Rebassa-Mansergas, A., and Southworth, J. (2008) *Mon. Not. R. Astron. Soc.*, **391**, L103.

Gladman, B. (1993) *Icarus*, **106**, 247.

Gladman, B.J. *et al.* (1997) *Science*, **277**, 197.

Goldreich, P. and Tremaine, S. (1982) *Annu. Rev. Astron. Astrophys.*, **20**, 249.

Graham, J.R., Matthews, K., Neugebauer, G., and Soifer, B.T. (1990) *Astrophys. J.*, **357**, 216.

Grigorieva, A., Thébault, P., Artymowicz, P., and Brandeker, A. (2007) *Astron. Astrophys.*, **475**, 755.

Heyl, J. (2007) *Mon. Not. R. Astron. Soc.*, **382**, 915.

Hill, G.W. (1886) *Acta Math.*, **8**, 1.

Holberg, J.B. and Bergeron, P. (2006) *Astron. J.*, **132**, 1221.

Juhász, A., Henning, T., Bouwman, J., Dullemond, C.P., Pascucci, I., and Apai, D. (2009) *Astrophys. J.*, **695**, 1024.

Jura, M. (2003) *Astrophys. J. Lett.*, **584**, L91.

Jura, M. (2004) *Astrophys. J.*, **603**, 729.

Jura, M. (2008) *Astron. J.*, **135**, 1785.

Jura, M., Farihi, J., and Zuckerman, B. (2007a) *Astrophys. J.*, **663**, 1285.

Jura, M., Farihi, J., Zuckerman, B., and Becklin, E.E. (2007b) *Astron. J.*, **133**, 1927.

Jura, M., Farihi, J., and Zuckerman, B. (2009) *Astron. J.*, **137**, 3191.

Kilic, M. and Redfield, S. (2007) *Astrophys. J.*, **660**, 641.

Kilic, M., Farihi, J., Nitta, A., and Leggett, S.K. (2008) *Astron. J.*, **136**, 111.

Kimura, H., Mann, I., Biesecker, D.A., and Jessberger, E.K. (2002) *Icarus*, **159**, 529.

King, A.R., Pringle, J.E., and Livio, M. (2007) *Mon. Not. R. Astron. Soc.*, **376**, 1740.
Koester, D. (2009) *Astron. Astrophys.*, **498**, 517.
Koester, D. and Wilken, D. (2006) *Astron. Astrophys.*, **453**, 1051.
Koester, D., Rollenhagen, K., Napiwotzki, R., Voss, B., Christlieb, N., Homeier, D., and Reimers, D. (2005) *Astron. Astrophys.*, **432**, 1025.
Laor, A. and Draine, B.T. (1993) *Astrophys. J.*, **402**, 441.
Maness, H.L. *et al.* (2009) *Astrophys. J.*, **707**, 1098.
Minton, D.A. and Malhotra, R. (2010) *Icarus*, **207**, 744.
Morbidelli, A. (1996) *Astron. J.*, **111**, 2453.
Murray, C.D. and Dermott, S.F. (2000) *Solar System Dynamics*, Cambridge University Press, UK.
Paquette, C., Pelletier, C., Fontaine, G., and Michaud, G. (1986) *Astrophys. J. (Suppl.)*, **61**, 197.
Parriott, J. and Alcock, C. (1998) *Astrophys. J.*, **501**, 357.
Rafikov, R.R. (2011) *Astrophys. J. Lett.*, **732**, L3.
Reach, W.T., Kuchner, M.J., von Hippel, T., Burrows, A., Mullally, F., Kilic, M., and Winget, D.E. (2005) *Astrophys. J. Lett.*, **635**, L161.
Reach, W.T., Lisse, C., von Hippel, T., and Mullally, F. (2009) *Astrophys. J.*, **693**, 697.
Saha, P. and Tremaine, S. (1992) *Astron. J.*, **104**, 1633.
Schröder, K.-P. and Connon Smith, R. (2008) *Mon. Not. R. Astron. Soc.*, **386**, 155.
Stern, S.A., Shull, J.M., and Brandt, J.C. (1990) *Nature*, **345**, 305.
Thompson, S.E. *et al.* (2010) *Astrophys. J.*, **714**, 296.
Tiscareno, M.S., Burns, J.A., Nicholson, P.D., Hedman, M.M., and Porco, C.C. (2007) *Icarus*, **189**, 14.
Villaver, E. and Livio, M. (2007) *Astrophys. J.*, **661**, 1192.
von Hippel, T., Kuchner, M.J., Kilic, M., Mullally, F., and Reach, W.T. (2007) *Astrophys. J.*, **662**, 544.
Wisdom, J. and Holman, M. (1991) *Astron. J.*, **102**, 1528.
Wyatt, M.C. (2005) *Astron. Astrophys.*, **433**, 1007.
Wyatt, M.C., Dermott, S.F., Telesco, C.M., Fisher, R.S., Grogan, K., Holmes, E.K., and Piña, R.K. (1999) *Astrophys. J.*, **527**, 918.
Zuckerman, B., Koester, D., Reid, I.N., and Hünsch, M. (2003) *Astrophys. J.*, **596**, 477.
Zuckerman, B., Koester, D., Melis, C., Hansen, B.M., and Jura, M. (2007) *Astrophys. J.*, **671**, 872.

7
Planetary Nebulae around White Dwarfs: Revelations from the Infrared

You-Hua Chu

7.1
Introduction: Expectations of Nebulae around White Dwarfs

Stars with initial masses greater than $0.5 M_\odot$ and less than 8–10$M_\odot$ end their evolution as white dwarfs. Within this mass range, the lower mass progenitor stars lose smaller amounts of mass over longer periods of time, while higher mass stars lose larger amounts of mass over shorter periods of time (Bloecker, 1995; Vassiliadis and Wood, 1993). For the former, the stellar winds have such low densities that the expelled mass simply disperses and may never form a visible nebula. For the latter, the mass loss during the asymptotic giant branch phase forms a circumstellar nebula. This circumstellar material is swept up by the subsequent fast wind to form an expanding shell, and is photo-ionized by the hot stellar core to become a visible planetary nebula (Kwok *et al.*, 1978).

The interiors of young planetary nebulae are filled with shocked fast stellar winds, which mix with the cool nebular shell material and have been observed to have plasma temperatures of $1–3\times10^6$ K (Chu *et al.*, 2001; Guerrero *et al.*, 2002; Kastner *et al.*, 2000). As the central star of a planetary nebula evolves, the fast stellar wind diminishes before the star morphs into a white dwarf. The hot gas in the planetary nebula interior cools, and as its thermal pressure drops, the cool nebular shell material back-fills. Evolved planetary nebulae around white dwarfs are expected to lose the sharp shell structures and eventually become centrally filled.

Planetary nebulae expand, fade, and merge into the interstellar medium (ISM). The time scale for a planetary nebula to merge into the ISM, τ, can be approximated by

$$\tau = 2 \times 10^6 \left(\frac{m}{1 M_\odot}\right)^{1/3} \left(\frac{n_0}{1\,\mathrm{cm}^{-3}}\right)^{-1/3} \left(\frac{V_{\mathrm{exp}}}{1\,\mathrm{km\,s}^{-1}}\right)^{-1} \mathrm{yr}\,, \tag{7.1}$$

where m is the total nebular mass, n_0 is the hydrogen density of the ambient ISM, and V_{exp} is the nebular expansion velocity. A nebula with a mass of a few solar masses and an expansion velocity of $10\,\mathrm{km\,s}^{-1}$ will merge into an ambient ISM of

White Dwarf Atmospheres and Circumstellar Environments. First Edition. Edited by D. W. Hoard.

density 1 H-atom cm^{-3} in less than a million years. This timescale is much shorter than the lifespan of white dwarfs; therefore, most white dwarfs no longer possess distinct circumstellar nebulae.

Even before a planetary nebula has merged into the ISM, the surface brightness of the nebula may have faded below the detection limit. For an ionized nebula, the surface brightness is proportional to the emission measure, defined as $\mathrm{EM} = n_e^2 L$, where n_e is the electron density in units of cm^{-3} and L is the line-of-sight length of emission in units of pc. For modern CCD cameras with a narrow Hα interference filter, the detection limit can reach an emission measure of $\sim 10\,\mathrm{cm}^{-6}\,\mathrm{pc}$ without much difficulty. By crudely assuming that the planetary nebula is back-filled, with its mass distributed uniformly in a sphere, the peak emission measure would be

$$\mathrm{EM} = 189 \left(\frac{m}{1 M_\odot}\right)^2 \left(\frac{r}{1\,\mathrm{pc}}\right)^{-5} \mathrm{cm}^{-6}\,\mathrm{pc}\,, \tag{7.2}$$

where m is the ionized nebular mass and r is the radius of the nebula in units of pc. For a nebular mass of $1 M_\odot$ and a detection limit of $\mathrm{EM} = 10\,\mathrm{cm}^{-6}\mathrm{pc}$, the nebula remains visible until its radius reaches ~ 2 pc, corresponding to an age of ~ 0.2 Myr. This short timescale indicates that only the youngest white dwarfs are surrounded by visible planetary nebulae.

Planetary nebulae move with their central stars through the ISM. If the relative motion is significant enough, then the ram pressure from the ISM will compress the leading edge of the planetary nebula and brake the nebula (Villaver *et al.*, 2003). This interaction is most pronounced for evolved planetary nebulae whose densities are not much higher than that of the ambient ISM; therefore, it is common for planetary nebulae around white dwarfs to show morphological indications of interaction with the ISM. If a hot white dwarf has already lost its circumstellar nebula and is surrounded by a dense ISM, then it can photo-ionize its surrounding ISM and produce a visible emission nebula. Such "H II regions" are often mis-identified as planetary nebulae.

7.2 Planetary Nebulae around White Dwarfs

About 40 white dwarfs in McCook and Sion's catalogue (McCook and Sion, 1999) are surrounded by planetary nebulae. Table 7.1 lists these white dwarfs, their planetary nebulae, spectral types, and nebular properties. As expected, a variety of nebular morphologies are seen, and these morphological characteristics can be associated with planetary nebula evolution and interactions with the ISM.

Planetary nebulae are broadly divided into two morphological groups: elliptical and bipolar. These morphologies are still preserved in some planetary nebulae around white dwarfs. Figure 7.1 shows four examples of elliptical planetary nebulae around white dwarfs; these nebulae still show limb-brightening indicating a shell structure. Figure 7.2 shows four examples of bipolar planetary nebulae around white dwarfs: NGC 650-1 shows a classical bipolar morphology, NGC 7293

Table 7.1 White Dwarfs in Planetary Nebulae.

WD name	PN name	Sp. type	Distance (pc)	Size (″)	Size (pc)	Shape[a]	Reference
WD 0029+571	Sh 2-176	DA	270	720	0.94	A, Int	1,2
WD 0044−121	NGC 246	PG 1159	630	273 × 224	0.83 × 0.68	E	3
WD 0103+732	EGB 1	DA.34	653	300 × 180	0.95 × 0.57	D, Irr	4
WD 0127+581	Sh 2-188	DAO.49	220	550	0.59	A, Int	5
WD 0139+513	NGC 650-1	DOZ.4	270	257	0.34	B	1
WD 0322+452	HDW 3	DAO.4	410	540	1.09	A, Int	1
WD 0342+374	HaWe 5	DA1.3	420	34	0.07	D, Irr	4
WD 0345+498	IsWe 1	DZQO.5	330	780	1.24	D, Int	1,2
WD 0439+466	Sh 2-216	DA.61	129	5760	3.6	E, D	6
WD 0500−156	Abell 7	DAO.51	705	760	2.59	E, D	2,4
WD 0533+555	HDW 4	DA1.1	246	105	0.125	D	4
WD 0556+106	WeDe 1	DA.36	968	1020 × 840	4.78 × 3.94	E, D	4
WD 0558−756	K1-27	DO.5	–	44	–	D	7
WD 0615+556	PuWe 1	DAO.54	695	1200	4.04	E, D	2,4
WD 0625−253	Abell 15	DA1	3656	36 × 32	0.63 × 0.57	E	8
WD 0713−468	Lo 3	PG 1159	–	80	–	E	–
WD 0726+133	Abell 21	PG 1159	240	685 × 530	0.80 × 0.62	A, Int	1
WD 0753+535	JnEr 1	DQZO.4	–	425 × 360	–	E/B	–
WD 0950+139	EGB 6	DA.46	460	834	1.86	E, Int	2
WD 1003−441	Lo 4	PG 1159	–	44	–	E, D	–
WD 1111+552	NGC 3587	DAO.54	1269	208 × 198	1.28 × 1.22	E	4
WD 1250−226	Abell 35	DAO.63	–	960 × 660	–	D, Irr	2
WD 1253+261	LoTr 5	sdO+G	–	210	–	D, Int	–
WD 1520+525	JavdSt 1	PG 1159.3	–	660	–	E, Int	9
WD 1625+280	Abell 39	DAO.43	1931	164	1.54	E	4
WD 1738+669	anon	DA.62	–	>3600	–	Irr	10
WD 1751+106	Abell 43	DAO.67	2649	85 × 76	1.2 × 1.0	E	4
WD 1822+008	Sh 2-68	DAOZ.7	1054	400	2.04	D, Irr	4
WD 1851−088	IC 1295	DAO.56	715	150 × 121	0.52 × 0.42	E, D	4
WD 1900+140	Sh 2-78	PG 1159	1420	600	4.13	B	1,2
WD 1909+304	NGC 6765	PG 1159	2300	40	0.45	B?, Irr	1
WD 1917+461	Abell 61	DAO.57	1380	187	1.25	E, D, Int	4
WD 1958+015	NGC 6852	PG 1159	2719	28	0.37	E	1
WD 2114+239	Abell 74	DAO.47	752	870 × 792	3.17 × 2.89	E/B, D	2,6
WD 2115+339	MWP 1	PG 1159.3	1400	780 × 540	5.3 × 3.7	B	2
WD 2134+125	NGC 7094	PG 1159	2246	94	1.02	E	4
WD 2212+656	IsWe 2	DA	260	960 × 840	1.21 × 1.06	E/A, Irr	1,2
WD 2218+706	DeHt 5	DAO.86	512	530	1.32	A, D	4
WD 2226−210	NGC 7293	DAO.47	216	1080	1.13	B	6
WD 2333+301	Jn 1	DOZ.3	710	330 × 295	1.13 × 1.02	E/B, D	1
WD 2342+806	Sh 2-174	DAO.78	556	900 × 600	2.0 × 1.2	Irr	2,11

a A = arc; B = bipolar; D = diffuse; E = elliptical; Int = interacting with the ISM; Irr = irregular.

1 Napiwotzki and Schoenberner (1995); 2 Tweedy and Kwitter (1996); 3 Perryman *et al.* (1997); 4 Napiwotzki (1999); 5 Wareing *et al.* (2006); 6 Harris *et al.* (2007); 7 Chu *et al.* (2009); 8 Cahn *et al.* (1992); 9 Jacoby and van de Steene (1995); 10 Tweedy and Kwitter (1994); 11 Napiwotzki (1993).

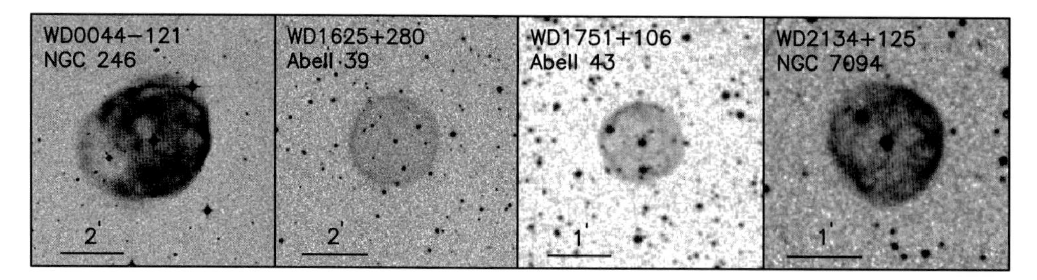

Figure 7.1 Examples of elliptical planetary nebulae around white dwarfs that still show limb-brightening, reproduced from Digitized Sky Survey 2 (DSS2) red images.

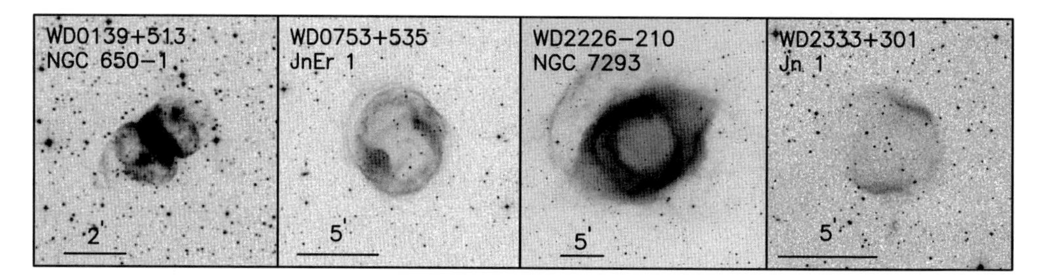

Figure 7.2 Examples of bipolar planetary nebulae around white dwarfs that still show limb-brightening, reproduced from DSS2 red images.

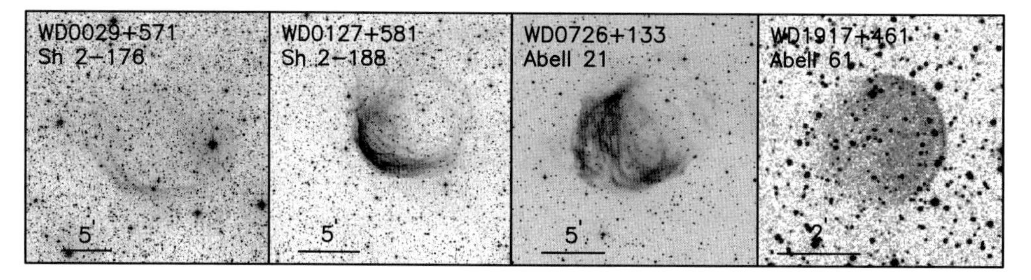

Figure 7.3 Examples of planetary nebulae around white dwarfs that show one-sided limb-brightening indicative of compression by ram pressure due to movement through the ISM, reproduced from DSS2 red images.

has a bipolar halo, while JnEr 1 and Jn 1 show only mildly bipolar morphologies that are sometimes called bi-nebular.

As a white dwarf and its planetary nebula move through the ISM, the ram pressure from the ISM compresses the leading side of the nebula and forms a sharp, bright edge (Tweedy and Kwitter, 1996). Figure 7.3 shows four examples of planetary nebulae exhibiting such one-sided brightening. The physical structure of Sh 2-188, associated with WD 0127+581, has been modeled by Wareing *et al.* (2006); its bright arc can be reproduced if the white dwarf and planetary nebula are moving at $125 \pm 25\,\mathrm{km\,s^{-1}}$ relative to the ambient ISM.

A prevalent morphological feature of planetary nebulae around white dwarfs is the lack of a prominent central cavity as a result of back-filling. NGC 3587 (the Owl Nebula), which is associated with WD 1111+552, has been shown to have a density profile exhibiting significant back-filling, although small cavities are still present (Guerrero *et al.*, 2003). Figure 7.4 shows three extreme examples of back-filled planetary nebulae whose surface brightnesses do not suggest any central cavities.

Some planetary nebulae around white dwarfs have irregular morphologies that can be caused by a combination of nebular back-fill and interactions with the ISM. Among the three examples of irregular planetary nebulae shown in Figure 7.5, NGC 6765 appears to be a distorted bipolar nebula and Abell 7 appears back-filled with a few bright spots, while Abell 35 interacts with the ambient ISM in a complex way (Tweedy and Kwitter, 1996).

Finally, some hot white dwarfs located in regions of dense ISM can photo-ionize the ambient medium and produce detectable emission nebulae. In principle, it is possible to distinguish between a planetary nebula and a photo-ionized interstellar nebula based on abundances; however, the surface brightnesses of nebulae

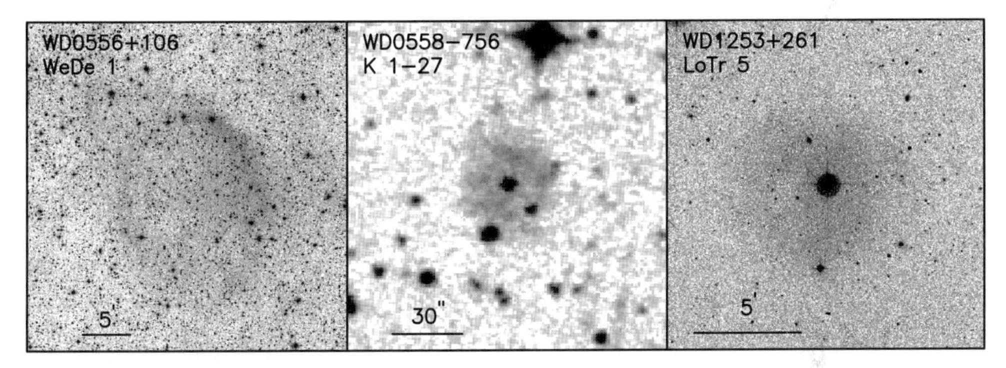

Figure 7.4 Examples of planetary nebulae around white dwarfs that have experienced back-filling and show almost uniform surface brightness, reproduced from DSS2 red images.

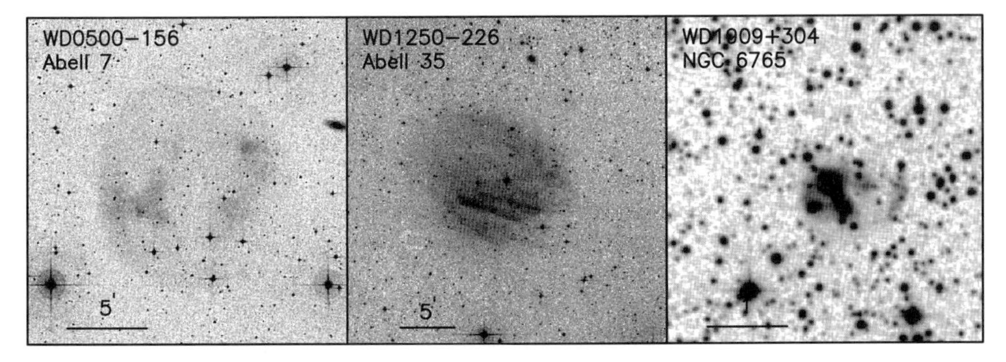

Figure 7.5 Examples of planetary nebulae around white dwarfs whose irregular morphologies may be caused by a combination of back-filling and interaction with the ISM, reproduced from DSS2 red images.

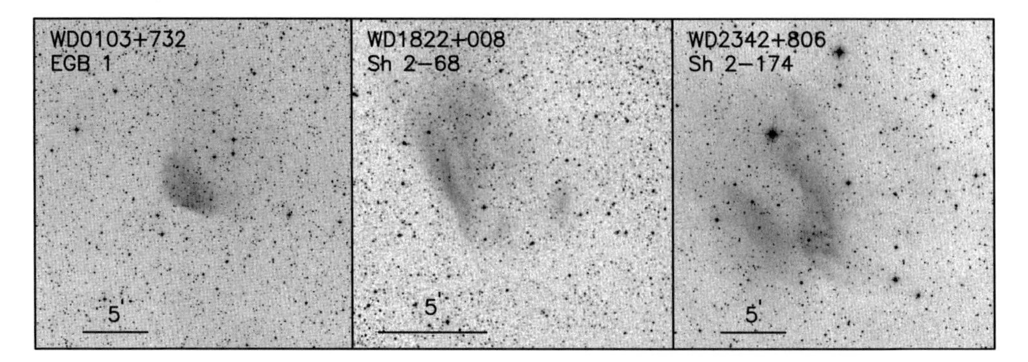

Figure 7.6 Examples of planetary nebulae around white dwarfs that are likely to be misclassified photo-ionized interstellar nebulae, reproduced from DSS2 red images.

around white dwarfs are so low that spectrophotometric measurements of emission lines of the nebulae are very difficult. It is easier to determine the interstellar origin of a white dwarf's nebula through radial velocity measurements that reveal whether the nebular velocity is similar to that of the background ISM or the stellar velocity. This method has been used to demonstrate that the emission nebula around WD 0005+511 (= KPD 0005+5106) and the alleged planetary nebula around PG 1034+001 (Hewett *et al.*, 2003) both consist of photo-ionized ISM (Chu *et al.*, 2004).

Figure 7.6 shows three emission nebulae around white dwarfs that have been reported to be planetary nebulae; however, their morphologies make this identification suspect. The nebula EGB 1 has a straight sharp southern edge that appears to be bordering a region of high extinction; such a structure is reminiscent of an H II region adjacent to a dark cloud, rather than an evolved planetary nebula interacting with the ISM. The planetary nebula classification of Sh 2-68 has been questioned by Tweedy and Kwitter (1996) based on the location of the star in the nebula and the straight bright eastern edge of the nebula. In the case of Sh 2-174 around WD 2342+806, the radial velocity of the ionized gas, $V_{LSR} = +4 \pm 1\,\mathrm{km\,s^{-1}}$ (Reynolds *et al.*, 2005), is remarkably similar to the interstellar H I velocity in this direction (Heiles and Habing, 1974); furthermore, the location of the white dwarf and the nebular morphology are not consistent with those expected from evolved planetary nebulae interacting with the ISM. Observations of the nebular velocities and stellar velocities are needed to test the interstellar origin of these three circumstellar nebulae of white dwarfs.

7.3 High-Excitation Nebulae around Hot White Dwarfs

White dwarfs that are associated with emission nebulae are the youngest and, thus, have high temperatures, commonly approaching 100 000 K or even higher.

Their stellar radiation peaks in the far ultraviolet or soft X-ray range. Consequently, their photo-ionized nebulae have high excitation. Highly ionized species, such as O VI, Ne V, N V, and C IV (with excitation potentials of 113.9, 97.12, 77.47, and 47.88 eV, respectively), have been detected through absorption or emission line observations. For example, O VI 1032 Å line emission from the interstellar nebula around WD 0005+511 has been detected in spectra obtained with the Far Ultraviolet Spectroscopic Explorer satellite (Otte *et al.*, 2004; Sankrit and Dixon, 2009). The [Ne V] 3426 Å line emission from the planetary nebula NGC 246, which is associated with WD 0044−121, has been detected in narrow-band imaging observations (Szentgyorgyi *et al.*, 2003).

The high excitation of nebulae around white dwarfs is caused by the energetic photons emitted by the star; the highly ionized species are produced by photo-ionization. This is different from those produced by collisional ionization in young planetary nebulae whose interiors are filled with X-ray-emitting plasma. Thermal conduction between hot plasma and cool nebular shell forms an interface layer at temperatures of a few $\times 10^5$ K, where thermal collisions can produce highly ionized species, such as O VI and N V (Weaver *et al.*, 1977). For example, the diffuse X-ray emission from the planetary nebula NGC 6543 indicates that its interior is filled with $\sim 2 \times 10^6$ K hot gas (Chu *et al.*, 2001); as its central star's effective temperature, 50 000 K, is not high enough to photo-ionize oxygen to O^{+5}, the detected O VI 1032 Å line emission must originate from the interface layer (Gruendl *et al.*, 2004). Hot white dwarfs have been frequently used as continuum sources to probe the intervening hot interstellar gas through absorption lines of the highly ionized species (e. g., Barstow *et al.*, 2010). In such work, care should be exercised to distinguish between the photo-ionized circumstellar component and the collisionally ionized interstellar component.

7.4
Mid-Infrared Emission from Circumstellar Nebulae of White Dwarfs

Planetary nebulae around hot white dwarfs usually have very low surface brightness in optical emission lines. Mid-infrared images of these planetary nebulae made with the Spitzer Space Telescope Multi-band Imaging Photometer for Spitzer (MIPS) camera (Rieke *et al.*, 2004) in the 24 μm band show that they are surprisingly bright and morphologically different from their optical counterparts (Chu *et al.*, 2009). To some extent, this is caused by the high sensitivity of the MIPS detector, but it is also caused by the inclusion of [Ne V] 24.31 μm and [O IV] 25.89 μm lines in the MIPS 24 μm band as well as the high excitation of planetary nebulae around hot white dwarfs.

Based on comparisons between Hα and 24 μm morphologies, planetary nebulae around hot white dwarfs have been divided into four groups. Twenty-two planetary nebulae with MIPS 24 μm observations are classified and listed in Table 7.2.

Group 1 planetary nebulae are characterized by the 24 μm band emission being more extended than the Hα emission. Figure 7.7 compares an Hα image of

Table 7.2 Hot White Dwarf Central Stars of Planetary Nebulae[a].

PN name	PNG number	CSPN name	T_{eff} [b] (10^3 K)	Size (arcsec)	Distance[c] (pc)	Size (pc)	HeII/Hβ flux[d]	24 µm SB (MJy sr^{-1})
			Group 1					
NGC 6852	042.5−14.5	WD 1958+015		28	2710	0.37	1.2	80.5
			Group 2					
Abell 39	047.0+42.4	WD 1625+280	117	164	1163	0.92	0.69	1.4
Abell 43	036.0+17.6	WD 1751+106	117	85 × 76	1619	0.67 × 0.60	0.93	11.3
Jn 1	104.2−29.6	WD 2333+301	150	330 × 295	709	1.13 × 1.02	0.5	0.9
Lo 4	274.3+09.1	WD 1003−441	120	44	–	–	0.9	8.8
MWP 1	080.3−10.4	WD 2115+339	170	300	–	–	–	4.4
NGC 246	118.8−74.7	WD 0044−121	150	273 × 224	470	0.62 × 0.51	1.2	36.6
			Group 3					
Abell 21	205.1+14.2	WD 0726+133	140	685 × 530	541	1.8 × 1.4	0.26	0.8
Abell 61	077.6+14.7	WD 1917+461	88	187	–	–	–	0.6
EGB 1	124.0+10.7	WD 0103+732	147	300 × 180	–	–	–	0.6
IC 1295	025.4−04.7	WD 1851−088	90	150 × 121	1024	0.74 × 0.60	0.5	16.1
JnEr 1	164.8+31.1	WD 0753+535	130	425 × 360	–	–	0.26	1.1
NGC 3587	148.4+57.0	WD 1111+552	94	208 × 198	615	0.63 × 0.60	0.14	4.5
Sh 2-188	128.0−04.1	WD 0127+581	102	550	–	–	–	0.6
Sh 2-216	158.5+00.7	WD 0439+466	83	5760	129	3.6	–	1.0
			Not Detected					
Abell 7	215.5−30.8	WD 0500−156	99	870 × 672	676	2.9 × 2.2	–	<0.2
Abell 74	072.7−17.1	WD 2114+239	108	870 × 792	752	1.0 × 0.9	–	<0.2
EGB 6	221.5+46.3	WD 0950+139	100	834	–	–	–	<0.2
HDW 3	149.4−09.2	WD 0322+452	125	540	–	–	–	<0.2
JavdSt 1	085.4+52.3	WD 1520+525	150	660	–	–	–	<0.2
PuWe 1	158.9+17.8	WD 0615+556	94	1200	365	2.12	–	<0.2
WeDe 1	197.4−06.4	WD 0556+106	141	1020 × 840	–	–	–	<0.2

a Data used in this table are from Chu *et al.* (2009).
b The stellar effective temperatures are either derived or compiled by Napiwotzki (1999), except those of Abell 21, Jn 1, and JnEr 1, which are from Werner and Herwig (2006).
c The distances are from Cahn *et al.* (1992), except those of Abell 7, Abell 21, Abell 74, PuWe 1, and Sh 2-216, which are from Harris *et al.* (2007). Some of these distances may be different from those given in Table 7.1 because they were reported by other researchers based on different methods. These different distances signify the uncertainties in distances to planetary nebulae.
d The He II λ4686/Hβ flux ratios are from Tylenda *et al.* (1994).

NGC 6852 to the MIPS 24 µm image. The surface brightness profiles clearly show that the 24 µm emission is more extended. If the 24 µm band emission were attributed to the [O IV] and [Ne V] lines, then the ratio of 24 µm to Hα would be expected to decrease radially outwards because of ionization stratification. Instead, the observed radial increase of this ratio indicates that the 24 µm emission originates from a dust continuum. The small size of NGC 6852, 0.37 pc in diameter, indicates that it is relatively young and dense. This planetary nebula is likely to be

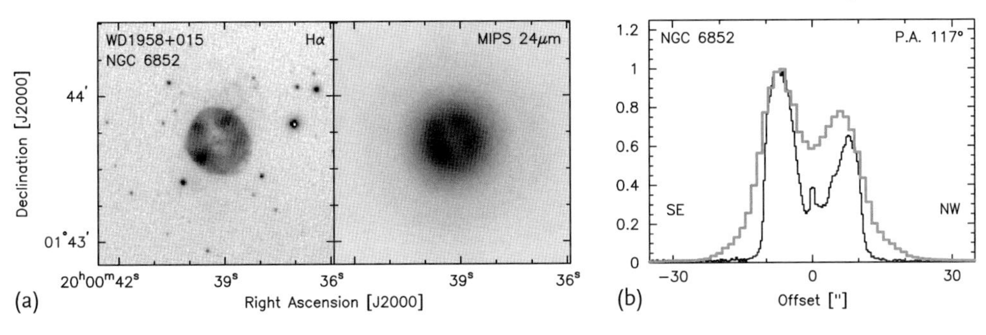

Figure 7.7 Images (a) and surface brightness profiles (b) of a Group 1 planetary nebula whose 24 μm emission is more extended than the Hα emission. The names of the central white dwarf and nebula are marked in the upper left corner and the passbands in the upper right corner of the image panels. Panel (b) shows the surface brightness profiles (Hα in black and 24 μm in gray) along the position angle marked in the upper right corner. From Chu *et al.* (2009), reproduced with the permission of the AAS.

ionization bounded and the extended 24 μm emission region may be dusty and neutral.

Group 2 planetary nebulae have 24 μm images resembling their Hα counterparts. Figure 7.8 shows three examples. Although their 24 μm sizes are similar to their Hα size, subtle morphological differences exist between the two passbands. NGC 246 around WD 0044−121 shows bright 24 μm knots that do not have Hα counterparts, and Jn 1 around WD 2333+301 is more filled-in in the 24 μm band than in Hα. The 24 μm band emission of these planetary nebulae is probably a combination of nebular line emission and dust continuum emission.

Group 3 planetary nebulae show 24 μm emission toward the central regions of the optical nebulae. Figure 7.9 shows four examples. The central regions of planetary nebulae are expected to have the lowest densities and highest excitation within the nebulae. It is more likely that high-ionization nebular lines, rather than dust continua, contribute to the MIPS 24 μm band emission. Indeed, Spitzer Infrared Spectrograph (IRS; Houck *et al.*, 2004) spectra of the diffuse 24 μm emission in the central region of Sh 2-188, associated with WD 0127+581, show predominantly [O IV] 25.89 μm emission within the MIPS 24 μm band (see Figure 7.10). Similar IRS spectra of the central regions of the Helix Nebula around WD 2226−210 (Su *et al.*, 2007), EGB 1 around WD 0103+732, and Sh 2-216 around WD 0439+466 all show prominent [O IV] lines.

The last group of planetary nebulae are not detected in MIPS 24 μm observations that are equally sensitive as those that made the above detections. As shown in Table 7.2, these nebulae are the largest. Therefore, their central densities must be the lowest and their surface brightnesses are below the detection limit.

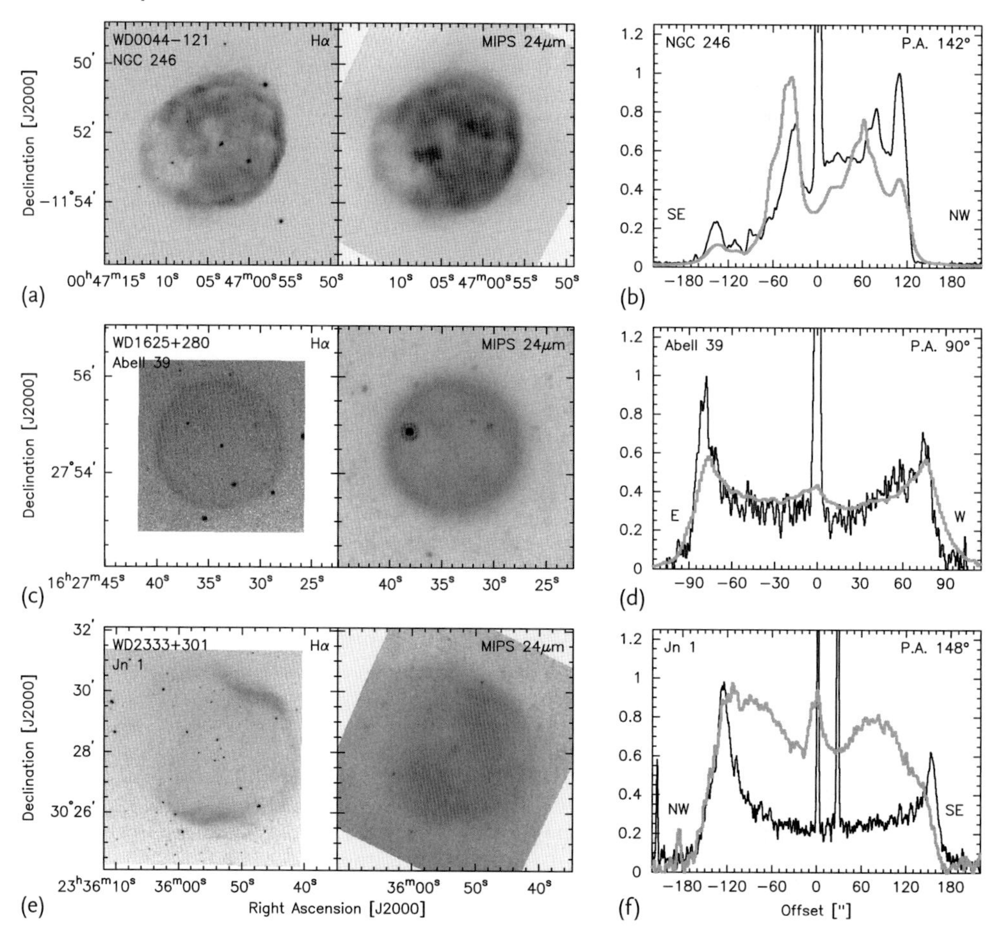

Figure 7.8 Same as Figure 7.7, but for Group 2 planetary nebulae whose 24 μm morphologies are similar to those in the Hα line. From Chu *et al.* (2009), reproduced with the permission of the AAS.

7.5 Conclusion

Planetary nebulae around white dwarfs represent the last visible stage of the nebular evolution before merging into the ISM. These evolved nebulae provide the best sites to study dynamical processes of the diffusion of nebular shell gas back to the central cavity (i. e., back-filling), and the interaction between planetary nebulae and the ISM. Planetary nebulae around white dwarfs can be compared with hydrodynamical simulations of late evolutionary stages of planetary nebulae to understand the pronounced morphological features produced by nebula–ISM interactions, and to probe the physical conditions and properties of the ISM.

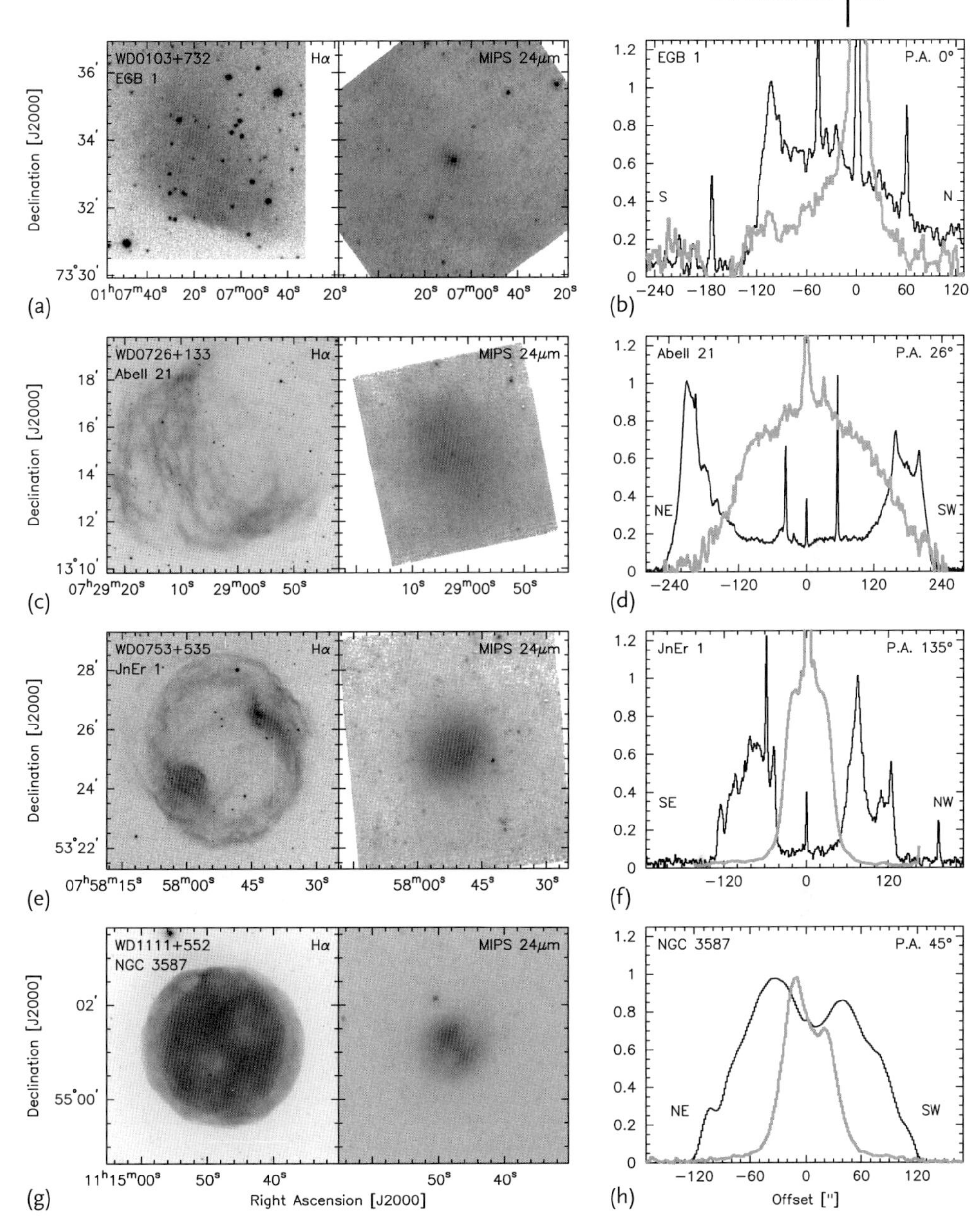

Figure 7.9 Same as Figure 7.7, but for Group 3 planetary nebulae whose 24 μm emission is distributed in the central regions around the white dwarfs. From Chu *et al.* (2009), reproduced with the permission of the AAS.

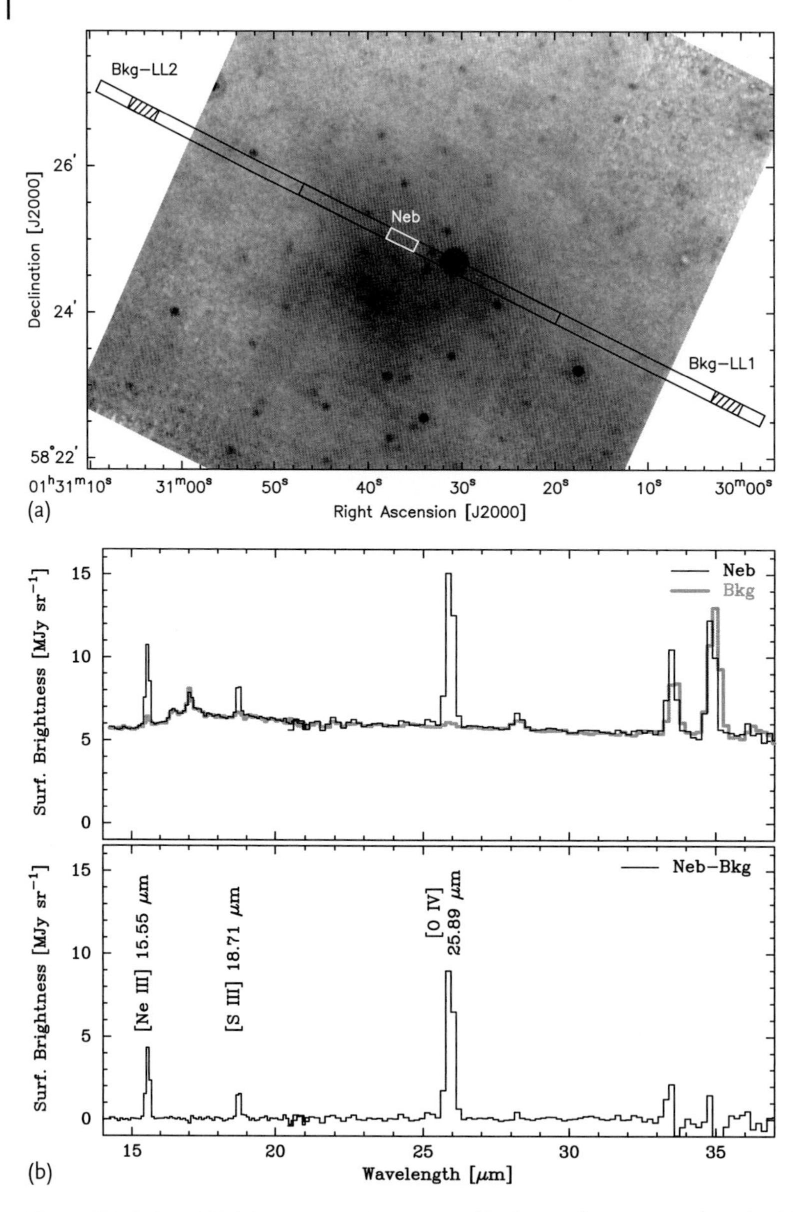

Figure 7.10 Spitzer MIPS 24 μm image (a) and IRS spectra of the diffuse 24 μm emission region and backgrounds around Sh 2-188 (b). The white box marked "Neb" in (a) shows the region where the nebular spectrum is extracted. The hatched areas at the two ends of the slit mark the background regions. The nebular and background spectra are plotted in black and gray, respectively, in the upper spectral panel, and the background-subtracted nebular spectrum is plotted in the lower spectral panel. From Chu *et al.* (2009), reproduced with the permission of the AAS.

Owing to the high temperatures of young white dwarfs, the stellar radiation field is extremely harsh and, thus, the photo-ionized planetary nebulae or ambient ISM have the highest excitation and ionization. Emission lines from highly ionized species, especially [Ne V] 24.31 μm and [O IV] 25.89 μm, have been detected by sensitive orbiting mid-infrared instruments, and provide a unique opportunity to probe the distribution and physical conditions of the low-density medium around hot white dwarfs.

Finally, it should be noted that it is also interesting to examine hot white dwarfs without visible planetary nebulae. Have they had planetary nebulae in the past at all? Have their planetary nebulae merged into the ISM or have the nebulae been stripped by the ISM due to ram pressure? To answer these questions, the evolutionary stage and the proper motion of the hot white dwarf need to be determined and a sensitive search for an ionized low-density circumstellar nebula is also needed. Regardless, it is fair to conclude that hot white dwarfs are fascinating probes of stellar evolution, nebular processes and ISM interaction whether or not they display associated planetary nebulae.

References

Barstow, M.A., Boyce, D.D., Welsh, B.Y., Lallement, R., Barstow, J.K., Forbes, A.E., and Preval, S. (2010) *Astrophys. J.*, **723**, 1762.

Bloecker, T. (1995) *Astron. Astrophys.*, **297**, 727.

Cahn, J.H., Kaler, J.B., and Stanghellini, L. (1992) *Astron. Astrophys. (Suppl.)*, **94**, 399.

Chu, Y.-H., Guerrero, M.A., Gruendl, R.A., Williams, R.M., and Kaler, J.B. (2001) *Astrophys. J. Lett.*, **553**, L69.

Chu, Y.-H., Gruendl, R.A., Williams, R.M., Gull, T.R., and Werner, K. (2004) *Astron. J.*, **128**, 2357.

Chu, Y.-H. *et al.* (2009) *Astron. J.*, **138**, 691.

Gruendl, R.A., Chu, Y.-H., and Guerrero, M.A. (2004) *Astrophys. J. Lett.*, **617**, L127.

Guerrero, M.A., Chu, Y.-H., Manchado, A., and Kwitter, K.B. (2003) *Astron. J.*, **125**, 3213.

Guerrero, M.A., Gruendl, R.A., and Chu, Y.-H. (2002) *Astron. Astrophys.*, **387**, L1.

Harris, H.C. *et al.* (2007) *Astron. J.*, **133**, 631.

Heiles, C. and Habing, H.J. (1974) *Astron. Astrophys. (Suppl.)*, **14**, 1.

Hewett, P.C., Irwin, M.J., Skillman, E.D., Foltz, C.B., Willis, J.P., Warren, S.J., and Walton, N.A. (2003) *Astrophys. J. Lett.*, **599**, L37.

Houck, J.R. *et al.* (2004) *Astrophys. J. (Suppl.)*, **154**, 18.

Jacoby, G.H. and van de Steene, G. (1995) *Astron. J.*, **110**, 1285.

Kastner, J.H., Soker, N., Vrtilek, S.D., and Dgani, R. (2000) *Astrophys. J. Lett.*, **545**, L57.

Kwok, S., Purton, C.R., and Fitzgerald, P.M. (1978) *Astrophys. J. Lett.*, **219**, L125.

McCook, G.P. and Sion, E.M. (1999) *Astrophys. J. (Suppl.)*, **121**, 1.

Napiwotzki, R. (1993) *Planetary Nebulae*. Proceedings of IAU Symposium 155 (eds R. Weinberger and A. Acker), Kluwer Academic Publishers, Dordrecht, p. 88.

Napiwotzki, R. (1999) *Astron. Astrophys.*, **350**, 101.

Napiwotzki, R. and Schoenberner, D. (1995) *Astron. Astrophys.*, **301**, 545.

Otte, B., Dixon, W.V.D., and Sankrit, R. (2004) *Astrophys. J. Lett.*, **606**, L143.

Perryman, M.A.C. *et al.* (1997) *Astron. Astrophys.*, **323**, L49.

Reynolds, R.J., Chaudhary, V., Madsen, G.J., and Haffner, L.M. (2005) *Astron. J.*, **129**, 927.

Rieke, G.H. *et al.* (2004) *Astrophys. J. (Suppl.)*, **154**, 25.

Sankrit, R. and Dixon, W.V.D. (2009) *Astrophys. J.*, **701**, 481.

Su, K.Y.L. *et al.* (2007) *Astrophys. J. Lett.*, **657**, L41.

Szentgyorgyi, A., Raymond, J., Franco, J., Villaver, E., and López-Martín, L. (2003) *Astrophys. J.*, **594**, 874.

Tweedy, R.W. and Kwitter, K.B. (1994) *Astrophys. J. Lett.*, **433**, L93.

Tweedy, R.W. and Kwitter, K.B. (1996) *Astrophys. J. (Suppl.)*, **107**, 255.

Tylenda, R., Stasińska, G., Acker, A., and Stenholm, B. (1994) *Astron. Astrophys. (Suppl.)*, **106**, 559.

Vassiliadis, E. and Wood, P.R. (1993) *Astrophys. J.*, **413**, 641.

Villaver, E., García-Segura, G., and Manchado, A. (2003) *Astrophys. J. Lett.*, **585**, L49.

Wareing, C.J., O'Brien, T.J., Zijlstra, A.A., Kwitter, K.B., Irwin, J., Wright, N., Greimel, R., and Drew, J.E. (2006) *Mon. Not. R. Astron. Soc.*, **366**, 387.

Weaver, R., McCray, R., Castor, J., Shapiro, P., and Moore, R. (1977) *Astrophys. J.*, **218**, 377.

Werner, K. and Herwig, F. (2006) *Publ. ASP*, **118**, 183.

Index

White Dwarf Atmospheres and Circumstellar Environments. First Edition. Edited by D. W. Hoard.

Object Index

White Dwarf Atmospheres and Circumstellar Environments. First Edition. Edited by D. W. Hoard.